IAEI
Soares Book on
Grounding

IAEI
Soares Book on
Grounding

Eighth Edition

International Association of Electrical Inspectors
Richardson, Texas

Notice to the Reader

This book has not been processed in accordance with NFPA Regulations Governing Committee Projects. Therefore, the text and commentary in it shall not be considered the official position of the NFPA or any of its committees and shall not be considered to be, nor relied upon as a formal interpretation of the meaning or intent of any specific provision or provisions of the 2002 edition of NFPA 70, *National Electrical Code.*®[1]

Publishers do not warrant or guarantee any of the products described herein or perform any independent analysis in connection with any of the product information contained herein. Publisher does not assume, and expressly disclaims, any obligation to obtain and include information referenced in this work.

The reader is expressly warned to consider carefully and adopt all safety precautions that might be indicated by the activities described herein and to avoid all potential hazards. By following the instructions contained herein, the reader willingly assumes all risks in connection with such instructions.

THE PUBLISHERS MAKE NO REPRESENTATIONS OR WARRANTIES OF ANY KIND, INCLUDING, BUT NOT LIMITED TO, THE IMPLIED WARRANTIES OF FITNESS FOR PARTICULAR PURPOSE, MERCHANTABILITY OR NON-INFRINGEMENT, NOR ARE ANY SUCH REPRESENTATIONS IMPLIED WITH RESPECT TO SUCH MATERIAL. THE PUBLISHERS SHALL NOT BE LIABLE FOR ANY SPECIAL, INCIDENTAL, CONSEQUENTIAL OR EXEMPLARY DAMAGES RESULTING, IN WHOLE OR IN PART, FROM THE READER'S USES OF OR RELIANCE UPON THIS MATERIAL.

[1]*National Electrical Code* and *NEC* are registered trademarks of the National Fire Protection Association, Inc., Quincy, MA 02269

Table of Contents

Preface

This book is dedicated to the memory of Eustace C. Soares, P.E., one of the most renowned experts in the history of the *National Electrical Code* in the area of grounding electrical systems. A wonderful teacher and man of great vision, Eustace foresaw the need for better definitions to clear up to the great mystery of grounding of electrical systems.

Eustace Soares' book, *Designing Electrical Distribution Systems for Safety* was originally published in 1966 and was based upon the 1965 edition of the *National Electrical Code*. Over the years, this book has become a classic.

A great majority of the recommendations contained in the original edition of his book have been accepted as part of Article 250 of the *National Electrical Code*. The grounding philosophies represented in the original edition are just as relevant today as they were then. To say that Eustace contributed more than any other man to solving some of the mysteries of grounding of electrical systems would not be an overstatement of fact.

Previous editions have been extensively revised both in format and in information. An effort has been made to bring this work into harmony with the 2002 edition of the *National Electrical Code* and retain the integrity of the technical information for which this work has been well known, at the same time adding additional information which may be more recent on the subject of grounding.

It is our intent to revise this work to complement each new *National Electrical Code* so this will be an on-going project. If the reader has suggestions for additional pertinent material or comments about how this work could be improved upon, they would be most welcome.

General

From the beginning, the use of electricity has presented many challenges ranging from how to install a safe electrical system to how to develop minimum *Code* requirements for safe electrical installations. These installations depend on several minimum requirements, many of which are covered in NFPA 70, *National Electrical Code*, Chapter 2, Wiring and Protection. Understanding the protection fundamentals and performance requirements in Chapter 2 is essential for electrical installation, design, and inspection. To truly understand how and why things work as they do, one must always start with the basics. Therefore, in this study of grounding and bonding, some basics, such as electrical circuits, are included. It is important that basic electrical circuits be understood, because grounding and bonding constitute an electrical circuit. The process of grounding and bonding is the building of safety circuits associated and working together with the electrical circuits and systems.

This book analyzes the how and why of these two functions of grounding and bonding and expresses their purpose in clear and concise language. It also examines grounding and bonding in virtually every article of the *Code* in addition to the major requirements of Article 250. Further, it provides information on grounding and bonding enhanced installations that exceed the minimum *NEC* requirements, such as for data processing facilities and sensitive electronic equipment installations. Chapter seventeen expands the information about those installations designed to exceed the *Code* requirements and establish an enhanced grounding electrode system or *earthing* system and install feeders and branch circuits in a fashion that works to reduce the levels of electrical or electro-magnetic interference (EMI) noise on the grounding circuits though insulation and isolation of the grounding circuit as it is routed to the original grounding point at the service or source. Some definitions of electrical terms that should be understood as they relate to the performance of grounding and bonding circuits are also included in this first chapter. This book emphasizes the use of the defined terms of both the electrical field and the *NEC* in order to develop a common language of communication.

Taking the Mystery Out of Grounding

For many years the subjects of grounding and bonding have been considered the most controversial and misunderstood concepts in the *National Electrical Code*. Yet there is no real reason why these subjects should be treated as mysteries and given so many different interpretations. Probably the single most effective method for clearing up the confusion is for one to review and clearly understand the definitions of the various elements of the grounding system. In addition, these terms should be used correctly during all discussions and instruction on the subject so that everyone will have a common understanding. For example, using the term *ground wire* to mean an "equipment grounding conductor" does no more to help a person understand what specific conductor is referenced than does the use of the term *vehicle* when one specifically means a "truck."

It is recommended that the reader carefully review the terms defined at the beginning of each chapter in order to develop a clear understanding of how those terms are used in regard to that particular aspect of the subject. Also, many of the terms associated with the overall grounding system are illustrated to give the reader a clearer understanding of their meaning.

This book is intended to assist the reader in establishing a strong understanding of the fundamentals of and reasons for the requirements of grounding and bonding to attain the highest level of electrical safety for persons and property. In Appendix A, the history and mystery of grounding and bonding are simplified and brought down-to-earth through the use of easy-to-read and understandable language. The publisher, International Association of Electrical Inspectors (IAEI), is committed to providing the highest quality information on grounding and bonding to the electrical industry and hopes that the reader profits immensely from this volume.

Definitions of Electrical Terms

Current (Amperes): Current, measured in amperes, consists of the movement or flow of electricity. In most cases, the current of a circuit consists of the motion of electrons, negatively charged particles of electricity.

Capacitance: A capacitor basically consists of two conductors that are separated by an insulator. A capacitor stores electrical stress. Capacitive reactance is the opposition to current flow due to capacitance of the circuit. The Institute of Electrical and Electronics Engineers (IEEE) defines capacitance as, "The property of systems of conductors and dielectrics which permits the storage of electricity when potential difference exists between the conductors."

Impedance: The term *resistance* is often used to define the opposition to the flow of current in both ac and dc systems. The correct term for opposition to current flow in ac systems is *impedance*. Resistance, inductive reactance, and capacitive reactance all offer opposition to the flow of current in alternating-current circuits. The three elements are added together vectorially, not directly. This results in the total impedance or opposition to the flow of current of an ac circuit. Impedance is measured in ohms.

Inductance: Inductance is caused by the magnetic field of an alternating-current circuit as a result of the alternating current changing directions. This causes the magnetic lines of force that surround the conductor to rise and fall. As the magnetic lines of force rise and fall, they work to oppose or cut the conductor and induce a voltage directly opposite the applied voltage. This induced voltage is called *counter-electromotive force* or counter EMF. Induction is the current effect of an ac circuit. Where there is an alternating magnetic field there will be induction. This induction will result in inductive reactance, which opposes the flow of current. Impedance is measured in ohms

> # The first and most vital element of a sound, safe structure is a solid footing or foundation on which to build the building.

Resistance: Resistance is the name given to the opposition to the flow of current which is offered by the internal structure of the different conductive materials to the movement of electricity through them, i.e., to the maintenance of current in them. This opposition results in the conversion of electrical energy into heat.

Voltage (Electromotive Force): A *volt* is the unit of measure of electromotive force (EMF). It is the unit of measure whereby the tendency to establish and maintain electric currents may be

measured. By international agreement it has been decided that 1 volt shall be taken as that EMF which will establish a current of 1 amp through a resistance of 1 ohm.

The Foundation of Grounding

The first and most vital element of a sound, safe structure is a solid footing or foundation on which to build the building. This foundation, usually consisting of concrete and reinforcing rods, must be adequate to support the weight of the building and provide a solid structural connection to the earth on which it sits. If the building or structure does not sit on a solid foundation, there may be continuous structural problems that might lead to unsafe conditions. Likewise, the grounding system serves as the foundation for an electrical service or distribution system supplying electrical energy to the structure. Often the grounding of a system or metal objects is referred to as *earthing*, being connected to the earth. When solidly grounded, the electrical system must be connected to a dependable grounding electrode or grounding electrode system. The grounding electrode(s) supports the entire grounding system and makes the earth connection. It must be effective and all grounding paths must be connected to it. This serves as the foundation of the electrical system. Chapter six covers the grounding electrodes, their functions, and their installations.

Electrical Circuitry Basics

Anyone who has been involved in the electrical field for any length of time has heard the phrase, "Electricity takes the path of least resistance." From the first-year apprentice to the seasoned veteran of the industry, the phrase is used to describe the path electrical current will take. The phrase is stated with pride, "Electricity takes the path of least resistance" or "Current takes the path of least resistance," and usually not much thought is given as to what is really going on. In reality, current will flow in all paths or circuits that are available. Where more than one path exists, current will

Voltage = Electromotive Force (E) or Pressure that pushes

Resistance = Opposition to Flow (R) or Resistance in Ohms

Current = Intensity (I) or Amperes that Flow

Impedance = Total Opposition (Z) or Total Opposition in AC Current

- Opposition to flow of current in a DC circuit is resistance
- Opposition to flow of current in a AC circuit is made up three components:

  R Resistance

  X_L Inductive Reactance

  X_c Capacitive Reactance

Ohms Law Basic

$$Z = \sqrt{R^2 + (X_L - X_c)^2}$$

Basic Impedance Formula

Figure 1-1. Basic Theory Terms

divide among the paths. As we will review later, current will divide in opposite proportion to the impedance. The lower impedance path or circuit will carry more current than the higher impedance path(s). The study of grounding and bonding is vital to applying basic rules relative to this important safety element of the electrical circuit. It is important to reivew some basic principles and the fundamental elements of electricity and how this current flow relates to electrical safety.

Ohm's Law in Review

First, in order for electrical current to flow, there needs to be a complete circuit (see figure 1-2). The amount of current that will flow in an electrical circuit depends on the characteristics of the circuit. Voltage or electromotive force (E) will push current or intensity (I) through a resistance (R). These are the basic components of Ohms Law (see Ohm's Law and its derivatives in Watt's Wheel below). Electrical current flow can be compared with water flowing through a water pipe. The bigger the pipe, the less the resistance is to the flow of water through the pipe. The smaller the pipe, the greater the resistance is to the flow of water through it. The same holds true for electrical current. Larger electrical conductors (paths) offer lower resistance to current flow. Smaller electrical conductors (paths) offer greater resistance to current flow. For electrical current to flow, there must be a complete circuit or path. This is true of both normal current flow and the flow of fault current.

Resistance as Compared to Impedance

Understanding the differences between the pure resistance of an electric circuit and the imped-

ance of a circuit is important in gaining a thorough understanding of the grounding or safety circuit. In Ohm's Law, resistance is the total opposition to the flow of current in a dc circuit. In an alternating-current circuit, the total opposition to the flow of current is the total of three components. The impedance (Z) of an ac circuit is the inductance, capacitance, and the resistance added together vectorially (see basic formula in figure 1-1). In a 60-cycle ac circuit, alternating current changes amplitude and direction 60 times per second as it flows, and develops a magnetic field that results from the inductive reactance of the circuit. Therefore, minimizing the amount of opposition to the flow of current in the grounding and bonding circuits of electrical systems is important. These circuits can be looked upon as silent servants, just waiting to perform the important function of carrying enough current so overcurrent protective devices can operate to clear a fault.

Current Flow in a Circuit

Where does the current flow? In any complete circuit or path that is available, current—be it normal current or fault current—will always try to return to its source. The statement on taking the path of least resistance is partially correct. Electrical current will take any and all available paths to return to its source. If several paths are available for current to flow, it will divide and the resistance or the impedance of each path will determine how much current will flow on that particular path. It can be concluded from the above that if there is no complete circuit, electrical current will not flow. Care must be given to the installation of ungrounded (hot) conductors so that the circuit will be complete for current to flow for normal operation. The same principles and fundamentals apply to the installation of grounding or safety circuits. The grounding (safety) circuit must be complete and must meet three important criteria: (1) the path for fault current must be permanent and continuous; (2) it must have adequate capacity to conduct safely any fault likely to be imposed on it; and (3) it must be of the lowest possible impedance. (See chapter

1 Ampere

1 Volt   1 Ohm

Ω = Ohm
1,000 Ohms = 1 Kilohm (K)
1,000,000 Ohms = 1 Megohm (Meg)
1,000 Kilohms (K) = 1 Megohm (Meg)

Figure 1-2. Watt's Wheel — Ohm's Law

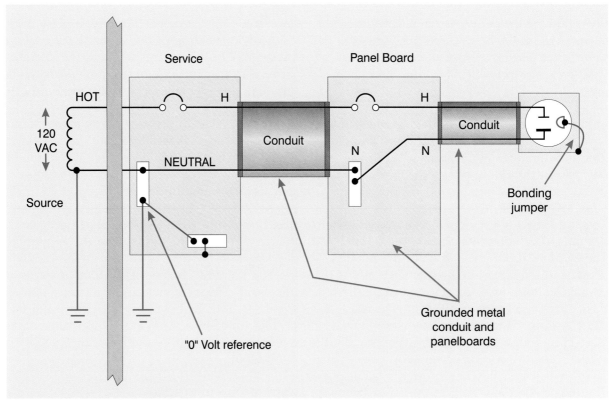

Figure 1-3. Overcurrent Device Operation

E = Electromotive force   R = Resistance   I = Intensity
E = Voltage   R = Ohms   I = Amperes

Voltage	=	*Resistance	x	**Current
120	=	60	x	2
120	=	40	x	3
120	=	20	x	6
120	=	10	x	12
120	=	5	x	24

* As the resistance or opposition to current flow increases, the current that will flow in the circuit decreases

** As the resistance or opposition to current flow decreases, the current that will flow in the circuit increases

Figure 1-4. Amperes Operate Overcurrent Devices

eleven for more specific information relative to clearing ground faults and short circuits.)

Article 250 mentions the term *low-impedance path* several times. As a quick overview, the opposition to current flow in a dc circuit is resistance. The total opposition to current flow in an ac circuit is impedance, which is made up of three elements: resistance, capacitance, and inductance. When *low-* *impedance path* is used in the *Code*, it is referring to a path that offers little opposition to current flow whether it is normal current or fault current. The key element is low opposition or impedance.

Overcurrent Device Operation

Overcurrent devices require current (amps) to operate. Generally speaking, the more current that flows through overcurrent devices the faster they open or operate; this is because they are "inverse time." The higher the impedance of the path, the lower the current that will flow through the overcurrent device. The lower the impedance of the path, the greater the current that will flow through the overcurrent device. Understanding these basic elements of electrical circuits helps one apply some important rules in Article 250. The following examples clearly demonstrate that amps operate overcurrent devices.

As with the electrical circuit built for normal current flow, the grounding (safety) circuit must also be built for current flow to ensure overcurrent device operation in ground-fault conditions. The grounding or safety circuit must be complete and constructed with as little impedance as practicable for quick, sure overcurrent device operation. Care must be taken when building electrical systems and circuits, including the grounding and bonding circuits of the system.

Proper Language of Communication

A common language of communication has been established to enable one to understand the requirements of the *NEC* in general and of grounding and bonding in particular. A common set of terms, defining and explaining the function of the

terms as used in the *Code* are included in Article 100 and in .2 of other articles. Two conductors of the grounding system should be mentioned and a brief story told about each: the grounded conductor, and the equipment grounding conductor.

Grounded and Grounded Conductor

The grounded conductor (usually a neutral) is generally a system conductor intended to carry current during normal operation of the circuit. The grounding of the grounded (often a neutral)

- In a series circuit there is only one path for current to travel on. It is important to remember that current will always try to seek out it's source. It does not matter if it is the fault current or normal current it will take any path available to it to get to the source.
- If there is no complete circuit current will not flow.

Source voltage

20 15 ohms R$_1$ 15 ohms R$_2$

120 volt circuit

$$I = E \div R = 120 \text{ volts} \div 30 \text{ ohms} = 4 \text{ amps}$$

Properly operating series circuit

Figure 1-5. Series Circuits (normal flow)

Source voltage

20 4 ohms R$_1$ 6 ohms R$_2$

120 volt circuit

.05 ohms at fault

$$I = E \div R$$

$$I = 120 \text{ volts} \div 4.05 \text{ ohms} = 29.6 \text{ amps}$$

Grounding electrode

Overcurrent device <u>will</u> operate!

Figure 1-6. Series Circuits with Low Resistance Fault in Circuit

conductor of a system is accomplished by a connection to ground through a grounding electrode conductor either at the service or at a separately derived system. Generally, it should be understood that the grounded conductor should not be used for grounding equipment on the load side of the grounding connection at the service or source of separately derived system. This separation between grounded conductors and equipment grounding conductors keeps the return cur-

$$I = E \div R$$

$$I = 120 \text{ volts} \div 17 \text{ ohms} = 7.058 \text{ amps}$$

Overcurrent device will _not_ operate!

Figure 1-7. Series Circuits with High Resistance Fault in Circuit

6.0 Amps
+ 4.9 Amps
─────────
10.9 Amps

24 Ohms + ¼ Ohm = 24 ¼ Ohms

120Volt / 24 ¼ Ohms = 4.9 Amps

Figure 1-8. Example of Ohm's Law With Human in the High Impedance Circuit

Basic Rule

1. Line-side of service only

2. No load - side connection

Grounded conductor

Line side service

Service equipment

Grounded electrode

Load side

Equipment grounding conductor

Figure 1-9. System grounding connections

rent on the neutral (grounded) conductor of the system flowing on the proper conductor back to its source. These principles are reinforced by requirements in 110.7 and 250.24(A)(5). *Code* rules and requirements for the grounded conductors are covered in depth in chapter three of this text.

Grounding and Equipment Grounding Conductor

As used in Article 250 and other articles, *grounding* is a process that is ongoing. The conductor to look at is the equipment grounding conductor. The action is ongoing through every electrical enclosure it is connected to all the way to the last outlet on the branch circuit. The equipment grounding conductor puts all metal enclosures at earth potential along the way, and also provides a low-impedance path for fault-current flow if a ground fault should occur in the system. So it is important that the equipment grounding conductor make a complete and reliable circuit back to the source. At the source or service is where the grounded (neutral) conductor and the equipment grounding conductor are required to be connected together through a main bonding jumper. In a separately derived system, this connection is made with a bonding jumper installed between

the grounded conductor and the equipment grounding conductor. These bonding jumpers complete the fault-current circuit back to the source. The rules and requirements for equipment grounding conductors are covered in depth in chapter nine.

Grounding as Compared to Bonding.

Defined in Article 100, both of these functions are essential for the complete safety anticipated by the rules in Article 250.

"**Grounded.** Connected to earth or to some conducting body that serves in place of the earth."

"**Bonding (Bonded).** The permanent joining of metallic parts to form an electrically conductive path that ensures electrical continuity and the capacity to conduct safely any current likely to be imposed."[1]

These are two separate functions with two different purposes. It is important that a clear understanding of the grounding (earthing) circuit and its purpose as compared to the equipment grounding conductors and bonding jumpers or connections be established. Section 250.4 has been broken down into grounded systems and ungrounded systems. Requirements in this section have been expanded and revised to include more descriptive perfor-

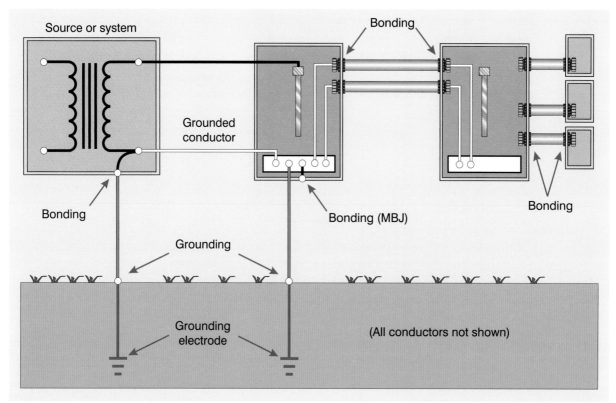

Figure 1-10. Grounding Compared to Bonding

mance requirements, and to establish the purposes served by each of these actions. Although the title remains simply "Article 250 Grounding," in reality the article contains an equally strong emphasis on bonding requirements. Chapter eight presents detailed information on bonding requirements.

The *National Electrical Code* Trend

The *NEC* in recent cycles has been revised to reduce the use of the grounded conductor for grounding equipment downstream from the main bonding jumper in a service, or downstream from the bonding connection at a separately derived system. As stated earlier, the reasons are elementary. Current, be it normal current or fault current, will take all the paths available to it to try to return to its source. If the grounded conductor (neutral) and equipment grounding conductors are connected at points downstream of the service or separately derived system, such as at sub-panels, multiple paths will be available on which the current will try to return to the source. This can lead to normal neutral current flowing on water piping systems, conduit, equipment grounding conductors, and any other electrically conductive path and can compromise overcurrent device operation in ground-fault condition.

In recent editions of the *NEC* (1996) electric range and dryer circuits were required to include an equipment grounding conductor in addition to an insulated grounded conductor. Existing range and dryer circuits are allowed to continue the use of the grounded conductor, or neutral, to ground the boxes at the outlet and the frames of the equipment. New installations, however, are required to maintain isolation between the grounded conductor and the equipment grounding conductor.

In the 1999 *NEC*, the rules covering the use of the grounded conductor for grounding purposes at a second building or structure were revised. Section 250.32(B)(1) requires that if an equipment grounding conductor is installed with the feeder supplying the second building or structure, separation between the grounded (neutral) conductor is to be maintained. There is an allowance in 250.32(B)(2) to utilize the grounded conductor of the feeder for grounding equipment under three specific and very restrictive conditions. First, an equipment grounding conductor is not included with the feeder supplying the building or structure. Second, there are no continuous metallic paths bonded to the grounding system in both buildings. Third, there

Parallel paths

Current flow

Series circuit path

For current to flow there must be a complete circuit
Current will always try to return to the source (fault current or normal current)
Current will flow back to the source on as many paths that are available to it
Amount of current flow on a particular path depends on the impedance of that path

Figure 1-11. Series and Parallel Paths for Current

is no ground-fault protection of equipment installed at the service. If all of these conditions are complied with, the grounded conductor must be used for grounding and be connected to the building or structure disconnecting means.

The grounded conductor is also required to be connected to a grounding electrode at the building or structure and installed in accordance with Part III of Article 250. This will serve as the grounding means and as the path for both normal current and fault current to clear overcurrent devices. Section 250.32(B)(2) requires that there be no continuous metallic paths bonded to the grounding system in each structure. This requirement encompasses all paths, not just wires or conduits. These paths could include items such as metal water pipes, other metal piping, steel members, and paths such as the shielding on a communications cable or a coaxial cable installed between the structures. If this connection were made and the service equipped with ground-fault protection in accordance with 230.95, these connections could desensitize the GFP device, possibly causing it not to operate properly when called upon to do so in ground-fault conditions because of multiple paths for current.

Current on the Proper Path
It is important that the basic elements of current flow be understood and carefully considered

while applying the rules of the *NEC*. Section 250.24(A)(5) states that a grounding connection to any grounded circuit conductor on the load side of the service disconnecting means shall not be made, unless otherwise permitted in the ar-

> It is important that the basic elements of current flow be understood and carefully considered while applying the rules of the *NEC*.

ticle. The fine print note (FPN) refers to three situations where this is acceptable but restrictive: for separately derived systems in 250.30(A); for separate buildings or structures in 250.32; and for grounding equipment under the limitations of 250.142. Installers and inspectors should be watchful to ensure there are no neutral-to-ground connections on the load side of the grounding connections at either a service disconnecting means or source of a separately derived system. In other words, isolate the neutrals and equipment grounding conductor connections. Give current, be it fault current or normal current, the

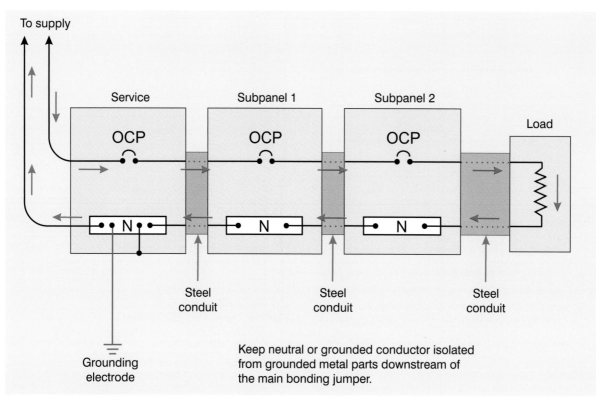

To supply

Service

OCP

N

Subpanel 1

OCP

N

Subpanel 2

OCP

N

Load

Steel
conduit

Steel
conduit

Steel
conduit

Grounding
electrode

Keep neutral or grounded conductor isolated
from grounded metal parts downstream of
the main bonding jumper.

Figure 1-12. Proper Current Flow

To supply

Service

OCP

N

Subpanel 1

OCP

N

Subpanel 2

OCP

N

Load
equipment

Steel conduit

Steel
conduit

Steel
conduit

Main bonding jumper

Legend

Normal
current

Circulating
current

Figure 1-13. Circulating Currents on all Paths Available

low-impedance path anticipated by the requirements of the *NEC*.

Grounding and bonding must be effective and include the following characteristics:

1. Be permanent and continuous path
2. Have adequate capacity for any fault currents likely to be imposed
3. Have lowest possible impedance of current flow in the fault-current path

The definitions of 250.2 describe three important terms used in Article 250. These three terms are:

Effective Ground-Fault Current Path. An intentionally constructed, permanent, low-impedance electrically conductive path designed and intended to carry current under ground-fault conditions from the point of a ground fault on a wiring system to the electrical supply source.

Ground Fault. An unintentional, electrically conducting connection between an ungrounded conductor of an electrical circuit and the normally non-current-carrying conductors, metallic enclosures, metallic raceways, metallic equipment, or earth.

Ground-Fault Current Path. An electrically conductive path from the point of a ground fault on a wiring system through normally non-current-carrying conductors, equipment, or the earth to the electrical supply source.

Overview of NFPA 70 Article 250

Proper grounding and bonding provide protection from electric shock hazards and facilitate operation of overcurrent devices to clear faulted circuits. Fast and effective operation of overcurrent devices, covered in Article 240, depends on the proper bonding and effective fault-current paths required by Article 250. Article 250 is located in Chapter 2, Wiring and Protection, of the *NEC* and is made up of ten parts. Part I includes general requirements and some performance requirements vital to the understanding and proper application of prescriptive requirements found in latter parts of the article. The parts of Article 250 are very much interlocking; that is, they must be used together in many instances to properly apply the rules. Figure 250.4 was new for the 1999 *NEC*, and serves as a built-in map for Article 250, which serves to enhance the usability issues in the *Code*. This figure is included here as figure 1-14 to assist the reader in understanding the arrangement of the article at a glance.

Article 250 includes general requirements for

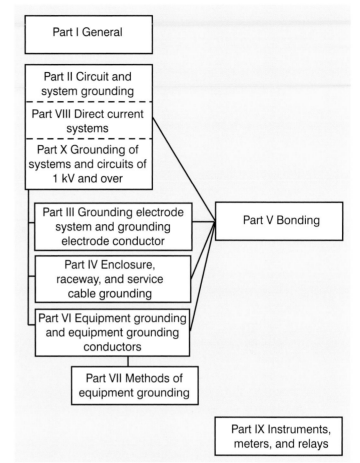

Figure 1-14. *NEC* **Figure 250.4 Grounding**

grounding and bonding of electrical installations and six specific requirements.

"1. Systems, circuits, and equipment required, permitted, or not permitted to be grounded
2. Circuit conductor to be grounded on grounded systems
3. Location of grounding connections
4. Types and sizes of grounding and bonding conductors and electrodes
5. Methods of grounding and bonding
6. Conditions under which guards, isolation, or insulation may be substituted for grounding"[2]

Alternatives to Grounding

Item six in 250.1 indicates that three conditions—guards, isolation, or insulation—may be substituted for grounding under certain conditions. Definitions of these terms, found in Article 100, help clarify the intent of the alternatives.

Guarded. Covered, shielded, fenced, enclosed, or otherwise protected by means of suitable covers, casings, barriers, rails, screens, mats, or platforms to remove the likelihood of approach or contact by persons or objects to a point of danger.

Isolated (as applied to location). Not readily accessible to persons unless special means for access are used.

Insulated. Equipment or materials that are covered with an insulating material. The insulation is of sufficient dielectric means to provide protection against electrically conductive parts that might likely become energized and present shock hazards to personnel. Double insulated power tools are an example of insulation being used as a substitute for grounding.[3]

Grounding of Electrical Systems

Some features of electrical safety are so fundamental they have appeared in some form in every edition of the *National Electrical Code*. These include requirements for suitable insulation for conductors; overcurrent protection for circuits, and grounding of electrical systems and equipment for safety. The grounding of equipment and conductor enclosures, as well as the grounding of one conductor of an electric power and light system, has been practiced in some quarters since the use of electricity began.

At first, there was no uniform standard for grounding (see the appendix of this text for information on the early history of grounding, including debates that took place on how best to approach the subject). However, it was not long before it became universal to ground one conductor on all 120-volt lighting circuits. Early editions of the *Code* firmly established the practice of grounding by making it mandatory to ground all such systems where the system can be grounded so that the voltage to ground does not exceed 150 volts. These editions also recommended that alternating current systems be grounded where the system voltage to ground did not exceed 300 volts, while at the same time stating that higher voltage systems were permitted to be ungrounded. Later, the *Code* made it mandatory to ground any system having a nominal voltage to ground of not more than 300 volts if the grounded service conductor (usually a neutral) was not insulated.

Grounded

The term *grounded* is defined in Article 100 as, "Connected to earth or to some conducting body that serves in place of the earth."[4] Conducting bodies that serve in place of the earth include conduit, boxes, enclosures, equipment grounding conductors and wiring devices. These are, in fact, an extension of the earth by being electrically connected to the earth by reliable electrical and mechanical means. Metal frames of buildings that are effectively grounded are acceptable as an extension of the earth so far as being a grounding electrode but are not permitted to be used as an equipment grounding conductor. Frames of vehicles are permitted to serve as a grounding electrode for vehicle-mounted generators and as a return path for low-voltage systems in recreational vehicles.

The earth as a whole is properly classed as a conductor. For convenience, its electric potential is assumed to be zero. Based upon the composition of the soil or earth, the resistance of segments of the earth can vary widely from one area to an-

Grounded: "Connected to earth or some conducting body that serves in place of the earth."

The earth as a conductor is assumed to have a potential of zero.

Figure 1-15. Grounded: Connected to Earth.

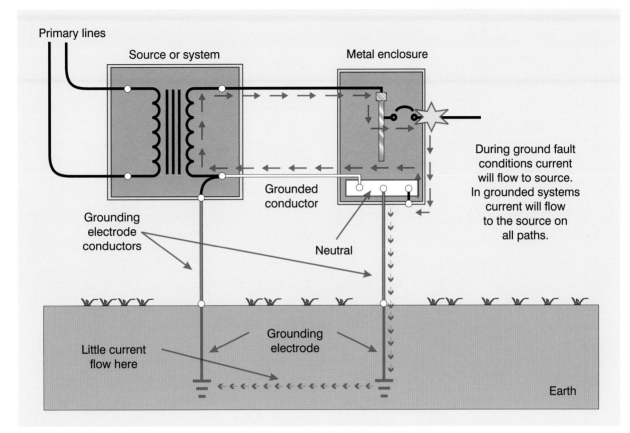

Primary lines

Source or system

Metal enclosure

During ground fault conditions current will flow to source. In grounded systems current will flow to the source on all paths.

Grounded conductor

Grounding electrode conductors

Neutral

Grounding electrode

Little current flow here

Earth

Figure 1-16. Grounding Fundamentals

other. The earth is composed of many different materials, some of which, especially when dry, are very poor conductors of electricity. Soil temperature, moisture content and chemical composition are factors that have a great influence on soil resistance. As a result of these factors, the capability of the earth to carry electrical current also varies widely (see chapter six and the appendix for more information on composition of the earth and effectiveness of grounding electrodes).

A metal object such as a box or other equipment enclosure that is grounded by connecting (bonding) it to the earth by means of a grounding electrode, grounding electrode conductor and/or equipment grounding conductor, is thereby forced theoretically to take the same zero potential as the earth. Slight differences in potential may exist due to differences in impedance of the materials or connections. Any attempt to raise or lower the potential of the grounded object results in current passing over the grounding path until the potential (voltage) of the object and the potential (voltage) of the earth (zero) are equalized. Usually, this potential above ground

is caused by a line-to-ground fault. Hence, grounding is a means for ensuring that the grounded object cannot take on a potential differing enough from earth potential to be hazardous. When the grounding conductor is broken, is inadequate in size, or has a poor connection, a hazardous, aboveground potential may be present from an abnormal condition.

Resistance and Impedance

Though a comprehensive discussion of the subject is beyond the scope of this text, a brief explanation of the terms *resistance* and *impedance* is offered.

For direct current (dc) systems and circuits, resistance (R) is properly used to describe the opposition to current flow. We are accustomed to using the term *ohms* to relate to the resistance of the circuit, such as 35 ohms. Ohm's Law can be summarized as follows: in a dc circuit, the current is directly proportional to the voltage and inversely proportional to the resistance. ($I = E/R$) This means, as the voltage is increased, the current will increase through a fixed resistance. As the resistance is reduced, the current will increase

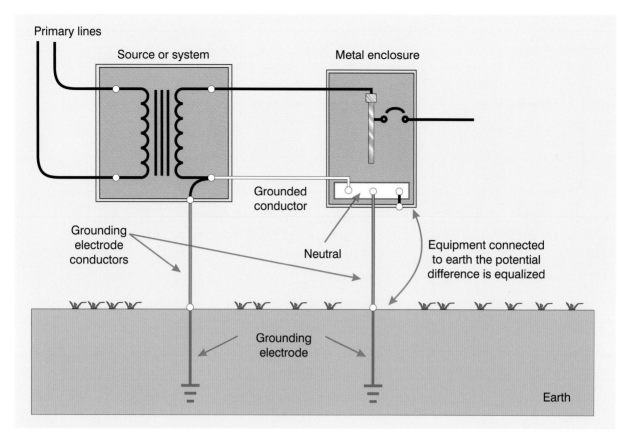

Figure 1-17. Grounding Fundamentals: Grounded and Bonded Properly

if the voltage stays the same. A pressure (voltage) of one volt will cause one ampere to flow through a resistance of one ohm.

For alternating current (ac) systems and circuits, *impedance* (Z) is the proper term to describe the total opposition to current flow. Impedance consists of three components: inductive reactance, capacitive reactance and resistance. Impedance, rather than resistance, is used most often throughout this text since, for the most part, ac electrical systems and circuits are being considered. See chapter eleven for a detailed discussion of the importance of keeping all conductors of the circuit, including the equipment grounding conductor, close together to keep the impedance as low as possible. Also, many excellent books on electrical theory cover the subject of impedance, inductive reactance, capacitive reactance and resistance in detail.

Equipment Supplied from a Grounded System

Where the electrical system is grounded, it is critical to provide a low-impedance path of adequate capacity from all the equipment supplied by the system back to the source of the system: (1) to maintain the metal equipment enclosures as close to earth potential as possible to reduce a shock hazard; and (2) to ensure the overcurrent protection will operate in the event of a line-to-ground fault. This is required by 250.4 and is emphasized in several portions of this text.

Electrical equipment supplied by a grounded system that is left ungrounded or is poorly grounded becomes a silent and often lethal source of electrical shock where a ground fault occurs. Another hazard of equal significance may occur where two pieces of electrical equipment supplied from a grounded system are within reach of a person. If one piece is not grounded or is poorly grounded and becomes energized through a failure of the insulation system (ground fault), the person making contact with the two (or more) pieces of equipment becomes the circuit (path) for current to pass through as it tries to find its way back to its source. In some cases, the person will receive a mild shock. In other cases, the shock may be fatal. Even though the impedance of the human body may be relatively high, approximately 200 milliamp (mA) of current flowing through it may be lethal. A milliamp is equal to one thousandth of an ampere.

Equipment Supplied from an Ungrounded System

Section 250.21 permits some electrical systems to be operated ungrounded. In this case, the electrical enclosures for the service, feeders, circuits and other equipment are connected to ground but the system itself is not grounded. In many parts of the country, the serving utility will provide only a grounded electrical system. From a practical standpoint, an ungrounded system exists only in theory or at the distribution transformers hanging on nonmetallic poles before the system is connected to the plant electrical system. Where the ungrounded system conductors are installed in grounded metal raceways or enclosures or connected to motors, the frame of which are grounded, the ungrounded system becomes capacitively coupled to ground (see chapter two for a more detailed discussion of this subject).

An ungrounded electrical system often becomes grounded through a line-to-ground fault. This essentially creates a grounded system. The other phases of the system then rise to a voltage to ground equal to the system voltage. For example, a 480-volt ungrounded delta system is installed in metal conduit and enclosures that are grounded. When checking the voltage to ground, it is found to be about 240 volts. If one of the phases becomes grounded anywhere on the system, the other phases will have a voltage to ground of approximately 480 volts. Obviously, this becomes a greater shock hazard to personnel who may be servicing the installation and adds greater stress on the conductor insulation, including transformer and motor windings.

Effect of Electricity on Humans

Like the bird on the electric wire, the human body is immune to electric shock as long as it is not a part of the electric circuit. The easiest way to avoid danger from electric shock is to keep one's body from becoming a part of the electric circuit. Due to the very common use of electric tools, equipment and appliances, the risk of being exposed to electric shock is multiplied by the numbers of these items the person is exposed to.

When a person becomes a path for electricity, he or she will experience an electrical shock. The

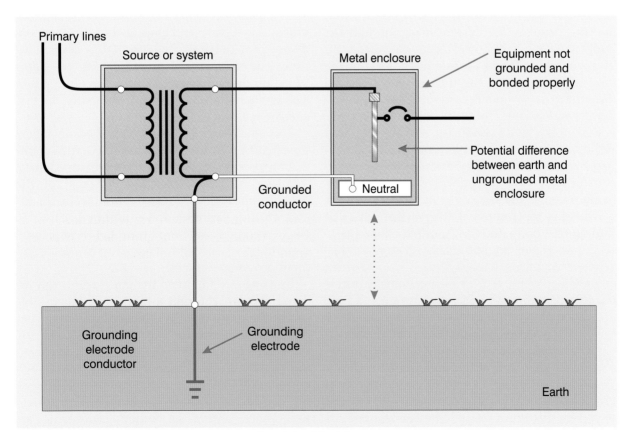

Figure 1-18. Potential Difference (Hazard): Not Properly Grounded and Bonded

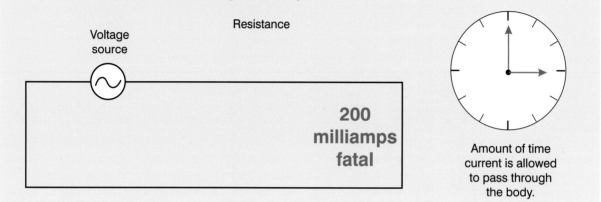

The severity of electric shock is determined by three elements. If the combination of these three elements is just right, the shock can be severe or lead to electrocution.

1. Amount of current that will flow.

2. Length of time that the current will flow.

3. Path of current through the body.

Voltage source

Resistance

200 milliamps fatal

Amount of time current is allowed to pass through the body.

Figure 1-19. Severity of Electric Shock

intensity and damage done to the person by the shock will be determined by the current level (the amount that flows through the person), how long the current flows (the duration), the person's size, and the pathway the current takes through the body. Speaking in electrical terms, the person can be thought of as a resistor or impedance in the circuit. The skin of the human body can be thought of as an insulation; however, its resistance is rather low. The inside of the human body is a relatively good conductor as it is composed of primarily fluids and salt. Once electrical current has entered the body through the skin, the resistance is very low, and severe damage or death is likely. The severity of the electrical shock is related to the amount of current that will flow through the body and the duration of the flow.

A person may become a pathway to ground or between conductors in one of two ways: in a series or a parallel circuit. In a series circuit, the person is the only path through which the current will flow in its attempt to return to its source. An example of this is when a person comes into contact with an electrical appliance that is energized through a line-to-ground fault and at the same time touches a grounded appliance like an electric range or a grounded kitchen sink. In general terms, the amount of current that flows through the person's body is determined by Ohm's Law by the voltage of the

circuit and the resistance offered by the person's body (see chapter fourteen for a greater discussion of the human body and current flow.)

In a parallel circuit, the human body and another path such as an equipment grounding conductor provide a path for current flow at the same time. The current will divide among the paths based on the impedance of the individual paths. The greater the impedance of the path, the less current will flow in that path.

The smallest risk of shock hazard occurs where the circuit protective device, usually a fuse or circuit breaker, opens the faulted circuit immediately. Another factor is the time it takes for the overcurrent device to clear the fault from the circuit. During this time, the equipment subject to the ground-fault will have a potential aboveground depending upon the voltage drop of the equipment grounding circuit.

In some installations, the equipment may be grounded but through a high impedance. This can be the result of an equipment grounding conductor that is too small or too long for its size; the connection has become loose; or it is routed improperly. In this situation, usually not enough current will flow through the equipment grounding conductor to cause the overcurrent protective device to clear the faulted equipment. Part of the fault current will then flow through the

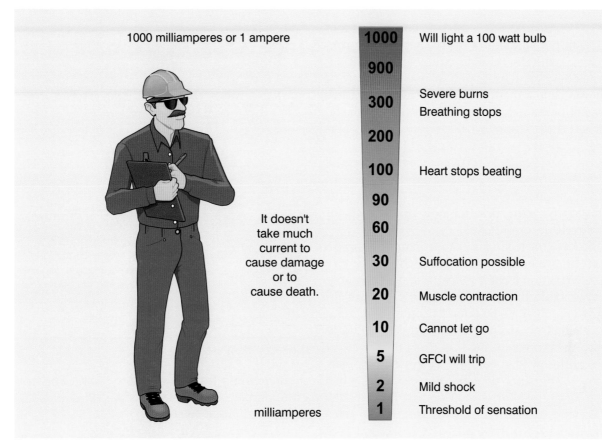

Figure 1-20. Effects of Electricity on Humans

person in contact with the energized equipment and part through the equipment grounding conductor path. The current will divide in opposite proportions to the resistance or impedance of the paths. The greatest amount of current will flow through the path providing the lowest resistance or impedance. Even though the human body may have an impedance that is greater than the other path(s), it only takes a small amount of current to cause serious injury or death.

The primary emphasis is to provide a path for ground-fault current that is permanent and continuous, is adequately sized and of low-impedance from all electrical equipment back to the source to facilitate the operation of the overcurrent device in a reasonable time. Protection of persons from electric shock is the reason ground-fault circuit devices (GFCIs) have become so popular over the recent years (see chapter fourteen for additional information on the subject).

Burns and Other Injuries
The most common shock-related injury is a burn. Burns suffered in electrical accidents may be of three types: electrical burns, arc burns, and thermal contact burns.

Electrical burns are the result of the current flowing through tissues, blood vessels or bone. Tissue damage is caused by the heat generated by the current flow through the body and is often immediately classified as a third degree burn. In many cases, the damage caused to the tissues becomes more apparent in the days or even months following the incident. Severe permanent damage can be caused to internal tissues or organs with little external indication until some time after the electric shock. Burns from electric shock are one of the most serious injuries that can be experienced and should be given immediate medical attention.

Arc or flash burns, on the other hand, are the result of high temperatures produced by an electric arc or explosion near the body. These burns can be of the more minor first-degree type, more severe second-degree, or most severe third-degree burns. They should also be attended to promptly and properly.

Finally, thermal contact burns are those normally experienced when the skin comes into contact with hot surfaces of overheated electric conductors, conduits, or other equipment. Additionally, clothing may be ignited in an electrical arc or explosion and a thermal burn will result. All three types of burns may be produced simultaneously.

The proper use of work procedures and personal protective equipment can minimize these injuries. (See NFPA 70E *The Standard for Electrical Safety Requirements for Employee Workplaces* for information about electrical safety requirements and safe working practice and procedures for employees in workplaces.)

Electric shock can also cause injuries of an indirect or secondary nature in which involuntary muscle reaction from the electric shock can cause bruises, bone fractures, and even death resulting from collisions or falls. In some cases, injuries caused by electric shock can contribute to delayed fatalities.

In addition to shock and burn hazards, electricity poses other dangers. For example, when an arcing type short circuit occurs, hazards can be created from the resulting arcs. If high-current is involved, these arcs can cause injury or start a fire. Extremely high-energy arcs can damage equipment, causing fragmented metal to fly in all directions, or melt steel, copper or aluminum. Even low-energy sparks can cause violent explosions in atmospheres that contain flammable gases, vapors, or combustible dusts.

Effect of Electricity on Animals

Some animals are especially sensitive to electricity. For example, past studies claimed that dairy cattle are so sensitive to electricity that a potential of as little as two volts between conductive portions of floors, walls, piping and stanchions caused behavior problems that resulted in loss of production. More recent studies claim the voltage difference to be about four volts. In other cases, severe health problems are attributed to the effects of electricity, which, if not corrected and treated may lead to death of the animal. (See chapter sixteen for methods of preventing and correcting these problems in agricultural structures.)

Purpose of Grounding and Bonding

The general requirements for grounding and bonding are contained in 250.4. The requirements include the grounding and bonding performance

Figure 1-21. Scope of Grounding

requirements for grounded systems and ungrounded systems as follows:

(A) Grounded System
(1) Electrical System Grounding.
(2) Grounding of Electrical Equipment.
(3) Bonding of Electrical Equipment
(4) Bonding of Electrically Conductive Materials and Other Equipment

Effective Ground-Fault Current Path
(B) Ungrounded Systems
(1) Grounding of Electrical Equipment.
(2) Bonding of Electrical Equipment.
(3) Bonding of Electrically Conductive Materials and Other Equipment.
(4) Path for Fault Current.

The requirements for grounded systems and ungrounded systems and the differences between them are covered in detail in chapter two of this text.

Grounding of Electrical Systems. Systems are solidly grounded to limit the voltage to ground during normal operation and to prevent excessive voltages due to lightning, line surges or unintentional contact with higher-voltage lines and to stabilize the voltage to ground during normal operation [see 250.4(A)(1)]. Several methods of grounding electrical systems are used depending on *National Electrical Code* requirements and system design and function. These methods of grounding electrical systems are covered in detail in chapter three.

Grounding of Electrical Equipment. Conductive materials enclosing electrical conductors or equipment, or that are a part of the equipment are grounded to limit the voltage to ground on these materials and bonded to facilitate the operation of overcurrent devices under ground-fault conditions [see 250.4(A)(2)]. Where the electrical system is grounded, the equipment grounding conductor is connected to the grounded system conductor (often a neutral) at the service or the source of a separately derived system. Where the electrical system is not grounded, the electrical equipment is connected to earth at the service to maintain the equipment at or near earth potential and the equipment is bonded together to provide a path for fault current. This occurs where a second ground fault occurs before the first one is cleared.

Bonding of Electrically Conductive Materials and Other Equipment. Electrically conductive materials, such as metal water piping, metal gas piping and structural steel members, that are likely to become energized, are bonded to provide a low-impedance path for clearing ground faults that otherwise would energize the equipment at a level above earth potential. For systems that are grounded, the equipment is bonded to the grounded system conductor (often a neutral) at the service or source of a separately derived system. Where the electrical system is not grounded, the electrical equipment is connected to earth at the service to maintain the equipment at or near earth potential and the equipment is bonded together to provide a path for fault current. Again, this occurs where a second ground fault occurs before the first one is cleared [see 250.4(A)(4)].

Effective Ground-fault Current Path. Section 250.4(A)(5), Effective Ground-Fault Current Path, provides requirements relative to one of the most critical elements of the grounding safety system. This section requires that the fault-current path: be permanent and electrically continuous; be capable of safely carrying the maximum fault current likely to be imposed on it; and have sufficiently low impedance to facilitate the operation of the overcurrent devices under fault conditions.

Each of these points is important and worthy of discussion. First, a permanent, reliable and electrically continuous grounding system is vital to the

Figure 1-22. Human Completing Grounding Circuit

Figure 1-23. Human Completing Grounding Circuit

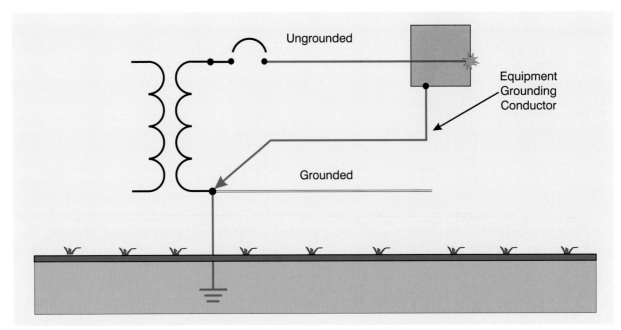

Figure 1-24. Purpose of Equipment Grounding Conductor

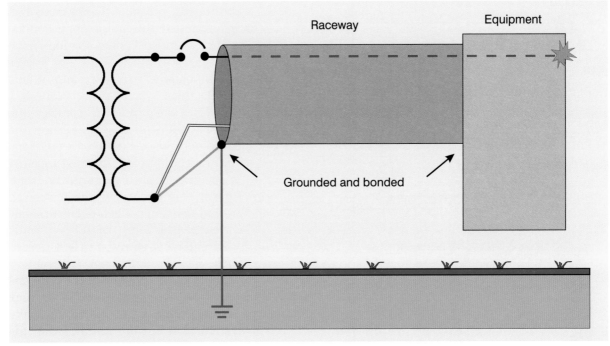

Figure 1-25. Purpose of Bonding Enclosures

overall safety of the electrical system. This includes both a stable voltage reference and provides an effective path for fault current due to abnormal conditions. Intermittent connections are like an unpredictable earthquake, waiting to wreak havoc on the unsuspecting. The grounding system, including all connections, whether a wire, conduit, equipment enclosure, or other element of the path, must be permanent, electrically con-

tinuous, and have all connections made up tightly.

If the insulation system of other conductors of the system, like ungrounded (hot) or grounded circuit conductors, fail, it usually is obvious since sometimes such as a piece of equipment or an appliance stops functioning. This is not true of the equipment grounding path. Usually, a failure of the ground-fault path is not known until someone is exposed to an electric shock. Also, the conductors of the equipment grounding conductor system normally only carry current in a fault situation. At this time, the equipment grounding conductor path will usually carry far more current than the ungrounded (hot) or grounded conductor (often a neutral) typically carry.

Providing adequate capacity for the fault current that will flow in the circuit is also of paramount importance. The minimum size system of the grounded conductor, bonding conductor and equipment grounding conductor are given in several locations, many of them in Article 250. Obviously, a grounding conductor that burns off (fuses) due to excessive current flow while it is carrying out its intended purpose is of little value in the safety system. The sizes of equipment grounding conductors given in Table 250.122 are the minimum size. There are cases where the withstand rating of conductors is exceeded by the available fault current and larger conductors are necessary (see chapter eleven for additional information on the subject).

The ground-fault path that is permanently connected and of adequate capacity is of little value if it does not have low impedance (measured in ohms). A high-impedance path provides resistance to the system, thus limiting the amount of current that can flow over the circuit. This allows a voltage above ground to be present on faulted equipment which may then present a shock hazard (E = I x R). The *NEC* stops short of specifying the maximum impedance acceptable for the ground-fault path, except to say that the path must have an impedance that is low enough to facilitate the operation of the overcurrent protective device. Every circuit has different characteristics associated with it that contribute to the impedance of the particular circuit and, thus, the impedance value of the grounding circuit will vary, but must be kept as low as possible in all grounding circuits (see chapter eleven for additional information on the subject).

The goal of good design of the ground-fault path is to provide a permanent and adequate path of low impedance so enough current will flow in the circuit to cause a circuit breaker to trip or a fuse to open. If the opening of the breaker or fuse on the line side of the faulted circuit does not take place rapidly, thus taking the faulted equipment

off the line, the grounding system will have failed to perform its critically important function. This failure may result in greater equipment damage, possibly fires or injury to personnel.

Electrical System Design

The ultimate goal of electrical system design is to prevent all faults, including ground faults, from occurring. This is accomplished by proper design, installation, operation and maintenance of electrical equipment and systems. System conductors that are ungrounded are separated from grounded conductors and grounded equipment by insulation. This insulation can be in the form of thermoplastic, thermosetting or other similar insulating medium applied to wires or busbars or by separating conductors having a potential (voltage) between them. This separation is usually accomplished by mounting busbars on insulators. The air then becomes the insulation between phases and a grounded enclosure. However, it is possible to

Section 110.7 requires that electrical wiring installations shall be free from short circuits and from grounds other than as required or permitted in Article 250.

have a failure or breakdown in insulation systems in any electrical installation. Therefore, ground-fault protection must be provided to safely clear line-to-ground faults which may occur.

Section 110.7 requires that electrical wiring installations shall be free from short circuits and from grounds other than as required or permitted in Article 250. Though not widely practiced, it is wise to test the insulation of all installations to be certain they are free from unintended short-circuits and/or ground-faults before the wiring is energized. This can be done with some means of continuity testing which may be a battery and a bell and leads (the origin of the term *ring-out*). An ohmmeter or megohm-meter (megger) can also be used. This is more than a simple continuity test. A current is passed through the conductor under test and the amount of leakage through the insulation is measured. Testing of the insulation of the system before it is energized is often required in specifications prepared by architects or engineers for commercial and industrial jobs.

Also, the impedance (rather than the resistance) of the equipment ground-fault return path can be verified by testing instruments. This is especially true for branch circuits where plug-in type testers are available and greatly simplify the testing process.

Designing Electrical Systems for Safety

Electrical systems need to be designed to be certain they are adequate for the loads to be served by them. The *NEC* requires that electrical services are to be adequate for the calculated load in accordance with Article 220 [see 230.42(A)]. Methods for calculating the minimum capacity for electrical services are found in Article 220 rather than in Article 230. The calculation method for determining the minimum size of the grounded service conductor (often a neutral) is found in 220.22. Sec-

> Measurements of insulation resistance can best be made with a megohm-meter insulation tester. These instruments are available from several manufacturers and vary in cost and features.

tion 250.24(B)(1) specifies the minimum size grounded service conductor.

The *Code* does not require that electrical systems be designed with any additional capacity for future load expansion [see 90.1(B)]. From a practical standpoint, some additional capacity should be provided for when the system is first installed. Some designers plan the electrical system for at least 25 percent spare capacity.

Electrical Systems Abused

Electrical systems are intended to provide reliable service for many years. That is generally the case, provided the system is designed for the load to be carried, and the system is installed in a proper manner. Overloading the system is one of the major abuses. With the expanded use of new or more modern electrical appliances, larger

electrical distribution systems are often required. It is not uncommon to find temporary power taps, extra wires placed under circuit breaker terminals, or splices being made in attics or in crawl spaces to add extra electrical equipment to the electrical system. Often, extra fuse or circuit breaker panels are added to existing systems without taking proper steps to ensure that the conductors and equipment supplying the loads are adequate.

In industrial or commercial installations, additional equipment is often added to existing electrical systems. At times, this is done without consideration for whether the existing system is adequate for the additional load.

In addition, electrical systems may be exposed to overvoltage from lightning or power-line crosses, insulation failure in high-to-low voltage transformers, and to short circuits and ground faults. In some cases, electrical systems are abused until they fail. The opening of a fuse or tripping of a circuit breaker often is the first indication of a failure in the electrical system.

Major Problems in Electrical Systems

The major cause of trouble in an electrical distribution system is insulation failure. The insulation may be air, such as in busways, switchboards and motor control centers, where clearance between uninsulated busbars and grounded metallic electrical equipment is maintained by air space. The clearance is maintained by insulators in the form of rubber, ceramic or thermoplastic, or insulation that is applied directly to the conductors. Insulation for wire or bus is usually rubber, thermoplastic or thermosetting material.

Insulation failure may result in two kinds of faults: line-to-line or line-to-ground. The least likely failure is line-to-line or between any two conductors of the system, that is, from one phase conductor to another or from one phase conductor to the neutral or grounded conductor. Experience has shown that most insulation failures (as high as 80 percent) are line-to-ground faults or from one phase conductor to the conductor enclosure or equipment.

While the entire *National Electrical Code*, and other safety electrical codes, like the *Canadian Electrical*

Code, are developed and updated to provide protection against electrical fires and shocks, they are not design specifications. However, by following the rules of these codes, one will have an installation that is essentially free from hazard, but not necessarily efficient, convenient, or adequate for good service or future expansion (see 90.1).

Insulation Resistance

Previous editions of the *Code* (e.g., up to the 1965 *NEC*) contained recommended values or results for testing insulation resistance. It was found that the resistance values were incomplete and not suf-

Photo 1-1. Insulation Resistance Tester.   **Courtesy of ETI**

ficiently accurate for use in modern installations, and the recommendation was to delete them from the *Code*. However, basic knowledge of and the need for insulation-resistance testing is important.

Measurements of insulation resistance can best be made with a megohm-meter insulation tester. These instruments are available from several manufacturers and vary in cost and features. As measured with such an instrument, insulation resistance is the resistance to the flow of direct current (usually at 500 or 1000 volts for systems of 600 volts or less) through or over the surface of the insulation in electrical equipment. The test is accomplished by energizing the conductor, wire or busbar with the potential and measuring the current that "leaks" through the insulation. By Ohm's Law, the insulation resistance is $R = E \div I$. The results are displayed in ohms or megohms, but insulation readings will be in the megohm range.

The insulation-resistance test is nondestructive, quite different from a high-voltage dielectrie insulation or breakdown test (often called a "hi-post" test). It is made with direct current rather than alternating current and is not a measure of dielectric strength as such. However, insulation resistance tests assist greatly in determining when and where not to apply high voltage.

In general, insulation resistance decreases with increased size of a machine or length of cable.

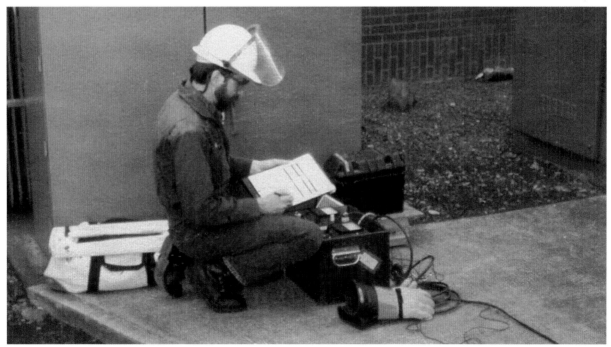

Photo 1-2. Insulation Resistance Testing Process   **Courtesy of Electro-Test Inc.**

This occurs because there is more insulating material in contact with the conductors and the frame, ground, or sheath. The greater volume of insulating material allows more total "leakage" and, therefore, lowers the overall insulation resistance reading.

Insulation-resistance usually increases with higher voltage rating of apparatus because of increased thickness of insulating material. Also different types of insulation, for example, air versus thermoplastic, will have different levels of insulation resistance.

Insulation-resistance readings are not only quantitative, but are relative or comparative as well, and since they are influenced by moisture, dirt, and deterioration of the insulation, they are reliable indicators of the presence of those conditions. Cable and conductor installations present a wide variation of conditions from the point of view of the resistance of the insulation. These conditions result from the many kinds of insulating materials used, the voltage rating or insulation thickness, and the length of the circuit involved in the measurement. Furthermore, such circuits usually extend over great distances and may be subject to wide variations in temperature, which will have an effect on the insulation resistance values obtained. The terminals of cables and conductors will also have an effect on the test values unless they are clean and dry, or guarded.

Records of insulation-resistance should be maintained so failures of conductors or equipment can be predicted. To make valid comparisons or to indicate trends, the recorded values must be corrected to a standard temperature, typically 20°C and differences in the relative humidity must be accounted for. Equipment or conductors can then be repaired or replaced to reduce plant downtime which often happens at inconvenient times causing expensive loss of production.

Differentiate Between a *Short Circuit* and a *Ground Fault*

It is common practice to call all faults or failures in the electrical system *short*s or *short circuits*. That can lead to misunderstanding, and has done so, because the terms are not used properly. A short circuit and a ground fault are different, although they both

stem from insulation failure. To end this confusion, the definitions of the two terms follow:

"Short circuit. A conducting connection, whether intentional or accidental, between any of the conductors of an electrical system whether it be from line-to-line or from line to the grounded conductor." The grounded conductor often is also the system or circuit neutral. The short circuit can be a solid or "bolted" connection or it may be an arcing fault completing the path through a short air space."

In the case of a short circuit, the failure may be from one phase conductor to another phase conductor or from one phase conductor to the grounded conductor or neutral. For either condition, the maximum value of fault current is dependent on the available capacity the system can deliver to the point of the fault. The maximum value of short-circuit current from line-to-neutral will vary depending upon the distance from the source

Figure 1-26. Short Circuit

Figure 1-27. Ground Fault

to the fault and the impedance of the path. The available short-circuit current is further limited by the impedance of the arc where one is established, plus the impedance of the conductors to the point of short circuit, (see figure 1-26.)

"Ground fault. A conducting connection, whether intentional or accidental, between any of the conductors of an electrical system and the grounding conductor or conducting material which encloses the conductors (such as conduit) or any conducting material that is grounded or that may become grounded."

In the case of a ground fault there is a fault from a hot or ungrounded phase conductor to the conductor enclosure or from the grounded phase conductor to the conductor enclosure (wire-to-conduit, wire-to-motor frame, and so forth). It is not common practice to refer to a conductor that is intentionally grounded, such as a grounded phase or system conductor or equipment grounding conductor, as being in a ground-fault condition. However, a grounded system conductor, such as a neutral, that is not permitted to be grounded again past the service can be considered a ground fault. This condition is particularly important with equipment ground-fault protection systems where grounding the system grounded conductor downstream from the point of protection will desensitize the protection system.

It is often the case that a fault may easily involve both conditions, short circuit and ground fault. The fault may start as a ground fault or line-to-ground fault and develop into a short circuit or phase-to-phase fault. A fault may also start as a short circuit and expand to a ground fault. For electrical systems of 120-volts-to-ground, the circuit protective devices will usually clear the fault or it will be extinguished when the alternating voltage passes through zero. For 277-volts-to-ground systems, destructive arcing faults can be easily maintained by the system voltage. Equipment ground-fault protection systems have been designed to address this problem. In some cases, the use of this equipment is required by the *NEC* (see chapter fifteen of this text for additional information on this subject).

NEC Requirements
Article 250 covers the subject of grounding.

Grounding is practiced for the protection of electrical installations, which in turn protects the buildings or structures in which the electrical sys-

> Electrical equipment intended to interrupt current at fault levels must be designed and installed so it has an adequate rating for the nominal circuit voltage and the fault current that is available at its line terminals.

tems are installed. Persons and animals that may come into contact with the electrical system or that are in these buildings or structures are also protected if the grounding system is installed and maintained properly. The *NEC* does not imply here that grounding is the only method that can be used for the protection of electrical installations, people or animals. Insulation, isolation and guarding are suitable alternatives under certain conditions.

Grounding of specific equipment is covered in several other articles. For example, health care facilities as covered in Article 517 and swimming pools in Article 680. Refer to the indices of this text, the *National Electrical Code*, or *Ferm's Fast Finder Index* for specific sections of the *Code* that apply to the equipment in question.

Circuit Impedance and Other Characteristics
Electrical equipment intended to interrupt current at fault levels must be designed and installed so it has an adequate rating for the nominal circuit voltage and the fault current that is available at its line terminals (see 110.9). Failure to comply with this important requirement can result in disastrous consequences. This includes the destruction of electrical equipment, such as circuit breakers or fuses that are themselves designed to protect the system.

Equipment such as time clocks, motor controllers, lighting contactors, and so forth, that are intended to interrupt current at only rated load must be suitable for the nominal circuit voltage and current that must be interrupted. These devices must also withstand the higher fault current levels until an upstream protective device, such as a fuse or circuit breaker, opens.

In addition, 110.10 requires that, "The overcurrent protective devices, the total impedance, the component short-circuit current ratings, and other characteristics of the circuit to be protected shall be selected and coordinated to permit the circuit-protective devices used to clear a fault to do so without extensive damage to the electrical components of the circuit. This fault shall be assumed to be either between two or more of the circuit conductors, or between any circuit conductor and the grounding conductor or enclosing metal raceway. Listed products applied in accordance with their listing shall be considered to meet the requirements of this section."

Listed equipment such as fuses, including the associated switch, and circuit breakers have nominal voltage and short-circuit current ratings. This is the maximum voltage and current for which the equipment is designed. In addition, other equipment such as busways may be marked similar to "Busways and associated fittings marked 'Short Circuit Current Rating(s) Maximum RMS Symmetrical Amperes _____ Volts _____' have been investigated for the rating indicated."[5] Electrical system components such as metering equipment, switchboards, panelboards, and motor control centers have short-circuit ratings that must not be exceeded.

Overcurrent protective devices such as circuit breakers and fuses should be selected and installed to ensure that the short-circuit current rating of components of the system will not be exceeded should a short-circuit or ground-fault occur. The overcurrent device that is selected must not only safely interrupt the fault current that is available at its line terminals, it also must limit the amount of energy that is let through to downstream equipment so as not to exceed its rating.

Electric utilities usually provide information on the short-circuit current (often referred to as the available fault current) that is available at the secondary terminals of the transformer or the service equipment. The short-circuit current must then be calculated to the point on the electrical system under consideration. Methods for calculating the available fault current at any point on

Equipment intended to interrupt current at fault levels shall have an interrupting rating sufficient for the nominal circuit voltage and the available short circuit current at the line terminals of the equipment.

Equipment intended to interrupt current at other than fault levels shall have an interrupting rating at nominal circuit voltage and current to be interrupted

Figure 1-28. Interrupting Ratings of Equipment

the electrical system can be found in literature from many circuit breaker or fuse manufacturers. See also the Institute of Electrical and Electronic Engineers (IEEE) "Buff Book" (IEEE 242)[6], Underwriters Laboratories product safety standard and "guide card information" for the equipment under consideration[7], and *Ferm's Fast Finder Index*[8] where, in addition to the point-to-point method, many pages of calculations are presented in an easy-to-use table form.

Other components of the electrical system, such as wire and cable, have published withstand ratings, in addition to allowable ampacity ratings, that must be considered for safety. Withstand ratings are not marked on the wire or cable but can be determined from the manufacturer's data, IEEE Standards, or calculated as described in chapter eleven.

As can be seen upon review of this literature, a conductor can safely carry much more current than allowed in the allowable ampacity table if the time the current flows is reduced. For insulated conductors, the ampacity of the conductor can be thought of as the conductor's long-time or continuous rating and the withstand-rating as the conductor's short-time rating. This is most important when considering the proper size of equipment grounding conductors to use, as they do not have overcurrent protection on their line side and must carry fault-current levels until an overcurrent device opens (see chapter eleven for additional information on this subject).

[1] NFPA 70, *National Electrical Code* 2002, Article 100, (National Fire Protection Association, Quincy, MA, 2002), p. 70-35.

[2] NFPA 70, Article 250, p. 70-95.

[3] NFPA 70, Article 100, p. 70-36.

[4] NFPA 70, Article 100, p. 70-35.

[5] General Information for Electrical Equipment Directory, "Busways and Associated Fittings (CWFT)," (Underwriters Laboratories Inc., Northbrook, IL, 2000), p. 6

[6] Available from IEEE Service Center, 445 Hoes Lane, P.O. Box 1331, Piscataway, NJ 08855-1331, (800)678-4333.

[7] Available from Publications Stock, Underwriters Laboratories, 333 Pfingsten Road, Northbrook, IL 60062-2096, (847)272-8800 Ext. 42612 or 42622.

[8] Available from International Association of Electrical Inspectors, P.O. Box 830848, Richardson, TX 75083-0848, (800) 786-4234, Fax (972) 235-6858.

• Overcurrent protective devices,
• The total impedance,
• The component short-circuit current rating,
• Other characteristics of circuit,

Selected and coordinated to permit circuit protective devices to clear fault without extensive damage to the electrical components of the circuit

Listed products applied in accordance with their listing are considered to meet requirements of Section 110.10.

Fault assumed to be:
• Between two or more of the circuit conductors
• Between any circuit conductor and the grounding conductor or enclosing metal raceway

Figure 1-29. Circuit Impedance and Other Characteristics

Chapter One: The questions included here were developed using material included in this chapter. The answers can be found by reviewing the text. It is also important that students make use of the 2002 *NEC*, where many of the answers can be found.

1. A grounded metal object is forced to take the same zero potential as the ____.
   a.   electrode
   b.   metal raceway
   c.   earth
   d.   system

2. Electrical systems are grounded to limit the voltage to ground during normal operation and to prevent excessive voltages because of ____, line surges or unintentional contact with higher voltage lines.
   a.   low voltage
   b.   overloads
   c.   loose connections
   d.   lightning

3. The effective ground-fault current path must be installed in a manner that creates a path that is ____.
   a.   permanent and electrically continuous
   b.   has the capacity to conduct safely the maximum ground-fault current likely to be imposed on it
   c.   has sufficiently low impedance to limit the voltage to ground and to facilitate the operation of the circuit protective devices in the circuit
   d.   all of the above

4. Insulation is considered to be air where conductors are mounted on insulators such as on poles, in busways or in equipment, or rubber or ____ material or wire or bus.
   a.   listed
   b.   approved
   c.   acceptable
   d.   thermoplastic

5. Measurements of insulation resistance can best be made with a ____ insulation tester.
   a.   voltage
   b.   megohm-meter
   c.   low voltage
   d.   high voltage

6. Insulation resistance is the resistance to the flow of direct current (usually at ____ or ____ volts for systems of 600 volts or less) through or over the surface of the insulation in electrical equipment megohm range.
   a.   500 or 1000
   b.   600 or 1500
   c.   800 or 1800
   d.   900 or 2000

7. An insulation-resistance test is made with direct-current rather than alternating-current, and is not a measure of ____ strength as such.
   a.   conductor
   b.   dielectric
   c.   terminal
   d.   equipment

8. A conducting connection, whether intentional or accidental, between any of the conductors of an electrical system whether it be from line-to-line or to the grounded conductor defines a ____.
   a.   phase fault
   b.   short circuit
   c.   overload
   d.   line disorder

9. A conducting connection, whether intentional or accidental, between any of the conductors of an electrical system and the conducting material which encloses the conductors or any conducting material that is grounded or that may become grounded is defined as a ____.
   a.   phase fault
   b.   overload
   c.   system ground
   d.   ground fault

10. In a dc electrical circuit, the total opposition to the flow of current due to primarily ____.
   a.   current
   b.   voltage
   c.   impedance
   d.   resistance

11. In an electrical circuit, the current flowing in the circuit will always attempt to return to which of the following:
   a.   the earth
   b.   the service
   c.   the source
   d.   the load

12. For current to flow there must be a complete ____.
   a.   grounding conductor
   b.   conduit system
   c.   circuit
   d.   overcurrent protective device

13. Current in an electrical circuit will take _____ to return to the source.
   a.   only the path of least resistance
   b.   any and all paths available
   c.   a high impedance path only
   d.   the path in the earth

14. The total opposition to the flow of current in an ac circuit is the_____ of the circuit.
   a.   resistance
   b.   impedance
   c.   capacitance
   d.   inductance

15. The severity of an electrical shock is dependant on the _____.
   a.   voltage of the circuit
   b.   resistance of the circuit
   c.   the amount of current and how long it flows through the body
   d.   the path for current back to the source.

16. The higher the impedance or resistance of a circuit, the _____ the amount of current flow in the circuit.
   a.   lower
   b.   higher
   c.   same
   d.   different

17. The NEC in recent cycles has continued to migrate away from the use of the _____ conductor for grounding of equipment downstream of the main bonding jumper at the service or bonding jumper at a separately derived system.
   a.   equipment grounding
   b.   grounded
   c.   bonding
   d.   grounding electrode

18. Alternatives for grounding include which of the following:
   a.   guarded
   b.   insulated
   c.   isolated
   d.   all of the above

19. A 480-volt, 3-phase circuit is connected to an electric heater rated at 480 volts, 3-phase/7500 watts. How many amperes will flow in the circuit?
   a.   13.6
   b.   9.03
   c.   15.7
   d.   27.9

To Ground or Not To Ground

When we refer to electrical systems, generally, we are considering the system as a unit of a specific voltage (potential) and often amperage (current or capacity). For example, a common system is 480Y/277-volts, 3-phase, 4-wire at some capacity such as 1600 amperes. Often, the premises are supplied by an electric utility. These systems are either over 600 volts (often referred to as *primary systems*) or 600 volts or less (often referred to as *secondary systems*). In addition, the system is referred to as single-phase or three-phase. Some two-phase systems exist but have fallen out of favor in recent years. Most systems are alternating current at 60 hertz although some direct-current systems are in use, primarily in industrial applications.

In addition, *systems* are produced at various voltages and phases on-site. The most common way to produce an electrical system is through a transformer or generator. In the case of a transformer, other than an autotransformer, one electrical system ends at the primary winding(s) and another system begins at the secondary winding(s).

An example is a plant that has a service at 480Y/277-volts and it is necessary to supply receptacle outlets at 120 volts. A single-phase, 480-volt to 240-volt transformer is installed with the winding on the secondary center-tapped to result in 120/240 volts single-phase. In this example, the 480-volt system ends at the primary windings and a new 120/240-volt system begins at the secondary winding. In other cases, a 208Y/120-volt system is derived by installing the appropriate transformer(s). Similar examples can be shown for generator-supplied systems, though, for these, we consider the system supplied as the new system. A new source is developed by the transformer or generator. For the purposes of grounding, the treatment of the grounded system conductor (often a neutral) determines whether the system is considered to be separately derived for grounding purposes (see chapter twelve for more information on this subject).

Definitions

Ground Fault. "An unintentional, electrically conducting connection between an ungrounded conductor of an electrical circuit and the normally non-current-carrying conductors, metallic enclosures, metallic raceways, metallic equipment, or earth."

Ground-Fault Current Path. "An electrically conductive path from the point of a ground fault on a wiring system through normally non-current-carrying conductors, equipment, or earth to the electrical supply source."

Examples of ground-fault current paths are of any combination of equipment grounding conductors, metallic raceways, metallic sheaths of cables, electrical equipment, and any other electrically conductive material such as metal water and gas piping, steel framing members, stucco mesh, metal ducting, reinforcing steel, shields of communications cables, and the earth itself.

Effective Ground-Fault Current Path. "An intentionally constructed, permanent, low-impedance, electrically conductive path designed and intended to carry current under ground-fault conditions from the point of a ground fault on a wiring system to the electrical supply source."[1]

Effective ground-fault current paths are created by effectively bonding together all of the electrically conductive materials that are likely to be energized by the wiring system. Effective bonding is accomplished through the use of equipment grounding conductors, bonding jumpers or bonding conductors, approved metallic raceways, connectors and couplings, approved metallic sheathed cable and cable fittings, and other approved devices. A ground-fault path is effective when it will carry the maximum ground-fault current likely to be imposed on it.

The 2002 *NEC* was revised to include the above definitions of the terms used in 250.4. These definitions work to contribute to a clearer understanding of some important perfor-

> Effective ground-fault current paths are created by effectively bonding together all of the electrically conductive materials that are likely to be energized by the wiring system.

mance characteristics of the ground faults and an effective ground-fault current path. These terms work to enhance the ability of proper application of the prescriptive grounding and bonding requirements in the later parts of Article 250.

Grounded Systems

Section 250.4 has been divided into two main parts. Part A includes the performance requirements for grounded systems, and Part B includes the performance requirements for ungrounded systems.

"**250.4 General Requirements for Grounding and Bonding.** The following requirements identify what grounding and bonding of electrical systems are required to accomplish. The prescriptive methods contained in Article 250 shall be followed to comply with the performance requirements of this section.

"(A) Grounded Systems.

(1) Electrical System Grounding. Electrical systems that are grounded shall be connected to earth in a manner that will limit the voltage imposed by lightning, line surges, or unintentional contact with higher-voltage lines and that will stabilize the voltage to earth during normal operation.

(2) Grounding of Electrical Equipment. Non-current-carrying conductive materials enclosing electrical conductors or equipment, or forming part of such equipment, shall be connected to earth so as to limit the voltage to ground on these materials.

(3) Bonding of Electrical Equipment. Non-current-carrying conductive materials enclosing electrical conductors or equipment, or forming part of such equipment, shall be connected together and to electrical supply source in a manner that establishes an effective ground-fault current path.

(4) Bonding of Electrically Conductive Materials and Other Equipment. Electrically conductive materials that are likely to become energized shall be connected together and to the electrical supply source in a manner that establishes an effective ground-fault current path.

(5) Effective Ground-Fault Current Path. Electrical equipment and wiring and other electrically conductive materials likely to become energized shall be installed in a manner that creates a permanent, low-impedance circuit capable of safely carrying the maximum ground-fault current likely to be imposed on it from any point on the wiring system where a ground fault may occur to the electrical supply source. The earth shall not be used as the sole equipment grounding conductor or effective ground-fault current path."[2]

Grounding Electrical Systems

Over the years, there have been great debates over the merits of grounding electrical systems versus leaving them ungrounded. This subject is covered at some length in the appendix of this text where the history of the development of the grounding rules in effect today is covered.

Many of the decisions about whether or not to ground electrical systems are made for us. The *National Electrical Code* requires that certain electrical systems falling within the parameters of 250.20 be grounded. Other electrical systems are permitted to be grounded, while some systems are not permitted to be grounded due to special conditions. Some systems today, primarily in the industrial or agricultural sector, are operated ungrounded. Common reasons for choosing to

Part A - Grounded System

Part B - Ungrounded System

The performance requirements of the fault-current paths for each system have been expanded and clarified

The new section 250.4 as revised will now be divided into two parts: (A) Grounded Systems and B) Ungrounded Systems and includes the performance criteria for each system.

Figure 2-1. 250.4: General Requirements for Grounding and Bonding operate an electrical system ungrounded are continuity of service for critical operations and minimizing downtime. These decisions are usually mutually agreed upon between the building owners and operators, the designers and engineering team, and others. The last sentence of 250.20 indicates that whether the system is required to be grounded, or is grounded by choice, the applicable grounding rules for grounded systems must be followed.

Systems That Must Be Grounded

In accordance with 250.20, the alternating-current systems that must be grounded are: (A) alternating-current systems of less than 50 volts, (B) alternating-current systems of 50 volts to 1000 volts, (C) alternating-current systems of 1 kV and over, and (D) separately derived systems. Each of these systems is discussed in the following sections.

(A) Less than 50-volt systems:

The following systems that are less than 50 volts, such as on the secondary of a transformer must be grounded:

(1) Where supplied by transformers if the supply voltage (primary) to the transformer exceeds 150 volts to ground.

(2) Where supplied by transformers if the transformer supply system (primary) is ungrounded.

(3) Where installed as overhead conductors outside of buildings.

(B) AC systems of 50 to 1000 volts:

Alternating-current systems of 50 to 1000 volts that supply premises wiring and premises wiring systems are required to be grounded under any of the following conditions:

(1) Where the system can be grounded so the maximum voltage to ground on the ungrounded conductors does not exceed 150 volts. Typical systems include:
- 120 volt, 1-phase,
- 120/240 volt, 1-phase, 3-wire, and
- 208Y/120 volt, 3-phase, 4-wire.

(2) Where the system is 3-phase, 4-wire wye connected and where the neutral is used as a circuit conductor. Typical systems include:
- 208Y/120 volt, 3-phase, 4-wire, and
- 480Y/277 volt, 3-phase, 4-wire.

(3) Where the system is 3-phase, 4-wire delta connected in which the midpoint of one

phase winding is used as a circuit conductor. Typical systems include:
- 120/240 volt, 3-phase, 4-wire, and
- 240/480 volt, 3-phase, 4-wire.

AC Systems of 1 kV and Over

These ac systems must be grounded if they supply mobile or portable equipment as covered in 250.188. Systems supplying portable or mobile high-voltage equipment, other than substations installed on a temporary basis, are required to comply with the rules in (A) through (F) below.

"**(A) Portable or Mobile Equipment.** Portable or mobile high-voltage equipment shall be supplied from a system that has its neutral conductor grounded through an impedance. Where a delta-connected high-voltage system is used to supply portable or mobile equipment, a system neutral shall be derived." This is usually accomplished by means of a zig-zag grounding transformer (see chapter three of this text for additional information on grounding systems by using zig-zag grounding transformers).

"**(B) Exposed Non-current-Carrying Metal Parts.** Exposed non-current-carrying metal parts of portable or mobile equipment shall be connected by an equipment grounding conductor to the point at which the system neutral impedance is grounded." This point may be at the service or source of a separately derived system.

"**(C) Ground-Fault Current.** The voltage developed between the portable or mobile equipment frame and ground by the flow of maximum ground-fault current shall not exceed 100 volts." This requirement, no doubt, necessitates an engineering study to determine the voltage drop across the equipment grounding conductor.

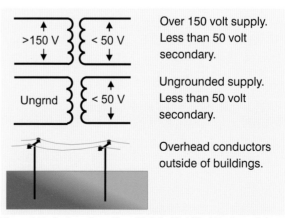

>150 V — < 50 V — Over 150 volt supply. Less than 50 volt secondary.

Ungrnd — < 50 V — Ungrounded supply. Less than 50 volt secondary.

Overhead conductors outside of buildings.

Figure 2-2. Systems Less Than 50 Volts That Must Be Grounded

Where the system can be grounded so the maximum voltage to ground does not exceed 150 volts.

Typical 120/240 volt 1-phase, 3-wire system.

3-phase, 4-wire wye systems where neutral is used as circuit conductor.

3-phase, 4-wire delta systems where the midpoint of one phase winding is used as circuit conductor.

Figure 2-3. Systems That Must Be Grounded 50 to 1000 Volts

"(D) Ground-Fault Detection and Relaying. Ground-fault detection and relaying must be provided to automatically de-energize any high-voltage system component that has developed a ground fault. The continuity of the equipment grounding conductor must be continuously monitored so as to de-energize automatically the high-voltage feeder circuit to the portable or mobile equipment upon loss of continuity of the equipment grounding conductor.

"(E) Isolation. The grounding electrode to which the portable or mobile equipment system neutral impedance is connected shall be isolated from and separated in the ground by at least 6.0 m (20 ft) from any other system or equipment grounding electrode, and there shall be no direct connection between the grounding electrodes, such as buried pipe, fence, and so forth.

"(F) Trailing Cable and Couplers. High-voltage trailing cable and couplers for interconnection of portable or mobile equipment must meet the requirements of Part III of Article 400 for cables and 490.55 for couplers."

This type of equipment is commonly found in mobile rock crushing plants and batch plants. Other applications are for open pit mining op-

erations. Note that self-propelled mobile surface mining machinery and its attendant electrical trailing cable are not covered by the *Code* [see 90.2(B)(2)]. Even though exempted from the *Code*, many requirements for this equipment are incorporated by regulations enforced by the Mine Safety and Health Administration (MSHA).

Other ac systems over 1000 volts are permitted but are not required to be grounded.

(D) Separately Derived Systems.

Electrical systems derived from a battery, a solar photovoltaic system, or from a generator, transformer, or converter windings that have no direct electrical connection, including a solidly connected grounded circuit conductor, to the supply conductors originating in another system, must be grounded if the system that is derived complies with the conditions in (A), (B), or (C) above. The separately derived system then must be grounded as specified in 250.30.

Examples of systems that are separately derived include:

(1) Inverters or batteries such as for uninterruptible power systems.

(2) Solar photovoltaic systems or fuel cells.

(3) Transformers with no direct electrical connection between the primary and secondary.

(4) Generator systems that supply power such as for carnivals, rock crushers or batch plants where the neutral is not connected to the utility system.

(5) Generator systems used for emergency, required standby or optional standby power that have all conductors, including a neutral, isolated from the neutral or grounded conductor of another system usually by a transfer switch.

(6) Ac or dc systems derived from inverters or rectifiers.

• Supplied from system having its neutral grounded through impedance

• Delta-connected system must have neutral derived

Eq. Gnd.

• Equipment grounded where system neutral impedance is grounded

• Max. 100 volts from fault current

• Ground-fault detection and relaying

• Grounding electrode isolated

Figure 2-4. AC Systems of 1 kV and Over Supplying Portable or Mobile Equipment

Transformers with no direct electrical connection between primary and secondary.

Generators with no direct electrical connection with another supply system.

AC or DC systems derived from converter windings.

[Ground per Section 250.30 if required to be grounded by Section 250.20(A) or (B)].

Figure 2-5. Separately Derived Premises Systems That Must Be Grounded

Figure 2-6 illustrates systems that are not separately derived and thus do not fall under the grounding requirements of 250.30. These systems include:

(1) A system supplied by an autotransformer since autotransformers by design have a conductor that is common to both primary and secondary.

(2) Systems from a generator that do not have all conductors, including the grounded system conductor (often a neutral), switched by a switching mechanism in a transfer switch.

The key to determining whether a generator supplied system is to be grounded as a separately derived system is often to examine electrical connections in the transfer switch.

If all the phases and the neutral or grounded conductor are switched by the transfer switch, the generator is a separately derived system that must be grounded in compliance with 250.30. If the neutral or grounded conductor is not switched, the system produced is not to be grounded as separately derived system and the neutral must not be grounded at the generator (see chapter twelve of this text for thorough information on grounding and bonding requirements for separately derived systems).

High-impedance Grounded Systems

High-impedance grounded systems should be considered where the plant operation cannot tolerate electrical system disruption from the first ground fault. This system has all the advantages of an ungrounded system, so far as operation of the plant or system with one phase faulted to ground is concerned, with none of the disadvantages of an ungrounded system. While the initial cost of the system can be significant, it can pay for itself many times over the installation cost by operational savings in more reliable plant operation (see chapter four of this text for additional information on high-impedance grounded neutral systems).

Ungrounded Systems

Ungrounded systems are derived electrical systems that have no conductor of the system grounded either solidly or through any resistance or impedance. Basically there is no potential between any of the system conductors and ground because there is not a connection of any conductor of the system to ground.

Circuits or Systems That Are Not Permitted To Be Grounded.

Only three circuits or systems are not permitted to be grounded by 250.22. They are as follows:

(1) Circuits for electric cranes that operate over combustible fibers in Class III locations (see 250.22(1) and 503.13). This action reduces the likelihood that sparks from faulted equipment will fall onto combustible fibers below the crane causing a fire.

(2) For health care facilities, those isolated power circuits in hazardous (classified) inhalation anesthetizing locations are required to be supplied by an isolation transformer or other ungrounded source [see 517.61(A)(1)]. In addition, receptacles and fixed equipment in "wet locations" of hospital patient care areas as defined in 517.3 must be protected by ground-fault circuit-interrupter devices where interruption of power to equipment under fault conditions can be tolerated. Where interruption of power under fault conditions cannot be tolerated, protection of these receptacles and fixed equipment is to be supplied from isolated, ungrounded sources by ungrounded electrical systems [see 250.22(2) and 517.20(A)].

(3) Circuits for electrolytic cells as provided in Article 668. Equipment located or used within the electrolytic cell line working zone or associated with the cell line dc power

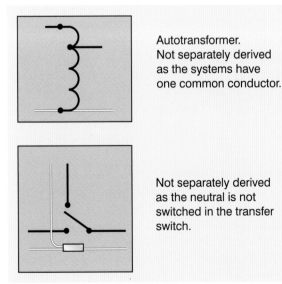

Autotransformer. Not separately derived as the systems have one common conductor.

Not separately derived as the neutral is not switched in the transfer switch.

Figure 2-6. Electrical Systems That Are Not Separately Derived Systems

circuits are not required to comply with Article 250 [see 250.22(3) and 668.3(C)(3)].

(4) Lighting systems as provided in 411.5(A). Article 411 covers lighting systems operating at 30 volts or less. The secondary circuits are required to be insulated from the branch circuit by an isolating transformer. Secondary circuits from these transformers are not permitted to be grounded.

Section 250.4(B) Ungrounded Systems includes four subsections (1) Grounding of Electrical Equipment (2) Bonding of Electrical Equipment (3) Bonding of Electrically Conductive Materials and Other Equipment, and (4) Path for Fault Current.

"**(B) Ungrounded Systems**

(1) Grounding of Electrical Equipment. Non-current-carrying conductive materials enclosing electrical conductors or equipment, or forming part of such equipment, shall be connected to earth in a manner that will limit the voltage imposed by lightning or unintentional contact with higher-voltage lines and limit the voltage to ground on these materials.

(2) Bonding of Electrical Equipment. Non-current-carrying conductive materials enclosing electrical conductors or equipment, or forming part of such equipment, shall be connected together and to the supply system grounded equipment in a manner that creates a permanent, low-impedance path for ground–fault current that is capable of carrying the maximum fault current likely to be imposed on it.

(3) Bonding of Electrically Conductive Materials and Other Equipment. Electrically conductive materials that are likely to become energized shall be connected together and to the supply system grounded equipment in a manner that creates a permanent, low-impedance path for ground-fault current that is capable of carrying the maximum fault current likely to be imposed on it.

(4) Path for Fault Current. Electrical equipment, wiring, and other electrically conductive material likely to become energized shall be installed in a manner that creates a permanent, low-impedance circuit from any point on the wiring system to the electrical supply source to facilitate the operation of overcurrent devices should a second fault occur on the wiring system. The earth shall not be used as the sole equipment grounding conductor or effective fault current path."[3]

Ungrounded systems are often utilized with industrial electric furnaces where continuity of the service is critical, and an outage due to a ground fault cannot be tolerated.

Ground detectors are usually installed on these systems to monitor the system for ground faults, that can be cleared prior to a second ground fault on the system.

Photo 2-1. Electrical Industrial Furnaces

FPN 1: A second fault that occurs through the equipment enclosures and bonding is considered a ground fault.[4]
FPN 2: See Figure 250.4 for information on the organization of Article 250.[5]

Electrical Systems Operated Ungrounded

Electrical systems that fall outside the requirements for grounding in 250.20(A), (B), (C) or (D) are often operated ungrounded. At times, the plant owner or engineer or combination will collectively choose to operate electrical systems ungrounded. These systems usually are found in industrial or agricultural applications and often are either 240-volt or 480-volt, three-phase, three-wire systems. Some higher voltage systems are also used in heavy-industrial applications. Where ungrounded systems are installed, the engineering decision is often based on an effort to obtain an additional degree of service continuity while providing equal and effective means for safety of equipment by the use of ground-fault indicator equipment.

Typical systems that are operated ungrounded include:
- 240 volt, 3-phase, 3-wire, delta connected
- 480 volt, 3-phase, 3-wire, delta connected
- 2300 volt, 3-phase, 3-wire, delta connected
- 4600 volt, 3-phase, 3-wire, delta connected
- 13,800 volt, 3-phase, 3-wire, delta connected

Since the system is ungrounded, the occurrence of the first ground fault (not a short circuit or line-to-line fault) on the system will not cause an overcurrent protective device for the service, feeder, or branch circuit to open. This fault does, however, ground the system but usually accidentally and through ineffective means (higher impedance) and in unspecified and uncontrolled lo-

cations. In essence, this system becomes a corner grounded delta system. Little, if any, current will flow in this first ground fault (see chapter three of this text for additional information on equipment and enclosure grounding requirements for these ungrounded systems).

When an ungrounded system with one ground fault experiences a second ground fault on a different phase, the result is a phase-to-phase fault on the system. This will usually cause one or more overcurrent protective devices to open provided adequate current flows in this path to cause the circuit protective devices to operate. A major concern for this type of system happens where the first and second faults may be located some distance apart. Often, these faults are from line-to-conduit or metallic enclosure such as wireway, pull boxes, busway or motor terminal housings in different parts of the plant. Where this occurs, a relatively high-impedance path for current flow is often established. In some cases, it has been found

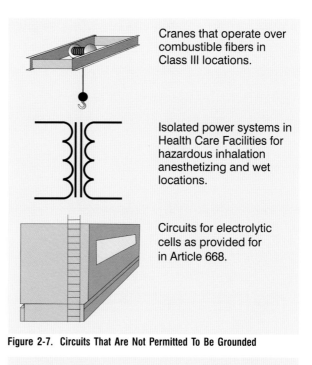

Cranes that operate over combustible fibers in Class III locations.

Isolated power systems in Health Care Facilities for hazardous inhalation anesthetizing and wet locations.

Circuits for electrolytic cells as provided for in Article 668.

Figure 2-7. Circuits That Are Not Permitted To Be Grounded

Typical Ungrounded Systems:
240 Volt, 3-Phase, 3-Wire
480 Volt, 3-Phase, 3-Wire
2300 Volt, 3-Phase, 3-Wire
4600 Volt, 3-Phase, 3-Wire
13.8 kV, 3-Phase, 3-Wire

Figure 2-8. Systems Permitted to be Ungrounded

that a great deal of heat along with arcing and sparking is produced along this fault path due to loose connections or inadequate bonding. Every conduit coupling and locknut connection to enclosures in the fault-current path must be tight to provide an adequate and low-impedance path and to reduce this arcing and sparking.

It is important for safety reasons, as well as for system continuity, that maintenance personnel locate and eliminate ground faults when first identified on ungrounded systems. This should be done as soon as practical and

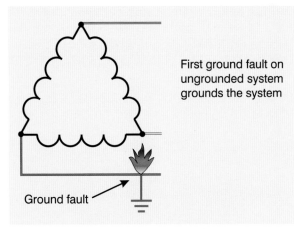

Figure 2-9. First Ground Fault on Ungrounded System

Figure 2-10. Second Ground Fault on Ungrounded System

especially before the second ground fault occurs on the system.

Ground-detection Indicator Systems

Commercially manufactured ground-detection equipment is available. This equipment, which can be located at the service equipment or in feeder distribution panels, can be set to operate an overcurrent relay or shunt-trip circuit breaker or to operate a visual or audible signaling system to indicate a ground-fault condition.

This monitoring equipment is not required by the electrical code but is a good practice from a plant management perspective. Successful operation of an ungrounded electrical system depends on good system maintenance and prompt elimination of the first ground fault. Sophisticated equipment is now available to identify the part of the electrical system where the fault is located while the system is energized. This significantly speeds detection and repair.

Ground Detector Lights

Often, ground detector lights or a neutralizer or potentializer plug are installed to indicate that a ground fault has occurred on the ungrounded system. The $7\frac{1}{2}$ watt indicator lights are connected to the lines through 18,000 ohm resistors. A tap is made to each resistor to give 120 volts to the lamp. The lamp burns until its phase goes to ground, at which time there is no or little potential across the lamp, and it stops glowing, thus identifying the faulted phase (see figure 2-11).

This indication is intended to alert the maintenance personnel to the problem so the ground-fault can be corrected during hours when the plant is not operating. The plant can continue to operate with one-phase grounded, thus preventing costly production down-times. In some cases, down-time in production plants can cost thousands of dollars per minute.

A wide variety of "homemade" systems have been installed over the years, some of which are downright dangerous. Some of these have consisted of nothing more than two "barn socket" lampholders with 240-volt lamps that are connected in series from phase-to-ground on 480-volt systems. Only proven and tested designs for ground detection systems should be used.

Figure 2-11. Ground Detector Lights

Ungrounded System Problems

An ungrounded system exists only in theory, in a laboratory or at the electrical distribution transformers hanging on the pole before connection to the plant electrical system. In the real world, ungrounded systems having insulated conductors installed in metallic enclosures are grounded to varying degrees through the distributed leakage capacitance of the system. Physically, a capacitor exists whenever an insulating material separates two conductors that have a difference of potential between them.

When any conductor is installed in close proximity to grounded metal, there is a capacitance between them that is increased as the distance between the conductors is reduced. In 600-volt systems, the two greatest sources of capacitance to ground are conductors in metal conduit and windings, such as for motors and transformers. In both cases, conductors are separated from grounded metal by fairly thin insulation. The capacitance to ground is known as the leakage capacitance, and the current flowing from the conductors to ground is known as the leakage current or charging cur-

rent. This capacitance is distributed throughout the electrical system but electrically acts like it is a single, lumped capacitance.

Disadvantages of operating systems ungrounded include but are not limited to the following:

1. Power system overvoltages are not controlled. In some cases, these overvoltages are passed through transformers into the premises wiring system. Some common sources of overvoltages include: lightning, switching surges and contact with a high-voltage system.
2. Transient overvoltages are not controlled, which, over time, may result in insulation degradation and failure.
3. System voltages above ground are not necessarily balanced or controlled.
4. Destructive arcing burndowns can result if a second fault occurs before the first fault is cleared.

Ungrounded systems have the characteristic that they are subject to relatively severe transient overvoltages. Such overvoltages may be owing to external disturbances as well as to internal faults in the wiring system and easily can reach a value of five to six times normal voltage. An actual case involved a 480-volt ungrounded system. Line-to-ground potentials in excess of 1200 volts were measured on a test meter. The source of the trouble was traced to an intermittent or sputtering (arcing) line-to-ground fault in a motor starting autotransformer.

These faults are not uncommon on 480-volt ungrounded systems. During the two-hour period that this arcing fault existed, between 40 and 50 motor windings had failed.

Circuit-switching operations can also be responsible for the creation of transient overvoltages in ungrounded systems. These generally are of short duration and typically only reach two to three times nominal system line-to-ground voltage.

Experience has shown that these overvoltages, that easily reach several times the system voltage, may cause failure of insulation at locations on the system other than at the point of the fault and may result in future system failures. This often occurs at a system weak point such as in a motor or transformer winding.

Locating the ground fault can be troublesome. While it is easy to spot a ground fault on a one-line diagram, locating it in a plant with a complex electrical system can be much more difficult unless sophisticated ground-fault detection equipment has been installed. The first step is to open the feeders one at a time and observe the ground detection indicator. After finding the feeder with the ground

> Where grounding of the electrical system is optional, the pros and cons of grounding must be carefully weighed by the plant owner or electrical designer so as to make the best decision.

fault, branch circuits are disconnected, one at a time, until the offending circuit is located. A significant loss of plant operation time can occur during this process. This is contrasted with a grounded system where only the offending equipment is taken off the line by the circuit protective devices.

Often, overcurrent devices are set above the current level of the fault in ungrounded systems. When arcing faults occur, destructive burndowns of electrical equipment can result. The fault arc releases a tremendous amount of energy such that conductors and metal enclosures in the vicinity are destroyed.

When the first ground fault occurs on the 480-volt ungrounded system, the other conductors of the system rise to a level of 480 volts to ground. This presents an additional risk of shock to operation and maintenance staff. This can be contrasted to a 480Y/277-volt grounded wye system where the voltage to ground does not exceed 277 volts while the phase-to-phase voltage is 480 volts, even under ground-fault conditions.

Systems Permitted but Not Required To Be Grounded

Five electrical systems are permitted to be grounded but are not required to be grounded. It should be noted that whether the system is required to be grounded or is grounded by choice, the grounding and bonding requirements for grounded systems must be followed (see

250.21). These alternating-current systems are:

"(1) Electric systems used exclusively to supply industrial electric furnaces for melting, refining, tempering, and the like

(2) Separately derived systems used exclusively for rectifiers that supply only adjustable-speed industrial drives

(3) Separately derived systems supplied by transformers that have a primary voltage rating less than 1000 volts, provided that all the following conditions are met:

  (a) The system is used for only control circuits.

  (b) ...only qualified persons service the installation.

  (c) Continuity of control power is required.

  (d) Ground detectors are installed on the control system."[6]

(4) Isolated power systems as permitted or required in Article 517 for health care facilities and in Article 668 for electrolytic cells commonly used for the production of aluminum, cadmium, chlorine, copper, and other products.

(5) Three-phase ac systems of 480 to 1000 volts that are high-impedance grounded-neutral systems. An impedance device, usually a resistor, limits the current of the first ground fault to a low value. All of the following conditions must be met before high-impedance grounded systems are permitted:

  (a) Only qualified persons will service the system.

  (b) Power continuity is required.

  (c) Ground detectors are installed.

  (d) Line-to-neutral loads are not supplied.

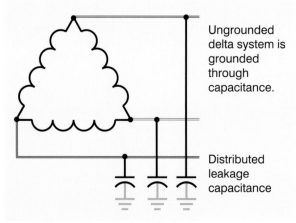

Ungrounded delta system is grounded through capacitance.

Distributed leakage capacitance

Figure 2-12. Actual Ungrounded Delta System

Factors to Consider Regarding System Grounding

Where grounding of the electrical system is optional, the pros and cons of grounding must be carefully weighed by the plant owner or electrical designer so as to make the best decision.

In the long run, greater service continuity may be obtained with grounded systems rather than ungrounded ones. Faults can be isolated to the feeder or circuit affected and cleared without disrupting the entire system. This is obviously a major consideration if the equipment or circuit affected is critical to the plant operation. This has to be balanced against the ungrounded system's tolerance of the first line-to-ground fault but with possible deterioration of conductor insulation from transient overvoltages and possible serious damage caused by a second ground fault on the system.

Bolted Faults

A common myth is that ground-faults are *bolted* or solidly connected and that a great deal of fault-current will flow to cause the overcurrent device to clear the fault. Bolted faults rarely occur, while sparking, intermittent or arcing faults are quite common. The higher impedance in the arc limits the total current, so standard overcurrent devices may be ineffective.

Arcing faults produce a great deal of heat in the vicinity of the fault and can lead to destructive burndowns of electrical switchboards and motor control centers.

This is the reason the *Code* requires equipment ground-fault protection systems for 3-phase, 4-wire, wye-connected systems of certain voltage and amperage services and feeders (see chapter fifteen of this text for additional information on this subject).

Bolted faults are usually utilized in testing laboratories under controlled conditions. A bolted fault, typically 3-phase bolted faults, must be used for the purposes of determining adequate interrupting ratings of overcurrent protective devices as well as bracing of busbars, and so forth. This is generally considered to be the "worst case" condition. For example, UL Standard 891 requires short-circuit current testing of equipment to determine a switchboard overall short-circuit current rating. For additional information, see UL 891 — *Standard for Switchboards and UL Standard 67 for Panelboards*.

[1] *NEC 2002 National Electrical Code*, Article 100, (Quincy, MA, National Fire Protection Association, 2002), p. 70-35-36.
[2] *NEC 2002*, 250.4, p. 70-97.
[3] *NEC 2002*, 250.4(B), p. 70-97.
[4] *NEC 2002*, 250.4(B), p. 70-97.
[5] *NEC 2002*, 250.4(B), p. 70-97.
[6] *NEC 2002*, 250.12, p. 70-99.

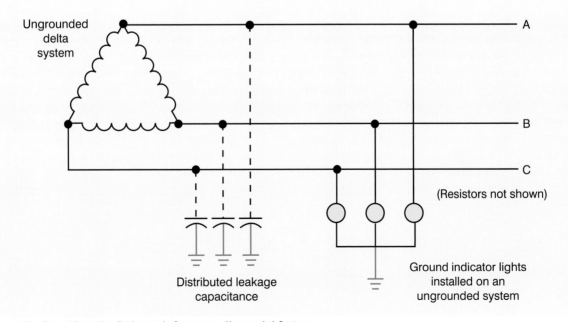

Figure 2-13. Ground Detection Equipment is Common on Ungrounded Systems

Chapter Two: The questions included here were developed using material included in this chapter. The answers can be found by reviewing the text. It is also important that students make use of the 2002 *NEC*, where many of the answers can be found.

1. Where operating at less than ____ volts, alternating-current systems are required to be grounded where supplied by transformers if the supply voltage to the transformer exceeds ____ volts to ground.
   a. 80 - 100
   b. 70 - 110
   c. 60 - 140
   d. 50 - 150

2. Where operating at less than 50 volts, ac systems are required to be grounded where supplied by transformers if the transformer supply system is ____.
   a. bonded
   b. ungrounded
   c. identified
   d. approved

3. Conductors installed on the outside of buildings where ac systems operate at less than 50 volts are required to be grounded when they are run as ____ conductors.
   a. overhead
   b. underground
   c. optical fiber
   d. Type IGS

4. Alternating-current system grounding is required for 50 to 1000-volt systems supplying premises wiring or premises wiring systems where the system can be grounded so the maximum voltage to ground on the ungrounded conductors does not exceed ____ volts.
   a. 180
   b. 150
   c. 240
   d. 208

5. Alternating-current system grounding is required for 50 to 1000-volt systems supplying premises wiring or premises wiring systems where the system is 3-phase, 4-wire, wye-connected in which the neutral is used as a ____ conductor.
   a. bonding
   b. circuit
   c. equipment grounding
   d. isolated

6. Alternating-current system grounding is required for 50 to 1000-volt systems supplying premises wiring, or premises wiring systems where the system operates at 3-phase, 4-wire, delta-connected in which the midpoint of one phase is used as a ____.
   a. bonding jumper
   b. equipment grounding conductor
   c. circuit conductor
   d. switch leg

7. Alternating-current system grounding is not required for 50 to 1000-volt electric systems used exclusively to supply industrial electric furnaces for ____.
   a. melting
   b. refining
   c. tempering
   d. all of the above

8. Alternating-current system grounding is not required for 50 to 1000-volt separately derived systems used exclusively for rectifiers supplying only ____.
   a. motor control centers
   b. adjustable speed industrial drives
   c. commercial buildings
   d. Class III hazardous (classified) locations

9. Alternating-current system grounding is not required for 50 to 1000-volt separately derived systems supplied by transformers that have a primary voltage rating less than 1000 volts if which of the following conditions is or are met. ____
   a. system is used for only control circuits.
   b. only qualified persons service installation, ground detectors are installed on the control system.
   c. continuity of power is required.
   d. all of the above

10. Alternating-current system grounding is not required for 50 to 1000-volt systems supplying premises wiring, or premises wiring systems where ____ are permitted or required for flammable anesthetizing systems in health care facilities.
   a. grounded power systems
   b. isolated power systems
   c. impedance grounded systems
   d. medium voltage power systems

11. Alternating-current system grounding is not required for 50 to 1000-volt systems supplying premises wiring, or premises wiring systems where three-phase ac systems of 480 to 1000 volts that are high-impedance grounded neutral systems if which one of the following conditions is or are met ____:
   a. only qualified persons will service the system.
   b. power continuity is required.
   c. ground detectors are installed, line-to-neutral loads are not supplied.
   d. all of the above

12. Ac systems of ____ and over must be grounded if they supply mobile or portable equipment as specified in 250.188.
   a. 250 volts
   b. 1000 volts
   c. 480 volts
   d. 600 volts

13. Ac systems operating at over ____ volts are permitted (not required) to be grounded where they do not supply mobile or portable equipment.
   a. 1000
   b. 300
   c. 277
   d. 480

14. Equipment operating at over ____ volts is commonly found in mobile rock crushing plants and batch plants. Other applications are for open pit mining operations.
   a. 277
   b. 300
   c. 600
   d. 1000

15. A ____ premises wiring system is one that is derived from a generator, transformer or converter windings that have no direct electrical connection to the supply conductors originating in another system.
   a. identified
   b. open neutral
   c. isolated power
   d. separately derived

16. Separately derived systems must be grounded as specified in ____.
   a. 250.24
   b. 250.26
   c. 250.30
   d. 250.32

17. Examples of separately derived systems include ____ with no direct electrical connection between the primary and secondary.
   a. phase converters
   b. transformers
   c. elevator motors
   d. Class 1 systems

18. Examples of separately derived systems include generator systems used for emergency, required standby or optional standby power that have all conductors including a neutral switched by a ____.
    a.   double pole
    b.   single pole
    c.   transfer switch
    d.   backfed device

19. Examples of separately derived systems include ac or dc systems derived from ____.
    a.   inverters
    b.   rectifiers
    c.   generators
    d.   all of the above

20. Certain systems are permitted to be operated ungrounded and usually are located in industrial or agricultural applications. Typical systems that are operated ungrounded include which of the following ____.
    a.   2300-volt, three-phase, three-wire, delta-connected.
    b.   4600-volt, three-phase, three-wire, delta-connected.
    c.   13,800-volt, three-phase, three-wire, delta-connected.
    d.   all of the above

21. Often ground detector lights or a potentializer plug are installed to indicate that a ground fault has occurred on the ____ system.
    a.   ungrounded
    b.   grounded
    c.   bonded
    d.   identified

22. For an ungrounded electrical system, the first ground fault
    a.   does nothing
    b.   causes a fuse to blow
    c.   grounds the system
    d.   none of the above

23. For an ungrounded electrical system, the second ground fault
    a.   does nothing
    b.   causes an overcurrent device to operate
    c.   grounds the system
    d.   none of the above

24. Disadvantages of operating an ungrounded electrical system include
    a.   power system voltages are not controlled
    b.   transient overvoltages are not controlled
    c.   destructive arcing burndowns can occur on a second ground fault
    d.   all of the above

25. Systems not permitted to be grounded include which of the following:
    a.   120-volt single-phase, three-wire system
    b.   High-impedance grounded neutral systems
    c.   A transformer secondary with a primary input of 460 volts
    d.   A low-voltage lighting system operating at 24 volts

Key Notes

Grounding Electrical Systems

OBJECTIVES

After studying this chapter, the reader will be able to understand

- Rules for system conductor to be grounded.
- Proper identification of grounded conductor.
- Methods of grounding electrical systems.
- Delta bank grounding.
- Grounding rules for ungrounded systems.
- Corner-grounded delta systems.

Section 250.20 requires that many electrical systems be grounded. It is important to understand that one is dealing with the electrical system and not the service equipment or disconnecting means at this point.

Electrical energy is delivered to the customer by the serving utility by either a grounded or ungrounded system or sometimes by both systems. Electrical utilities have tariffs and standards that dictate whether or not they will deliver a system of a certain voltage level and phase configuration as grounded or ungrounded systems. Many utility companies require that all low voltage (600 volts and under) systems be grounded. Others will supply 3-phase, 480-volt delta-connected systems ungrounded, while they insist on furnishing wye systems only in a 480Y/277-volt grounded wye-configuration.

Large industrial plants may purchase power at medium or high voltage levels such as 69,000-volt, 115,000-volt, or 230,000-volt levels and own and maintain their electrical distribution systems. Transformers, capacitor banks, controls, overcurrent devices and relaying systems are then installed at customer-owned switchyards. Power is distributed to utilization points on the premises where transformers are installed as needed for the utilization voltages desired. Generally, electrical systems at the utilization level are grounded at the voltage levels and configurations required by the *NEC*.

Definitions

"**Ground.** A conducting connection, whether intentional or accidental, between an electrical circuit or equipment and the earth, or to some conducting body that serves in the place of the earth.

"**Grounded.** Connected to earth or to some conducting body that serves in place of the earth.

"**Grounded Conductor.** A system or circuit conductor that is intentionally grounded."[1]

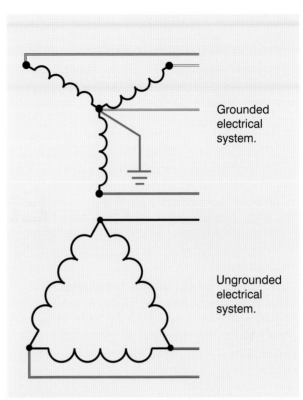

Grounded electrical system.

Ungrounded electrical system.

Figure 3-1. Electrical Systems

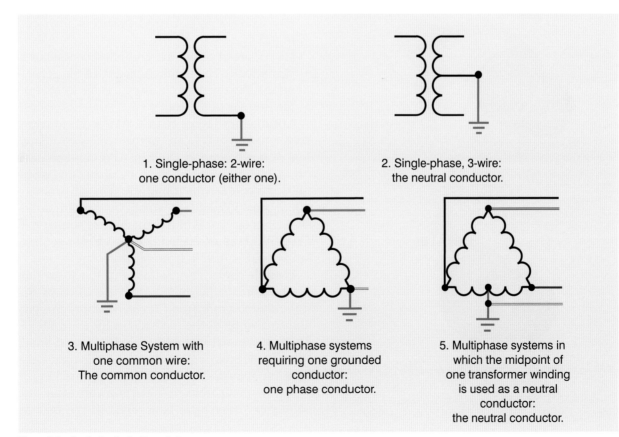

1. Single-phase: 2-wire: one conductor (either one).

2. Single-phase, 3-wire: the neutral conductor.

3. Multiphase System with one common wire: The common conductor.

4. Multiphase systems requiring one grounded conductor: one phase conductor.

5. Multiphase systems in which the midpoint of one transformer winding is used as a neutral conductor: the neutral conductor.

Figure 3-2. Conductor To Be Grounded

Conductor To Be Grounded

Where the electrical system is grounded, 250.26 specifies which conductor in an alternating-current system shall be grounded. For alternating-current premises wiring systems, the conductor required to be grounded is specified below:

(1) Single-phase, 2-wire: one conductor (either one).

(2) Single-phase, 3-wire: the neutral conductor.

(3) Multiphase systems having one wire common to all phases: the common conductor.

(4) Multiphase systems requiring one grounded phase: one phase conductor.

(5) Multiphase systems in which one phase is used as in (2) above: the neutral conductor.

Identification of Grounded Conductor

Grounded conductors are required to be identified by the means specified in 200.6. Rules are provided for identification of grounded conductors of sizes 6 AWG or smaller, conductors larger than 6 AWG, flexible cords, grounded conduc-

For conductors 6 AWG and smaller, the grounded conductor is now permitted to be identified by "three continuous white stripes on other than green insulation along its entire length." This is in addition

For sizes larger than 6 AWG, the grounded conductor is permitted to be identified as indicated above or by "distinctive white marking at its terminations." This marking must encircle the conductor or insulation.

Figure 3-3. Identification of Grounded Conductor

tors of different systems, and grounded conductors of multiconductor cables.

6 AWG and smaller. Three general means of identification of 6 AWG and smaller insulated grounded conductors are provided which include:

- a continuous white outer finish, or
- gray outer finish, or
- three continuous white stripes on other than green insulation.

This identification must be along the entire length of the conductor. It is generally not permitted to paint or phase-tape grounded conductors of 6 AWG and smaller.

For specific conductor assemblies or applications, the following grounded conductor identification is provided regardless of size:

(1) For Type MI cable, the grounded conductor is permitted to be identified by distinctive marking at its termination at the time of installation.

(2) For single-conductor, sunlight-resistant, outdoor-rated cable for solar photovoltaic systems, the grounded conductor is permitted to be identified at terminations by distinctive white markings. This will usually be made with white adhesive vinyl marking tape.

(3) Fixture wire can be identified as provided in 402.8.

(4) For aerial cable, the grounded conductor is permitted to be identified by any of the general means identified above or by a ridge located on the exterior of the cable.

Conductors larger than 6 AWG. Conductors larger than 6 AWG are permitted to be identified either like conductors 6 AWG and smaller or, at the time of installation, by distinctive white marking at each termination. This distinctive white marking usually consists of white adhesive vinyl tape or white paint. Where so marked, the marking must encircle the conductor or insulation so it is visible from all sides.

Flexible cords. The grounded conductor within flexible cords is permitted to be identified by the three methods included for conductors 6 AWG and smaller or by the methods included in 400.22. These additional methods include: colored braid, tracer in the braid, colored insulation, colored separator, tinned conductors, and surface marking.

A fine print note in 200.6 indicates that the color gray may have been used in the past as ungrounded conductors.

Grounded conductors of different systems. In

larger electrical installations, it is common to have more than one electrical system installed in the same enclosure, such as raceways like conduits and wireways, pull boxes and cables. For example, grounded conductors from both a 480Y/277-volt, three-phase, four-wire system and a 120/240-volt, single-phase system may be in the same raceway or other enclosure. Where this happens, one of the grounded conductors is required to be identified, as covered above, for conductors smaller or larger than 6 AWG. Each other system grounded conductor is required to be identified differently by one of the means provided for conductors smaller or larger than 6 AWG or by white insulation with a readily distinguishable different colored stripe that is not green and runs along the insulation.

Grounded conductors of multiconductor cables. Where only qualified persons will service and maintain the installation, grounded conductors of multiconductor cables are permitted to be permanently identified at terminations by a distinctive white marking or by other effective means.[2]

Use of conductors with white insulation. The previous permission to install conductors with white insulation in a conduit or other raceway and phase-tape them and use them as ungrounded conductors has been removed from the *NEC* (see 200.7 for additional requirements or restrictions on the use of conductors with white, gray, or three white stripes on other than green insulation).[3]

Methods of Grounding Electrical Systems
A variety of methods are used to ground electrical systems. The method chosen will vary, depending upon the system voltage, code requirements, plant owner specifications or engineer's philosophy. The various methods commonly used are shown in figure 3-5 and are as follows:

(1) **Solidly grounded:** no intentional grounding impedance. Solidly grounded refers to the connection of the electrical system or of a separately derived system of a generator, power transformer, or grounding transformer directly to the station ground or grounding elec-

Grounded conductors 6 AWG and smaller are permitted to be identified as white or gray finish or three continuous white stripes on other than green insulation.

Grounded conductors larger than 6 AWG either identified as white or gray or three continuous white stripes on insulation other than green or by distinctive white markings at the termination that encircles the conductor

New FPN cautions users that the color gray may have been used in the past as an ungrounded conductor.

Figure 3-4. Identifying Grounded Conductors

trode system without intentionally introducing an impedance.

(2) **Effectively grounded:** to permit the use of reduced rated (80 percent) surge arresters. Grounding of surge arresters rated near, but not less than, 80 percent of line-to-line voltage constitutes an effectively grounded system. This will carry with it a line-to-ground circuit current of at least 60 percent of the three-phase short-circuit value.

High-resistance grounding maintains control of transient overvoltages, but may not furnish sufficient current for ground-fault relaying.

(3) **Reactance grounded:** the system is grounded through a reactor. This grounding is accomplished by a reactor (grounding transformer), a device that introduces impedance to the circuit, the principal element of which is inductive reactance.

(4) **Resistance grounded:** intentional insertion of resistance into the system grounding connection. For resistance grounding, the system is grounded by connecting the system neutral to ground through a resistor. This grounding is accomplished by a resistor which introduces impedance to the circuit, the principal element of which is resistance.

(5) **High-resistance grounded:** the insertion of nearly the highest permissible resistance into the grounding connection. This grounding is accomplished by a resistor which introduces impedance to the circuit, the principal element of which is resistance. For high-resistance grounding, the system is grounded by the system neutral through a resistor (may be a grounding transformer) that typically limits the ground-fault current to 10-amperes or less. High-resistance grounding maintains control of transient overvoltages, but may not furnish sufficient current for ground-fault relaying. The protective scheme usually associated with high-resistance grounding is detection and alarm rather than immediate tripout (see

250.36 and chapter four for additional information). High-resistance grounding is typically applied to 3-phase systems such as 480-volt systems and also higher voltage wye-connected systems, such as 12.47 kV systems.

(6) **Ungrounded:** no intentional system grounding connection. Ungrounded systems are used in industrial plants and where desired for manufacturing processes to give an additional degree of service continuity. While a ground fault on one phase of an ungrounded system generally does not cause a service interruption, the occurrence of a second ground fault on a different phase before the first fault is cleared does result in an outage.

1. Solidly grounded. No intentional grounding impedance.

2. Effectively grounded. Grounded through surge arresters.

3. Reactance grounded. Grounded through an inductor.

4. Resistance grounded. Grounded through a resistor. Typically 50 to 600 amperes ground fault current.

5. High-Resistance grounded. Grounded through the highest permissible resistance. Typically 1 to 10 amperes ground-fault current.

6. Ungrounded. No intentional system grounding connection

Figure 3-5. Methods of System Grounding

A ground detection system is recommended, and adequate maintenance and repair will minimize interruptions.

Low-voltage systems, 600 volts and below, that are grounded are almost always solidly or high-impedance grounded; medium-voltage systems are usually either solidly or resistance grounded; and high-voltage systems above 34.5 kV are nearly always effectively grounded or ungrounded.

Delta Bank Grounding

Where three transformers that have center taps are connected in a delta bank, only one transformer may have its midpoint grounded. Any attempt to use a second transformer of the delta bank to supply a second 3-wire, single-phase source would require grounding the midpoint of a second transformer. However, since there will be a difference of potential between the midpoints of the different transformers, there would be an abnormal flow of current through both grounded neutrals. This would sooner or later show up in heating or a short circuit, depending on how both neutrals were connected.

With midpoint grounding of one phase in a delta bank, one of the phase conductors will operate at a higher voltage to ground than the other two. In practice, this *high leg* is identified as required by 110.15, 215.8, 230.56 and 408.3(E) by orange color-coding or other effective means (see figure 3-6).

The voltage of the phase conductor with the higher voltage to ground is determined by the following formula: 1/2 of phase-to-phase voltage x 1.732 = high-leg voltage to ground. For example, a 240-volt, 3-phase system center tapped to establish a neutral: 1/2 of 240 = 120 x 1.732 = 208 V high-leg voltage to ground.

Grounding of Enclosures and Equipment for Ungrounded Systems

In some cases, it is desirable to ground electrical systems that originally were installed ungrounded or are permitted to be, but are not required to be grounded (see chapter two for a thorough discussion of the subject). There are four methods commonly in use for grounding of ungrounded systems of 600 volts or less. Solid grounding is used in all the methods. They are as follows:

(1) Grounding the neutral of wye-connected secondary windings of a transformer bank. As shown in figure 3-7, this is the most universal and commonly used method of grounding a system. Standard voltages are 208Y/120, 480Y/277, 575Y/332 and 600Y/346 volts. The first two voltage systems, 208Y/120 and 480Y/277 are in most common use today in the United States with the growth tending in favor of the 480Y/277-volt system owing to the better economies of that system. In general, the primary windings of the transformers serving those systems are delta connected.

(2) Grounding a delta bank with a zigzag grounding transformer. This method is best adapted to an existing 3-wire, 3-phase delta-connected distribution system which is ungrounded and where it is desired to ground the system to obtain the advantages gained through operating the system grounded. For new systems, the use of transformers with wye-connected secondaries is advisable to obtain a grounded system. The neutral derived by the zigzag transformer can be used as a system current-carrying conductor if the zigzag grounding transformers are sized for the maximum unbalanced current (see 220.22 for

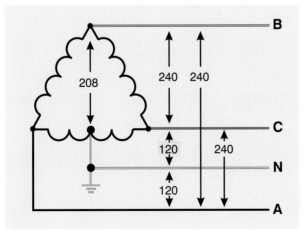

Figure 3-6. 240 Delta Bank Voltages

Grounded by connecting midpoint to grounding electrode.

Common voltages:
• 208Y/120
• 480Y/277
• 575Y/332
• 600Y/346

Figure 3-7. Grounding Wye-Connected Bank

the method for calculating the maximum un-balanced current on the grounded conductor).

A zigzag transformer, shown applied to a delta system in figure 3-8, obtains its name from the manner in which the windings are installed and connected. Windings for each phase are on the same core leg. All windings have the same number of turns but each pair of windings on the same core leg is wound in opposite directions. The impedance of the transformer to 3-phase currents is so high that under normal conditions only a small magnetizing current will flow. If a ground fault develops on one phase of the system, the transformer impedance to ground current is so low that a high ground current will flow. The ground current will divide into three equal parts in the three phases of the grounding transformer.

Such a 3-phase zigzag transformer has no secondary winding. This type of grounding transformer is required to carry rated current for only a short time (the duration of the ground fault). The short-time kVA rating of such a grounding transformer may thus be equal to the rating of a regular 3-phase transformer yet be only about 10 percent of the physical size.

Electrically, a zigzag grounding transformer connection appears to superimpose a wye-connected system inside the delta-connected transformers. This wye-type connection does permit the transformer to be used like a wye transformer

with phase-to-neutral loads to be utilized. In this case, it is important for the zigzag transformer to be sized for the calculated or connected load.

A wye-delta grounding transformer may also be used to provide a neutral for an existing delta-connected ungrounded system, but the use of the zigzag grounding transformer is more practical and economical.

The grounding transformer may be connected to the system, as shown in figure 3-9, directly on the power transformer secondaries. Where a grounding transformer is so connected, one grounding transformer is required for each delta-connected power transformer bank. In this manner, the switching out of any one transformer bank will not disturb the secondary system ground. The grounding transformer and the power transformer are considered a single unit, both being protected by the overcurrent protective means provided for the main power transformer.

The grounding transformer also may be connected directly to the main bus through its own overcurrent protection means. In that event, an alarm should be provided to indicate the system is operating ungrounded if the grounding transformer should be disconnected from the line.

Where a grounding transformer is used on low-voltage systems (600 volts and below), it is important that the equipment grounding conductor be connected to the neutral of the grounding transformer in such a way as to provide a low-impedance path for ground-fault currents to flow back to the system. The same grounding electrode should be used for the neutral of the grounding transformer and the equipment grounding conductor and a common grounding conductor used for both the equipment grounding conductor and the neutral.

The same disconnecting means and overcurrent

Figure 3-8. Grounding Delta Bank with Zigzag Transformer

Figure 3-9. Relationship of Zigzag Transformer to Delta System

protection means for the system would be used as described under method (1).

(3) Grounding the midpoint of one of the transformers in a delta-connected system. This method of grounding is commonly used, especially in smaller distribution systems, to provide a three-phase power source and a three-wire single-phase lighting source from the same transformer bank. This then becomes a three-phase, four-wire delta-connected system. At the same time, the advantages of a neutral grounded system are obtained. It is common for the serving utility to use three single-phase transformers connected in a delta configuration for this system. In such a system, the single-phase transformer which supplies the lighting load usually is sized larger than the other two single-phase transformers which supply the power load only.

Since the conductors supplying the 3-phase power will now be from a grounded source, the grounded conductor must be run to the service equipment. There it must be connected to the equipment grounding conductor of the power system and the grounding electrode. This provides a ground-fault current return path of low impedance. Such a connection enhances the safety of the power system. If the power service and the lighting service go to the same building, which is usually the case, the grounding electrode conductor from the power service, as well as the grounding electrode conductor from the lighting service, must be connected to the same grounding electrode (see chapter four for additional information on this subject).

(4) Grounding one corner of a delta system. In the past, most three-phase ungrounded delta distribution systems comprised three single-phase transformers which were connected delta-delta. The main reason was to be able to continue operating if one transformer failed by disconnecting the faulted transformer and operating the bank open-delta, although at reduced capacity.

When the advantages of grounding became very apparent, it first was the practice to ground one corner of the delta. This is shown in figure 3-10. The grounded leg must be positively identified throughout the system in accordance with 200.6.

Where the grounded conductor is disconnected by the switch or circuit breaker, it is important that a bonding connection be made on the line side of the disconnect or circuit breaker. If this is

not done, the electrical equipment will become electrically isolated from the grounded conductor when the switch or circuit breaker is in the open position. In this case, any fault that occurs would cause the enclosure to raise to a hazardous potential above ground equal to the system voltage.

Generally, fuses are not permitted to be installed

Grounding midpoint of one transformer.

(If three, single-phase transformers connected in Delta-bank configuration.)

Figure 3-10. Grounding Delta Transformer Banks (High-Leg)

in the grounded conductor. However, a two-pole fused switch may be used with fuses in the ungrounded phases. Also, a three-pole switch may be used with a solid neutral in the grounded phase. For services, 230.90(B) does not allow an overcurrent device, other than a circuit breaker that opens all conductors, to be inserted in the grounded service conductor. Section 240.22 generally prohibits connecting an overcurrent device in series with any conductor that is intentionally grounded. Subsection (1) permits an overcurrent device to be used that "opens all conductors of the circuit, including the grounded conductor, and is designed so that no pole can be operated independently."

This requirement means that three-pole circuit breakers can be used for this purpose while a three-pole switch with a fuse in series with each phase could not as the switch is not the overcurrent device, the fuse is. Figure 3-11 shows an acceptable use of a fused switch for corner-grounded systems.

Section 430.36 permits a fuse to be inserted in the grounded conductor for the purpose of providing motor overload protection. However, based on the examination of *Code* requirements discussed in the previous paragraph, this application would be limited to being located downstream from the service equipment.

Keep in mind that 250.24(B) requires that the grounded system conductor be run to each service disconnecting means and be bonded to the service disconnecting means enclosure. This is usually accomplished by connecting it to a terminal bar that is mounted inside the service enclosure. Nothing permits this bonding connection to be interrupted by a switch or circuit breaker.

Where it is desirable to disconnect the grounded system conductor from the feeder, it

From the service on throughout the system, the grounded conductor must be insulated from equipment and enclosures [see 250.24(A)(5)].

is acceptable to route the conductor from the terminal bar through the switch or circuit breaker. It is not uncommon to see three-pole switches with a copper bar or other suitable slug inserted in the grounded leg.

For purposes of installing or grounding a cor-

ner-grounded delta system, it is helpful to think of it as a single-phase system. This is illustrated in figures 3-11 and 3-12. The system is grounded at one corner of the delta. Three conductors are taken to the service where the two ungrounded conductors connect to the disconnecting means or circuit breaker. The grounded service conductor is connected to the neutral terminal bar where it is bonded to the enclosure and connected to the grounding electrode conductor.

Where circuit breakers are used as the disconnecting means for corner-grounded systems, they must be marked with a voltage rating suitable for the system voltage and will have a "straight voltage rating" such as 240V or 480V. Two-pole circuit breakers that are suitable for a corner-grounded delta system are marked "1-phase/3-phase" (see 240.85).

From the service on throughout the system, the grounded conductor must be insulated from equipment and enclosures [see 250.24(A)(5)]. An equipment grounding conductor is carried with the circuit-to-ground equipment and enclosures such as disconnecting means, motor controllers, and other non-current-carrying equipment that are required to be grounded.

---

[1] NFPA 70, *National Electrical Code* 2002, Article 100, (Quincy, MA, National Fire Protection Association, 2002), p. 70-35.
[2] NFPA 70, 200.6, p. 70-49.
[3] NFPA 70, 200.7, p. 70-50.

Figure 3-11. Fusible Switch for Corner Grounded Delta System

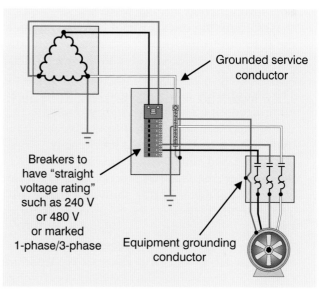

Figure 3-12. Corner-Grounded Delta System

Chapter Three: The questions included here were developed using material included in this chapter. The answers can be found by reviewing the text. It is also important that students make use of the 2002 *NEC*, where many of the answers can be found.

1. A conducting connection, whether intentional or accidental, between an electrical circuit or equipment and the earth, or to some conducting body that serves in the place of the earth is defined as a ____.
   a. grounded conductor
   b. ground
   c. grounding electrode
   d. bonded conductor

2. Connected to earth or to some conducting body that serves in place of the earth is defined as ____.
   a. grounded
   b. a grounded conductor
   c. being identified
   d. being bonded

3. A system or circuit conductor that is intentionally grounded is defined as one that is ____.
   a. identified
   b. bonded
   c. a grounded conductor
   d. a grounding conductor

4. ____ AWG and smaller insulated grounded conductors are required to be identified by a continuous white or gray outer finish.
   a. 6
   b. 4
   c. 2
   d. 1

5. For installations that will be serviced by qualified persons only, grounded conductors of ____ cables are permitted to be marked at their terminations.
   a. Type IGS
   b. Type TFFN
   c. multiconductor
   d. single

6. Conductors ____ than 6 AWG are permitted to be identified either like conductors 6 AWG and smaller, or at the time of installation, by distinctive white marking at their terminations.
   a. of aluminum or smaller
   b. of copper or smaller
   c. smaller
   d. larger

7. Insulated conductors 6 AWG or smaller are permitted to be identified by which of the following methods?
   a. a continuous white insulation
   b. a continuous gray insulation
   c. three white stripes on other than green insulation
   d. all of the above

8. In practice, a "high leg" is required to be identified by a (an) ____ color coding or other effective means.
   a. green
   b. brown
   c. orange
   d. yellow

9. Grounding of surge arresters rated near, but not less than, ____ percent of line-to-line voltage constitutes an effectively grounded system.
   a. 70
   b. 60
   c. 50
   d. 80

10. For resistance grounding, the system is grounded by connecting the system neutral to ground through a ____.
   a. resistor
   b. terminal
   c. ground rod
   d. generator

11. The insertion of nearly the highest permissible resistance into the grounding connection results in a system that is ____.
    a.   extra low-resistance grounded
    b.   medium-resistance grounded
    c.   low-resistance grounded
    d.   high-resistance grounded

12. High-resistance grounding maintains control of transient overvoltages, but may not furnish sufficient current for ground-fault ____.
    a.   detection
    b.   relaying
    c.   alarms
    d.   conditions

13. Low-voltage systems operating at 600 volts and below are almost always solidly grounded; medium-voltage systems are usually either solidly or resistance grounded; and high-voltage systems above ____ kV are nearly always effectively grounded or ungrounded.
    a.   12.7
    b.   10.4
    c.   15.3
    d.   34.5

14. A wye-delta grounding transformer may also be used to provide a neutral for an existing delta-connected ungrounded system, but the use of the ____ grounding transformer is considered as being more practical and economical.
    a.   converter
    b.   autotransformer
    c.   zigzag
    d.   scott connect

15. Where a grounding transformer is used on low-voltage systems operating at 600 volts and below, it is important that the equipment grounding conductor be connected to the neutral of the grounding transformer in such a way as to provide a ____ path for ground-fault currents to flow back to the system.
    a.   high-impedance
    b.   low-impedance
    c.   current limiting
    d.   straight

16. The same grounding electrode should be used for the ____ of the grounding transformer and the equipment grounding conductor and a common grounding conductor used for both the equipment grounding conductor and the neutral.
    a.   unidentified conductor
    b.   equipment grounding conductor
    c.   bonding jumper
    d.   neutral

17. In the past, most 3-phase ungrounded delta distribution systems comprised three single-phase transformers that were connected ____.
    a.   wye-delta
    b.   delta-wye
    c.   delta-delta
    d.   delta

18. For purposes of installing or grounding a corner-grounded delta system, it is best to think of it as a ____ system.
    a.   three phase
    b.   single phase
    c.   5-wire
    d.   4-wire

19. Circuit breakers used for corner-grounded delta systems must be marked
    a.   with a voltage rating such as 120/240
    b.   for at least 600 volt operation
    c.   with a straight voltage rating such 240 V or 480 V
    d.   none of the above

20. Fuses are permitted to be inserted in a grounded conductor
    a.   as desired
    b.   for grounded systems only
    c.   for ungrounded systems only
    d.   for motor overload protection

Key Notes

Grounding Electrical Services

OBJECTIVES

After studying this chapter, the reader will be able to understand

- Important elements for grounding of electrical services.
- Proper location of service grounding connection.
- Rules for low-impedance grounding electrode connections.
- Dwelling unit services and feeders.
- Proper sizing of grounded service conductor.
- Rules for parallel service conductors.
- Rules for multiple services to one building.
- Rules for high-impedance grounded systems.
- Grounding requirements for instrument transformers, relays, etc.
- Hazards of operating an ungrounded services from grounded systems.

Electrical services are furnished to the premises by the serving utility as either grounded or ungrounded. At the service disconnecting means, the system is either solidly grounded, left ungrounded, or may be resistance or reactance grounded. How the services are treated regarding grounding depends on the type of system installed, design criteria, and *Code* rules.

Grounded Electrical Systems

One of the most important requirements for grounded systems and services is contained in 250.24(B). This section requires that:

"(1) Where an ac system operating at less than 1000 volts is grounded at any point, the grounded conductor shall be run to each service disconnecting means.

(2) It must be bonded to each disconnecting means enclosure.

(3) The grounded conductor must be routed with the phase conductors.

(4) It must not be smaller than the required grounding electrode conductor specified in Table 250.66.

(5) For service-entrance phase conductors larger than 1100 kcmil copper or 1750 kcmil aluminum, the grounded conductor shall not be smaller than 12 1/2 percent (0.125) of the area of the largest service-entrance phase conductor."

(6) Where the service-entrance phase conductors are installed in parallel, the minimum size of the grounded service conductor must be based on the equivalent area of the ungrounded parallel service conductors.[1]

Three exceptions to the rule for running the grounded service conductor to the service or sizing the conductor are provided.

"(1) Where more than one service disconnecting means are located in an assembly listed for use as service equipment such as a switchboard, the grounded conductor (conductors in the case of parallel runs) is only required to be run to the assembly, not to each disconnecting means or switch. It (or they in the case of parallel conductors) must be bonded to the assembly enclosure. Since the sections of the switchboard are bolted together to form the assembly, the grounded service conductor is required to be run to only one section. In addition, it is common to have an internal equipment grounding bus connected to each section.

(2) The grounded conductor is not required to be larger than the largest ungrounded service-entrance phase conductor.

(3) See 250.36 for high-impedance grounded neutral systems grounding connection requirements.

In addition to providing the return portion of

Figure 4-2. Grounded Conductor Run to Each Service Disconnecting Means Enclosure

Grounded service conductor must be:

1. Run to each service disconnecting means;

2. Bonded to service disconnecting means enclosure;

3. Routed with phase conductors;

4. Not smaller than grounding electrode conductor;

5. At least 12-1/2 percent of area of conductors where larger than given in Table 250.66;

6. Based on equivalent area of ungrounded parallel service-entrance conductors.

Figure 4-1. Grounded Electrical System

the circuit for unbalanced loads, this grounded service conductor provides the vital low-impedance path for ground-fault current to return to the source. The grounded system conductor must be installed to each service disconnecting means and be bonded to the enclosure regardless of whether or not it is needed for the service or is used to supply a load.

Figure 4-4 illustrates a 3-phase power service taken from a grounded system. The grounded conductor is not needed since only 3-phase or phase-to-phase connected loads are supplied. If the grounded conductor is not run to the service, as required in 250.24(B), a ground-fault circuit of high impedance is present as shown by the dashed lines.

Figure 4-3. Grounded Conductor to Service Equipment

High-Impedance Path
Grounded conductor not installed.
Only high-impedance path for ground-fault current to return to grounded source.

High-Impedance Path
Grounded conductor installed. Both high- and low-impedance paths for return of ground-fault current to source.

Figure 4-4. High- and Low-Impedance Paths

This is a violation of 250.4(A)(5) and it becomes virtually impossible to clear a ground fault, thus introducing an unnecessary hazard in the system.

However, the installation complies with 250.24(B) if the grounded conductor is run to service. A low-impedance ground-fault path is provided as required by 250.4(A)(5). In the latter case, the safety of the service is improved immeasurably. Since the grounded conductor is run only to provide a ground-fault path, the size of the conductor to be run will depend on the size of the phase conductors in the service. This conductor must "not be smaller than the required grounding electrode conductor specified in Table 250.66."[2]

Figure 4-5 represents the relationship between the service and equipment ground in a 3-phase, 4-wire delta system with the midpoint of one transformer grounded. Such systems are intended to supply, economically, both a 3-phase, 3-wire power service and a single-phase, 3-wire lighting service from one transformer bank. The lighting transformer must be big enough to supply all the lighting load plus the three-phase load. The other transformers are sized to carry the 3-phase, 3-wire load. Though two separate panelboards are shown for illustration purposes, all loads can be supplied from a single 3-phase panelboard. Where this is done, the electrician needs to exercise caution in making connections for 120-volt loads. These loads must be connected to only the A and C phases as the B phase will have a voltage to ground of approximately 208 volts which is enough to severely damage equipment designed to operate at 120 volts.

3-Phase Service and loads

Grounded service conductor run to both service disconnects.

1-Phase. Service and loads

Figure 4-5. Power and Lighting Service for 3-Phase System

In the case of the 3-phase, 3-wire power service, there is no need for the neutral (or grounded service conductor), as far as meeting voltage or phase requirements on the power service is concerned. This is true where no line-to-neutral loads are supplied. However, it is required to run the grounded conductor of the system to the three-phase power service equipment and to use the main bonding jumper to satisfactorily clear a ground fault that may develop in the service equipment or in equipment that is supplied by the service.

Location of Service Grounding Connection.

Section 250.24(A) requires that a grounding electrode conductor be used to connect the grounded service conductor to a grounding electrode. The connection to the grounded service conductor must be "at an accessible point from the load end of the service drop or service lateral to and including the terminal or bus...at the service disconnecting means [see 250.24(A)(1)]. These locations include current-transformer enclosures, meter enclosures, pull and junction boxes, busways, auxiliary gutters and wireways as well as switchboards, panelboards or motor control centers.

Many inspection authorities and serving utilities will not permit the grounding electrode connection to the system grounded conductor to be within current-transformer cans, meter enclosures, or other enclosures that are sealed by the utility. Some utility policies prohibit connection within metering equipment, and many inspection authorities prohibit that connection location as they interpret the connection to be no longer accessible, as utilities seal the metering equipment

Figure 4-6. Service Grounding Connection To Be Accessible

to prevent unauthorized access. It is important that serving utilities are consulted before beginning the project to be certain their system grounding policies are complied with.

The most practical and commonly accepted location for the grounding electrode conductor connection to the grounded service conductor is within the service disconnecting means or within a wireway that is not sealed at the service equipment location. The connection is usually made to the neutral bus.

Low-impedance Grounding Electrode Conductor Connections

To obtain sufficiently low impedance, the point of connection of the grounding electrode conductor must be within the same metallic enclosure as the phase conductors and the grounded conductor. The most practical point is within the service equipment.

The connection may be at one of the locations shown in figure 4-6. Locations indicated are within the service panelboard, to the grounded conductor terminal block inside the current transformer cabinet, to the grounded service conductor terminal block inside the service wireway, or within the meter base. See Figure 4-6.

It is also permissible to make the service grounding connections within each service disconnecting means rather than within the wireway. Means and methods of connecting and sizing these conductors are covered in chapter seven.

Where the service is supplied by a transformer located outside the building, an additional grounding connection to an electrode must be made outside the building, usually at the transformer. This connection is usually provided by the electric utility.

Section 250.24(A)(5) prohibits grounding of the grounded circuit conductor (often a neutral) at any point beyond the service. This prevents neutral current from flowing on unintended paths such as metal piping, cable trays, cable sheaths, and so forth. Three exceptions to this requirement are provided in the FPN that follows 250.24(A)(5), several of which will be covered in later chapters. An additional grounding connection beyond the service is permitted: (1) for separately derived systems, (2) where more than one building or structure on the same premises is supplied by the same service, and (3) for existing circuits for electric ranges and dryers. The use of the words *grounded circuit conductors* is intended to cover all such conductors whether they are feeders or branch circuits.

Sizing of Grounded Service Conductor

The size of the grounded service conductor (often the neutral) varies with the load as calculated in accordance with 220.22. This becomes the minimum size grounded service conductor unless modified by 250.24(B)(1).

First, a load calculation should be performed in accordance with 220.22. Then, the minimum size of conductor must be compared with Table 250.66. Section 250.24(B)(1) requires the grounded conductor to be no smaller than the grounding electrode conductor as determined in Table 250.66. If the load calculation permits a conductor smaller than provided for in Table 250.66, the conductor must be increased in size to be not smaller than provided for in the table. If the load calculation requires a conductor larger than required by Table 250.66, the larger conductor must be used.

For example, if a 400-ampere service is to be

Grounded service conductor based on size of ungrounded service conductors.

Figure 4-7. Minimum Size of Grounded Service Conductor

installed from a grounded system, 500 kcmil copper conductors are selected for the ungrounded service conductors. Table 250.66 shows that the grounded service conductor can be no smaller than 1/0 copper or 3/0 aluminum. This conductor must be routed with the phase conductors and be bonded to the disconnecting means enclosure.

This is the minimum size grounded service conductor permitted and must be installed as noted even though there is no neutral load on the system or the calculated neutral load would permit a smaller conductor. Follow a similar procedure for other service installations.

It is important to note that the minimum size of the grounded service conductor that must be run to the service disconnecting means is based on the size of the ungrounded (hot) service-entrance conductors and not on the rating of the circuit breaker or fuse that is installed in the service. It is helpful to remember that Table 250.66 generally applies up to the service, and Table 250.122 applies beyond the service.

Dwelling Unit Services and Feeders

Special rules or provisions for sizing service-entrance conductors and feeders are provided in 310.15(B)(6) and Table 310.15(B)(6). These rules apply only to 120/240-volt, 3-wire, single-phase dwelling services and feeders. By following the conditions of the section and table, the specified size of service-entrance or feeder conductors shown in the table are per-

Size of Largest Ungrounded Service-Entrance Conductor or Equivalent Area for Parallel Conductors[a] (AWG/kcmil)		Size of Grounding Electrode Conductor (AWG/kcmil)	
Copper	Aluminum or Copper-Clad Aluminum	Copper	Aluminum or Copper-Clad Aluminum[b]
2 or smaller	1/0 or smaller	8	6
1 or 1/0	2/0 or 3/0	6	4
2/0 or 3/0	4/0 or 250	4	2
Over 3/0 through 350	Over 250 through 500	1/0	3/0
Over 350 through 600	Over 500 through 900	2/0	4/0
Over 600 through 1100	Over 900 through 1750		
Over 1100	Over 1750	3/0	250

Notes:
1. Where multiple sets of service-entrance conductors are used as permitted in 230.40, Exception No. 2, the equivalent size of the largest service-entrance conductor shall be determined by the largest sum of the areas of the corresponding conductors of each set.
2. Where no service-entrance conductors, the grounding electrode conductor size shall be determined by the equivalent size of the largest service-entrance conductor required for the load to be served.
[a] This table also applies to the derived conductors of separately derived ac systems
[b] See installation restrictions in 250.64(A)

Table 250.66. Grounding Electrode Conductor for Alternating-Current Systems

mitted to be used based upon the service or feeder rating.

The grounded conductor (often a neutral) is permitted to be smaller than the ungrounded (hot) conductors provided the rules of 215.2, 220.22, and 230.42 are met.

Section 215.2 provides that the feeder neutral must be adequate for the load, must be a minimum size for certain loads, and does not have to be larger than the service-entrance conductors that supply them.

Section 220.22 provides the method for calculating the feeder neutral load. The basic requirement is that the neutral must carry the maximum unbalanced load from the ungrounded conductors.

Section 230.42 generally requires that service-entrance conductors be of sufficient size to carry the loads as computed by Article 220. For grounded conductors in 230.42(C), the conductor cannot be smaller than required by 250.24(B).

Note that for dwelling unit services and feeders, the previous permission to size the grounded conductor not more than two sizes smaller than the ungrounded conductors given in 310.15(B)(6) has been deleted from the *Code*. As a result, the grounded system conductor now has to be sized according the above rules.

Parallel Service Conductors

Where parallel service conductors as permitted by 310.4 are installed, the minimum size grounded service conductor is determined by multiplying the circular mil area of the ungrounded conductors installed by the number of conductors installed in parallel. Installing conductors in parallel means connecting two or more conductors together at each end to form one larger conductor. This is usually done for economic and practical reasons. As can be seen in

Table 310.16, the ampacity of conductors does not increase in proportion to their size.

Example No. 1.

Given: Three, 4/0 AWG copper conductors per phase are installed in parallel. Before multiplying the conductor size (4/0), it must be converted from the American Wire Gauge (AWG) 4/0 designation to the circular mil area.

- Refer to *NEC* Chapter 9, Table 8 to determine the circular mil area of the 4/0 AWG conductors. There we find the area to be 211,600 circular mils.
- Three times 211,600 = 634,800 circular mils.
- By referring to Table 250.66, we find the minimum size grounded service conductor is 2/0 AWG copper or 4/0 AWG aluminum.

This example assumes that all of the conductors are installed in the same raceway. This may not be practical due to the requirement that the derating or ampacity adjustment factors of Table 310.15(B)(2)(a) be applied. This obviously has the effect of requiring larger conductors to be installed than if the individual sets of service-entrance conductors were installed in individual conduits or raceways. Where conductors are installed in sheet-metal wireways, derating may be avoided if the number of conductors does not exceed 30 and the conductors do not fill more than 20 percent of the square-inch area (see 376.22 for additional information).

Where installed in separate metal raceways, 300.3(B) requires that the neutral conductor must

Parallel conductors must be:
1. Same length.
2. Same conductor material.
3. Same size in circular mil area.
4. Same insulation type.
5. Terminated in the same manner.

Figure 4-8. Parallel Service Conductors

Where all service-entrance conductors are in the same raceway, the grounded conductor is to be not less than given in Table 250.66. For conductors larger than Table 250.66, to be not less than 12.5 percent of total area of largest set of ungrounded conductors. See 250.24(B)(1).

Where installed in two or more raceways, the grounded conductor size is based on the size of the ungrounded conductors in the raceway (based on Table 250.66), but not smaller than 1/0 AWG.

Figure 4-9. Parallel Service Conductors

be installed in each raceway. In this case, 310.4 requires that the paralleled conductors be not smaller than 1/0 AWG, so the minimum size grounded conductor permitted in parallel is 1/0 AWG.

Example No. 2.

If six 4/0 AWG copper conductors are installed parallel, the minimum size grounded service conductor is determined as follows:

As explained above, the area of 4/0 AWG conductors (211,600 circular mils) in *NEC* Chapter 9, Table 8 is used.

- Six times 211,600 = 1,269,600 circular mils. Since the total conductor area exceeds the 1100 kcmils for copper conductors given in Table 250.66, the rule in 250.24(B)(1) must be followed. There we find a requirement that the grounded service conductor be not smaller than 12 $^1/_2$ percent (0.125) of the equivalent area for parallel conductors.
- 1,269,600 times 0.125 = 158,700 circular mils.
- By again referring to Chapter 9, Table 8, the conductor that is the next size larger than 158,700 circular mils is a 3/0 AWG conductor which has a circular mil area of 167,800.

This example also assumes that all the service-entrance conductors are installed in the same raceway, which, due to derating requirements, may not be too practical.

Again, where installed in separate metal raceways, a grounded service conductor must be installed in each raceway and must not be smaller than 1/0 AWG. The grounded service conductor must also comply with 230.42(A). This section requires that the conductor be adequate to carry the load as determined by Article 220.

Section 220.22 requires that the neutral be sized for the maximum unbalance of the load. Examples of neutral conductor load calculations are found in Annex D of the *National Electrical Code*. In addition, where the length of run of the grounded conductor from the transformer to the service equipment is long, the size of grounded conductor should be increased.

Underground Parallel Service Conductors

As illustrated in figure 4-10, for underground installations in nonmetallic raceways, it is permitted by 300.5(I) Exception No. 2 to install all

Figure 4-10. Underground Parallel Conductors

the conductors of each phase in the same raceway. This is also permitted in 300.3(B)(1) Exception. To comply with these sections, all the ungrounded conductors of phase A are installed in one raceway, phase B in another, C in the third and the grounded service conductors in another. This method is often chosen to allow phase conductors to readily line up with bus terminations in bottom-fed switchboards. As can be seen in the illustration, this reduces the "rats nest" in the bottom of these enclosures caused by many conductors criss-crossing each other for termination.

Another advantage is it is much easier to comply with the requirement that parallel conductors be the same length. When this type of installation is made, care must be taken to eliminate the inductive heating of metal enclosures with magnetic properties by cutting a slot in the metal between the conduit entries or arranging for the manufacturer of the equipment to install a nonmetallic plate in the bottom of the equipment where the conduits terminate [see 300.20(A) and (B)]. Usually, a slot cut with a single hacksaw blade is adequate to provide the desired relief. Another option is to terminate these conduits above the floor in the compartment of an open-bottom, floor-standing switchboard.

See 408.10 where the conduit or raceways, including their end fittings, are not permitted to rise more than three inches above the bottom of the enclosure.

400 Ampere Service   100 Ampere Service

Size of grounded service conductor to each service based upon the size of ungrounded service conductor to the individual service.

Figure 4-11. Two Services to One Building from Grounded System

Neutral must be a fully insulated conductor.

Not smaller than current rating of impedance device, minimum 8 AWG copper or 6 AWG aluminum.

Neutral connects to impedance device.

Equipment grounds connected below impedance device.

Figure 4-12. High Impedance Grounded Systems

Multiple Services to One Building

Section 230.2 permits several services to one building under one of several conditions given. Each service that is supplied from a grounded system must be provided with a grounded service-entrance conductor. The size of the ungrounded service-entrance conductor for each service determines the minimum size of grounded service conductor for that service. Each service is considered individually for sizing the grounded service conductor to it.

For example, a building has a 400-ampere, 480-volt 3-phase and a 100 ampere 120/240-volt service. The minimum size of grounded service conductor is determined as follows:

400-ampere service
• 750 kcmil THW aluminum ungrounded service conductors
• Table 250.66 = 1/0 AWG copper or 3/0 AWG aluminum grounded service conductor

100-ampere service
• 2 AWG copper ungrounded service conductors
• Table 250.66 = 8 AWG copper or 6 AWG aluminum grounded service conductor

Again, it must be emphasized that this method determines the minimum size of grounded service conductor to comply with 250.24(B). A larger conductor may be required to carry the maxi-mum unbalanced load on the neutral conductor as determined by 220.22.

High-impedance Grounded Systems

Industrial plants having a continuous industrial process often need uninterrupted electrical power and systems. It is quite common to see these plants located near a power company substation and to have more than one high-voltage feeder to improve system reliability.

Another step that is commonly taken to improve system reliability is to install high-impedance grounded systems rather than solidly grounded systems. Advantages include improved reliability, the ability to have ground-fault relaying and fault-indicating alarms, as well as fewer problems to the system from transient overvoltages.

Four conditions must be met before the *Code* will permit high-resistance grounded systems to be installed. They are as follows:

1. Qualified persons must be available to service and maintain the system.
2. Continuity of power is required.
3. Ground detectors must be installed to indicate an insulation failure.
4. Line-to-neutral loads are not served.

Specific rules are provided in 250.36 for installing these systems. The grounding impedance, usually a resistor, is installed between the transformer supplied grounded service-entrance conductor and the grounding electrode. Usually, the impedance device is sized to a value just greater than the capacitive charging current of the system. For 480-volt systems, this is usually about 10 amperes.

A fully insulated grounded service-entrance conductor must be run to the impedance device. This conductor must have an ampacity not less than the maximum current rating of the grounding impedance. The minimum size grounded service conductor cannot be smaller than 8 AWG copper or 6 AWG aluminum or copper-clad aluminum. Since the grounding impedance will limit the fault current of the first line-to-ground fault to a low value, usually

10 amperes or less, the minimum size of neutral conductor is primarily related to mechanical strength, not to its current-carrying capabilities.

It is not required that the grounded service conductor be routed with the ungrounded service-entrance conductors since it will carry very little current in the event of a fault [see 250.36(D)]. At the service, it is connected to the impedance device which may be located outside the service enclosure to dissipate heat. An unspliced equipment bonding jumper is installed from the load side of the impedance device to the service enclosure where a terminal bar is installed for connection of equipment grounding conductors [see 250.36(E)]. The grounding electrode conductor is permitted to be connected to any point from the grounded side of the grounding impedance to the equipment grounding connection at the service equipment or the first system disconnecting means [see 250.36(F)].

This system is designed to limit the fault current on the first ground fault that might occur on the system. As can be seen in figure 4-12, the impedance device is in series with the first ground fault. The electrical system will continue to function normally with the first ground fault present on the system. The ground-detection system will indicate the presence of the faulted condition by either a visual or audible signal or both. This is intended to alert maintenance personnel of the ground-fault condition so corrective action can be taken, hopefully during a pe-

Resistor for high impedance grounded neutral system.

Control panel for high impedance grounded neutral system.

Photo 4-1. Equipment for High Resistance Grounded Neutral System

High-impedance path through metal enclosures, loose raceway connections, etc.

Ground fault 1   Ground fault 2

Figure 4-13. Second Ground Fault on High-Impedance Grounded System

riod the plant is not planned to be in operation (see photo 4-1).

One difference from ground-detection on an ungrounded system is that the faulted circuit can be identified without shutting the plant down. A second ground fault (illustrated in figure 4-13) that occurs before the first fault is cleared will be a line-to-line or phase-to-phase fault that would be cleared by the service or feeder overcurrent device which will result in a power outage. This can, and at times has, involved two pieces of equipment in separate parts of the plant supplied by different feeders. In this situation, a great deal of current can flow in the circuit between the faults and can cause extensive damage to electrical equipment. Every metal conduit locknut and fitting connection must be made wrench tight to avoid arcing.

Instrument Transformers, Relays, etc.

Part IX of Article 250 deals with grounding requirements for instrument transformers, relays, and so forth. The *Code* requires the grounding of secondary circuits of current and potential instrument transformers where the primary windings are connected to circuits of 300 volts or more to ground and, where on switchboards, shall be grounded irrespective of voltage (see 250.170). If a switchboard has no live parts or wiring exposed or accessible to other than qualified persons, such circuits need not be grounded where the primary windings are connected to circuits of 1000 volts or less.

Instrument transformer cases or frames are required to be grounded where accessible to other than qualified persons. Such cases or frames of current transformers need not be grounded where the primaries are not over 150 volts to ground and the current transformers are used exclusively to supply current to meters (see 250.172).

The cases of instruments, meters and relays, where operated with windings or working parts at 1000 volts or less, are required to be grounded. If such instrument cases and relays are not on switchboards and are accessible to other than qualified persons, the cases or other exposed metal parts shall be grounded if they operate at 300 volts or more to ground [see 250.174(A)].

For dead-front switchboards having no live parts on the front of the panel, such instrument cases shall be grounded [see 250.174(B)].

For live-front switchboards which have exposed live parts on the front of the panel, such instrument cases are not required to be grounded. However, the *Code* recommends the use of an insulating mat for the protection of the operator where the voltage to ground exceeds 150 volts [see 250.74(C)].

If the operating voltage exceeds 1000 volts, the cases of instruments shall not be grounded but shall be isolated by elevation or protected by suitable barriers, grounded metal or insulating covers or guards. In the case of electrostatic ground detectors, the internal ground segments of the instruments are to be connected to the instrument case and grounded. The ground detector is to be isolated by elevation (see 250.176).

Size of Instrument Grounding Conductor

The *Code* specifies that the grounding conductor for circuits of instrument transformers and for instrument cases shall not be smaller than 12 AWG copper or 10 AWG aluminum. However, if cases are mounted directly on grounded metal surfaces of enclosures, they shall be considered to be grounded (see 250.178).

Ungrounded Systems

Ungrounded systems are subject to relatively severe transient overvoltages which can reach several times normal voltage to ground (see chapter two for more information on this subject). Such abnormal voltages become potential hazards and often cause insulation failure and equipment breakdowns. If a system has one conductor

Base on Ungrounded Conductors
Size per Section 250.66

Connect service disconnect for ungrounded
system to grounding electrode at
building or structure.

Figure 4-14. Grounding of Ungrounded Systems

grounded and is thus a grounded system, the value of such transient overvoltages as they develop is greatly reduced.

An ungrounded system must have its conductor and equipment enclosures connected to a grounding electrode system at the building or structure served. This keeps such enclosures as near to ground potential as possible and reduces shock hazards to a minimum. These service equipment enclosures are grounded by connecting them to a grounding electrode system per 250.24(D). Grounding electrodes and grounding electrode systems are covered extensively in chapter six.

Hazard of Operating a Service Ungrounded from a Grounded System

Figure 4-15 illustrates the hazard of operating an ungrounded service from a grounded system. The original ungrounded service on the right in the figure was installed some time before the service on the left. The first and original service was supplied by an ungrounded utility system. The service and feeder shown supplying equipment were protected by large overcurrent devices. Some time later, the service on the left which included a grounded service conductor was installed.

At that time, the serving utility grounded the transformer bank and extended a grounded system conductor to the new service weatherhead. However, the grounded service conductor was not extended nor connected to the older, existing service disconnecting means. The older, existing service, which supplied only 3-phase equipment loads continued to supply power to a portion of the building, though without a grounded conductor connection.

The two services were connected with properly sized grounding electrode conductors to a common grounding electrode system which included a metal water piping system.

Some time later, a ground fault occurred in the equipment being supplied from the original ungrounded service equipment. Since the system supplying the equipment was now grounded, current attempted to find its way back to the source of power to complete the circuit. Current flowed over the grounding electrode conductor at the ungrounded service, to the grounding elec-

trode, through the grounding electrode to the grounded service and returned to the transformer bank over the grounded service conductor. In this and similar situations, the current will seek a path back to the system through all available paths and will divide among the paths based upon the impedance of the paths that are available.

A great deal of current flowed over the grounding electrode conductors that were not designed or installed for that amount of current which caused the conductors to get extremely hot. The resulting fire burned a great deal of the industrial plant to the ground.

The action necessary to prevent the fire was to run a properly sized grounded system conductor to the existing service when the new service from a grounded supply was installed. That conductor should have been extended to the service disconnecting means and bonded to it. The grounded service conductor would have provided a low-impedance path back to the transformer bank and would have allowed the overcurrent devices to clear the ground fault. Methods of clearing ground faults and short circuits are discussed in detail in chapter eleven.

---

[1] NFPA 70, *National Electrical Code* 2002, 250.24(B)(1) & (2), (Quincy, MA, National Fire Protection Association, 2002), p. 70-100.
[2] NFPA 70, 250.24(B)(1), p. 70-100.

Figure 4-15. Hazard of Supplying an Ungrounded Service from Grounded System

Chapter Four: The questions included here were developed using material included in this chapter. The answers can be found by reviewing the text. It is also important that students make use of the 2002 *NEC*, where many of the answers can be found.

1. The grounded conductor is required to be run to each ac service disconnecting means where they operate at less than ____ volts and are grounded at any point.
    a.   2000
    b.   1500
    c.   1000
    d.   1200

2. Where an ac system operating at less than ____ volts is grounded at any point, the grounded conductor is required to be bonded to each disconnecting means enclosure.
    a.   1500
    b.   1200
    c.   2000
    d.   1000

3. Where an ac system operating at less than ____ volts is grounded at any point, the grounded conductor is required to be routed with the phase conductors.
    a.   1000
    b.   1200
    c.   1300
    d.   2000

4. Where an ac system operating at less than 1000 volts is grounded at any point the grounded conductor cannot be smaller than the required _____ conductor specified in Table 250.66.
    a.   service-entrance
    b.   grounding-electrode
    c.   equipment grounding
    d.   ungrounded service

5. For service-entrance phase conductors larger than 1100 kcmil copper or 1750 kcmil aluminum, the grounded conductor cannot be smaller than ____ percent of the area of the largest service-entrance phase conductor.
    a. $12\frac{1}{2}$
    b. $14\frac{1}{2}$
    c. $13\frac{3}{4}$
    d.   12

6. Where the service-entrance phase conductors are installed in ____, the minimum size of the grounded service conductor is required to be based on the equivalent area of the ungrounded service conductors.
    a.   a raceway
    b.   a trench
    c.   parallel
    d.   a cable tray

7. Where more than one service disconnecting means is located in an assembly listed for use as service equipment, only one grounded conductor is required to be run to the assembly. It is required to ____ to the assembly enclosure.
    a.   be sized at 8 AWG and bonded
    b.   not be bonded
    c.   be bonded
    d.   be sized at 6 AWG and bonded

8. Some utility policies prohibit connection within metering equipment, and many inspection authorities prohibit that connection location because they interpret the connection to be no longer ____.
    a.   accessible
    b.   identified
    c.   serviceable
    d.   visible

9. The grounding electrode conductor connection to the grounded service conductor is most often made within the service disconnecting means and is connected to the ____ bus.
    a.   unidentified
    b.   isolated ground
    c.   floating neutral
    d.   neutral or grounded conductor

10. Where the transformer supplying the service is located outside the building an additional grounding connection to an electrode is required to be made outside the building, usually at the ____.
    a.   transformer
    b.   pole
    c.   meter
    d.   grid

11. Where copper conductors sized at 500 kcmil are selected for the ungrounded service conductors supplying a 400-ampere service, and are installed from a grounded system, the size of the copper grounding electrode conductor cannot be less than ____ AWG.
   a.   8
   b.   6
   c.   4
   d.   1/0

12. Section 310.15(B)(6) requires or permits the grounded service conductor for dwelling services and feeders to be:
   a.   two sizes smaller than the ungrounded conductors.
   b.   used for 60-ampere feeders.
   c.   sized to meet the requirements of 215.2, 220.22, and 230.42.
   d.   Type TW conductors.

13. Industrial plants having a continuous industrial process often have a need for ____ electrical power and systems.
   a.   isolated
   b.   larger
   c.   two-phase, five-wire
   d.   uninterrupted

14. Certain conditions must be met before the *NEC* will permit high-resistance grounded systems to be installed. They include ____.
   a.   qualified persons must be available to service and maintain the system.
   b.   continuity of power is required, line-to-neutral loads are not served.
   c.   ground detectors must be installed to indicate an insulation failure.
   d.   all of the above.

15. Secondary circuits of current and potential instrument transformers where the primary windings are connected to circuits of ____ volts or more to ground and, where on switchboards, are required to be grounded irrespective of voltage.
   a.   300
   b.   100
   c.   200
   d.   150

16. Where a switchboard has no live parts or wiring exposed or accessible to other than qualified persons, such instrument transformer secondary circuits are not required to be grounded where the primary windings are connected to circuits of ____ volts or less.
   a.   480
   b.   1000
   c.   277
   d.   150

17. Instrument transformer cases or frames are required to be grounded where accessible to other than qualified persons. Such cases or frames of current transformers are not required to be grounded where the primaries are not over ____ volts to ground, and the current transformers are used exclusively to supply current to meters.
   a.   100
   b.   120
   c.   130
   d.   150

Key Notes

Service and Main Bonding Jumpers

OBJECTIVES

After studying this chapter, the reader will be able to understand

- Definitions of bonding and bonding jumpers.
- Functions of the main bonding jumper.
- Sizing of main and equipment bonding jumpers.
- Methods for bonding at service equipment.
- Use of neutral for bonding on line side of service.
- Requirements for grounding and bonding of remote metering.

The main bonding jumper is one of the most critical elements in the safety grounding system. This conductor is the link between the grounded service conductor, the equipment grounding conductor and, in some cases, the grounding electrode conductor. The primary purpose of the main bonding jumper is to carry the ground-fault current from the service enclosure and from the equipment grounding system that is returning to the source. In addition, where the grounding electrode conductor is connected directly to the grounded service conductor bus, the main bonding jumper ensures that the equipment grounding bus is at the same potential as the earth.

Definitions:

"Bonding (Bonded): The permanent joining of metallic parts to form an electrically conductive path that ensures electrical continuity and the capacity to conduct safely any current likely to be imposed."

"Bonding jumper, Main: The connection between the grounded circuit conductor and the equipment grounding conductor at the service."[1]

Main Bonding Jumper

For a grounded system, 250.28 requires that "an unspliced main bonding jumper be used to connect the equipment grounding conductor(s) and the service-disconnect enclosure to the grounded conductor" of the electrical system. This connection must be made within the enclosure for each service disconnect.

An example of this is two or more service disconnecting means in individual enclosures that are grouped at one location. This type of installation often is made with a wireway or a short section of busway installed downstream from the metering equipment. In other cases, a wireway or short section of busway is installed ahead of metering and is supplied by a service lateral or service-entrance conductors. Sets of service-entrance conductors supply each of the service disconnecting means. Service disconnecting means are installed from the wireway or auxiliary gutter. [If there are nipples between the disconnecting means and the metal or nonmetallic trough, the trough meets the definition of a wireway from Article 376 (metal wireways) or 378 (nonmetallic wireways) rather than an auxiliary gutter from Article 366]. Section 250.28 requires a main bonding jumper be installed in each service-disconnect enclosure. As previously mentioned, 250.24(B) requires that the grounded service conductor be brought to each service disconnecting means and be bonded to the enclosure. The main bonding jumper is the means to accomplish this requirement.

The rules are a little different where more than one service disconnecting means is in a common enclosure. This equipment usually consists of listed switchboards, panelboards, or motor control centers. "Where more than one service disconnecting means is located in an assembly listed for use as service equipment," 250.28 Exception No. 1 permits the grounded service conductors to be run to a single grounded conductor bus in the enclosure and then be bonded to the assembly enclosure. This means that only one main bonding jumper connection is required to be installed from the common grounded conductor bus to the assembly enclosure. The sections of the assembly are bonded together by means of an equipment grounding conductor bus and by being bolted together.

Exception No. 2 to 250.28 permits alternate means for bonding of impedance grounded neutral systems (see chapter four for methods and requirements for grounding high-impedance grounded

Figure 5-1. Main Bonding Jumper

- Bus for neutral or grounded conductor
- Main bonding jumper may be wire, bus or screw, green finish if screw
- Equipment grounding bus bonded to enclosure

(All conductors not shown)

Figure 5-2. Main Bonding Jumper – Multiple Enclosures

Main bonding jumper in each enclosure

Assembly listed for use as service equipment. 250.28 Exc. No. 1

Grounded Service Conductors Terminal Bar Main Bonding Jumper Equipment Grounding Bus

Figure 5-3. Main Bonding Jumper for Listed Assembly

neutral systems). Also, see 250.36 and 250.186 for the specific requirements and allowances.

The main bonding jumper is permitted to consist of a wire, bus, screw, or other suitable conductor. It must be fabricated of copper or other corrosion-resistant material. Aluminum alloys are permitted where the environment is acceptable. In addition, where the main bonding jumper consists of a screw, it must have a green finish that is visible with the screw installed. This green finish assists in identifying the bonding-jumper screw from the other screws that are on or near the neutral bus [see 250.28(A) and (B)].

Functions of Main Bonding Jumper

The main bonding jumper performs three major functions:

1. Connects the grounded service conductor to the equipment grounding bus or conductor and the service enclosure.

Photo 5-1. Main Bonding Jumper in Service Disconnect

Figure 5-4. Major Functions of Main Bonding Jumper

2. Provides the low-impedance path for the return of ground-fault currents to the grounded service conductor. The main bonding jumper completes the ground-fault return circuit from the equipment through the service to the source as is illustrated in figure 5-4.

3. Connects the grounded service conductor to the grounding electrode conductor. Under certain conditions given in 250.24(A)(4), it is permitted to connect the grounding electrode conductor to the equipment grounding terminal bar rather than the terminal or bus for the grounded service conductor. This scheme is common on larger switchboards and service equipment and is necessary for proper operation of certain types of ground-fault protection equipment (see chapter fifteen for additional information on this subject).

Size of Main Bonding Jumper in Listed Enclosures

Where listed service equipment consisting of a switchboard, panelboard or motor control center is installed, the main bonding jumper that is provided with the equipment is rated for the size of conductors that would normally be used for the service. The method for sizing of the main bonding jumper in listed service equipment is found in Underwriters Laboratories Safety Standard for the equipment under consideration and is verified by the listing agency. Therefore, if a main bonding jumper that is a busbar, strap, conductor, or screw is furnished by the manufacturer as part of the listed equipment, it may be used without calculating its adequacy. Section 408.3(C) requires the equipment manufacturer to provide the main bonding jumper. If equipment is listed as service equipment, it will be marked with a label that identifies it as being "Suitable for use as service equipment." This equipment has been evaluated for use as service equipment and will include appropriate provisions for connecting the grounded conductor to the equipment grounding conductor or enclosure. This connection means is usually one of the forms specified in 250.28(A) or (B). Equipment that is marked "Suitable for use Only as Service equipment" usually is manufactured with the grounded conductor bus or terminal bar electrically and mechanically bolted to the enclosure.

This type of equipment may be installed only in the service position. There is no main bonding

Figure 5-5. **Main Bonding Jumper at Single Disconnect**

Three 250 kcmil conductors per set
3 x 250 - 750 kcmil
Refer to Table 250.66
Select 1/0 AWG copper or 3/0 AWG aluminum

Figure 5-6. **Main Bonding Jumper for Parallel Runs**

jumper in equipment that is suitable for use only as service equipment, but there may be provisions in larger equipment for disconnecting the grounded conductor bus from the enclosure for testing purposes only. More information on these two types of equipment can be found in UL 67 *Standard for Panelboards* and UL 891 *Standard for Switchboards*.

Size of Main Bonding Jumper at Single Service Disconnect or Enclosure

Since the main bonding jumper must carry the full ground-fault current of the system back to the grounded service conductor (which may be a neutral), its size must relate to the rating of the service conductors which supply the service. The minimum size of the main bonding jumper is found in Table 250.66 as required by 250.28(D). This relationship is based on the conductor's ability to carry the expected amount of fault current for the period needed for the overcurrent device to open and stop the flow of current.

For example, where 250 kcmil aluminum service-entrance conductors are installed, the main bonding jumper is found to be 4 AWG copper or 2 AWG aluminum by reference to Table 250.66.

The size of the main bonding jumper does not directly relate to the rating of the service overcurrent device. Do not attempt to use Table 250.122 for this purpose. Table 250.122 gives the minimum size of equipment grounding conductors for feeders and circuits on the load side of the service.

Sizing of Main Bonding Jumper for Parallel Service Conductors

Where service conductors are installed in parallel (connected together at each end to form a larger conductor), the total circular mil area of the conductors connected in parallel for one phase are added together to determine the minimum size main bonding jumper required [see 250.28(D)]. For example, where three 250 kcmil conductors are connected in parallel per phase, they are treated as a single 750 kcmil conductor. By reference to Table 250.66 the main bonding jumper, if aluminum service-entrance conductors are used, is 1/0 AWG copper or 3/0 AWG aluminum.

Where the service-entrance conductors are larger than the maximum given in Table 250.66, Section 250.28(D) requires the main bonding jumper to be not less than $12^{1}/_{2}$ percent (0.125) of the area of the largest phase conductors.

This is illustrated by the following example:
Three 500 kcmil copper conductors are installed in parallel as service-entrance conductors.
3 x 500 kcmil = 1500 kcmil.
1500 x .125 = 187,500 circular mils.

Since a 187,500 circular mil conductor is not a standard size, refer to *NEC* Chapter 9, Table 8, to find the area of conductors.

The next conductor exceeding 187,500 circular mils is a 4/0 AWG conductor which has an area of 211,600 circular mils. It is always necessary to go to the next larger size conductor since the $12\frac{1}{2}$ percent size is the minimum size permitted. Follow a similar procedure for determining the minimum size main bonding jumper required for other sizes of parallel service-entrance conductors.

Bonding of Service-conductor Enclosures

Special rules are provided for bonding enclosures on the line side of the service disconnecting means. This is because this equipment does not have overcurrent protection on its line side such as feeders and branch circuits have. Fault current of sufficient magnitude must flow during a short period to allow the fuse on the line side of the utility transformer to open. The level of fault current and particularly the duration the current may flow could be much larger than would flow in a feeder or branch circuit, as there is not an overcurrent device in series with the conductor.

The basic rule is that all metallic enclosures that contain a service conductor must be bonded together. The bonding ensures that none of the equipment enclosures can become isolated elec-

Figure 5-7. Size of Equipment Bonding Jumper on Line Side of Service

Figure 5-8. Size of Equipment Bonding Jumper on Line Side of Service

trically and become a shock hazard should a line-to-ground fault occur. The bonding also provides a low-impedance path for fault current to flow in so the fuse or circuit breaker on the line side of the electric utility transformer will open.

Sizing of Equipment Bonding Jumper on Line (Supply) Side of Service

Equipment bonding jumpers on the line side of the service and main bonding jumper must be sized to comply with Table 250.66. This bonding jumper sizing is required by 250.102(C). For example, where 250-kcmil copper conductors are installed as service-entrance conductors, Table 250.66 requires a 2 AWG copper or 1/0 AWG aluminum bonding jumper.

Where the sum of the circular mil area of the service-entrance phase conductors exceeds 1100-kcmil copper or 1750-kcmil aluminum, the equipment grounding conductor must be not less than $12\frac{1}{2}$ percent (0.125) of the area of the ungrounded phase conductors.

Sizing of Equipment Bonding Jumper for Parallel Conductors

Two methods are provided for bonding service raceways that are installed in parallel. The first method is to add the circular mil area of the service-entrance conductors per phase together and treat them as a single conductor. The bonding jumper size is determined from Table 250.66 and is connected to each conduit bonding bushing in a *daisy-chain fashion*. This method often results in an equipment bonding jumper that is quite large and difficult to work with.

For example, if five 250-kcmil copper conductors are installed in parallel for a phase, the equipment bonding jumper for bonding the metal raceways must not be smaller than 3/0 copper.

This is determined as follows:
Five x 250 kcmil = 1250 kcmil
1250 kcmil x .125 = 156,250 circular mils
The next larger conductor found in Chapter 9, Table 8, is 3/0 with an area of 167,800 circular mils.

In this case, a 3/0 copper equipment bonding conductor must be connected from the grounded service conductor or equipment grounding bus to each metal raceway in series (daisy-chain fashion from one raceway to another) [see figure 5-7].

A more practical method of performing the

bonding for services supplied by multiple raceways may be to connect an individual bonding conductor between each raceway and the grounded service conductor terminal bar or equipment grounding bus. This is permitted by 250.102(C). This will usually result in a smaller equipment bonding conductor which is easier to install.

Again, using the example above and referring to Table 250.66, the minimum size equipment bonding conductor for the individual raceways containing 250-kcmil copper service-entrance conductors is 2 AWG copper or 1/0 AWG aluminum. A properly sized equipment bonding jumper is installed from the terminal bar for the grounded service conductor or from the equipment grounding terminal bar to each conduit individually (see figure 5-8).

Bonding Service Equipment Enclosures

The *Code* requires that electrical continuity of service equipment and enclosures that contain service conductors be established and maintained by bonding. The items required to be bonded together are stated as follows in 250.92(A):

(1) The service raceways, cable-trays, cablebus framework, auxiliary gutters, or service cable armor or sheath.

(2) All service equipment enclosures containing service conductors, including meter fittings, boxes or the like, interposed in the service raceway or armor.

(3) Any metallic raceway or armor enclosing the grounding electrode conductor as provided in 250.64(B). (This subject is covered in detail in chapter seven.)

An exception to this requirement for bonding

• Service raceways and cable armor or sheath

• Service equipment enclosures including meter fittings

• Metallic raceway or armor enclosing GEC

Figure 5-9. Bonding Service Equipment Enclosures

at service equipment is mentioned in 250.92(A)(1). It refers to 250.84 which has rules on underground service cables that are metallically connected to the underground service conduit. The *Code* points out that if a service cable contains a metal armor, and if the service cable also contains an uninsulated grounded service conductor which is in continuous electrical contact with its metallic armor, then the metal covering of the cable is considered to be adequately grounded.

Use of Neutral for Bonding on Line Side of Service

Section 250.92(B) permits the use of the grounded service conductor (may be the neutral) for grounding and bonding equipment on the line side of the service disconnecting means. This is also permitted by 250.142(A)(1). (Two other applications of this method of bonding are explored in later chapters of this book.) Often, connecting the grounded service conductor to equipment such as meter bases, current transformer enclosures, wireways or auxiliary gutters is the most practical method of bonding these enclosures.

Usually self-contained meter sockets and meter-main combination equipment are produced with the grounded conductor terminals or bus (often a neutral) bonded directly to the enclosure. The meter enclosure is then effectively bonded when the grounded conductor is connected to these terminals within the meter socket enclosure. No additional bonding conductor connection to the meter enclosure is required. Current from a ground fault to the meter-main enclosure will return to the source by the grounded service conductor (may be a neutral) and, hopefully, will allow enough current to flow in the circuit to operate the overcurrent protection on the line side of the utility or other transformer.

In addition, meter enclosures installed on the

Figure 5-10. Use of Neutral for Bonding on Line Side of Service

load side of the service disconnecting means are permitted to be grounded (bonded) to the grounded service conductor provided that:

(a) service ground-fault protection equipment is not installed; and

(b) the meter enclosures are located near the service disconnecting means, [note: no distance is given to clarify what is meant by the word *near*], and

(c) the size of the grounded circuit conductor

Threaded couplings or bosses on enclosures made up wrenchtight.

Conduit hub furnished in many trade conduit sizes as accessory by equipment manufacturer. Install according to manufacturer's instructions.

Threadless couplings and connectors made up tight for rigid metal conduit, intermediate metal conduit and electrical metallic tubing.

Figure 5-11. Methods of Bonding Service Equipment

is not smaller than the size specified in Table 250.122 for equipment grounding conductors.

Method of Bonding at Service Equipment

The methods for bonding at service equipment are outlined in 250.92(B). These requirements for bonding are more restrictive at services than downstream from the service disconnect. The reason this is so important is service equipment and enclosures may be called upon to carry heavy fault currents in the event of a line-to-ground fault.

The service conductors in these enclosures have only short-circuit protection provided by the overcurrent device on the line side of the utility transformer. Only overload protection is provided at the load end of the service conductor by the overcurrent device. This is one of the reasons the *Code* limits the length of service conductors inside a building.

Bonding of these enclosures is to be done by one or more of the following methods specified in 250.92(B):

(1) Bonding to the grounded service conductor through the use of exothermic welding, listed pressure connectors such as lugs, listed clamps, or other listed means. These connections cannot depend solely upon solder.

(2) Threaded couplings and threaded bosses in a rigid or intermediate metal conduit system where the joints are made up wrench-tight. Threaded bosses include hubs that are either formed as a part of the enclosure or are supplied as an accessory and installed according to the manufacturer's instructions.

(3) Threadless couplings and connectors are permitted where they are made up tight for rigid and intermediate metal conduit and elec-

Photo 5-2. Bonding Bushing at Service Disconnect

Weatherproof conduit hub for field application.

Provided with "O" ring seal.

Boss is provided to accept lug for bonding

Photo 5-3. Weatherproof Conduit Hub

trical metallic tubing and metal-clad cables.
(4) Other approved devices such as bonding wedges and bonding-type locknuts and bushings.

Bonding jumpers are required to be used around concentric or eccentric knockouts that are punched or otherwise formed to impair an ad-

Bonding Locknut - Used where no concentric or eccentric knockouts remain. Standard locknut opposite side.

Bonding Wedge - Use with bonding jumper around concentric or eccentric knockouts; with or without bonding jumper where no concentric or eccentric knockouts. Standard locknut opposite side.

Figure 5-12. Bonding Fittings

Conduit insulated grounding bushing under test. 1½ in. bushing with 4 AWG copper conductor.

Bushing and wire during test with 2,450 amperes.

Conduit bushing at end of test. Conductor failed.

Photo 5-4. Conduit Bonding Bushing

equate electrical path for ground-fault current. It is important to recognize that concentric and eccentric knockouts in enclosures such as panelboards, wireways, and auxiliary gutters have not been investigated for their ability to carry fault current. Where any of these knockout rings remain at the conduit connection to the enclosure, they must always be bonded around to ensure an adequate fault-current path.

The *Code* states that "Standard locknuts or bushings shall not be the sole means for the bonding required by this section."[2]

This statement does not intend to prevent the use of "standard" locknuts and bushings, it is just that they cannot be relied upon as the sole means for the bonding that is required by this section. Standard locknuts are commonly used outside the enclosure on conduit that is bonded with a bonding bushing or bonding locknut inside the enclosure. Standard locknuts are used to make a good, reliable mechanical connection as required by 300.10. They must be tightened wrench tight to prevent arcing during the flow of fault current.

Different Conductor Material

Section 250.28(D) provides instructions on sizing the main bonding jumper or equipment bonding jumper on the supply side of the service where different conductor materials are used for the service-entrance conductors and the bonding jumper. The procedure involves assuming the phase conductors are of the same material (copper or aluminum) as the bonding jumper and that they have an equivalent ampacity to the conductors that are installed. This is illustrated as follows:

- Assume aluminum phase conductors and a copper bonding jumper are installed.
- Three 750-kcmil Type THW aluminum conductors are installed.
- Determine the ampacity of the phase conductors from Table 310.16: 385 amperes x 3 = 1155 amperes. The smallest type THW copper conductor that has an equivalent rating is 600 kcmil with an ampacity of 420.
- Next, determine the total circular mil area of the copper conductors.
  Three x 600 kcmil = 1800 kcmil.
  1800 kcmil x .125 = 225 kcmil.

The next standard size is 250 kcmil copper which is the minimum size bonding jumper permitted to bond equipment at or ahead of the service equipment in this example.

Parallel Bonding Conductors

Section 250.102(C) requires that where service-entrance conductors are paralleled in two or more raceways or cables and the equipment bonding

These requirements for bonding are more restrictive at services than downstream from the service.

jumper is routed with the raceways or cables, the equipment bonding jumper must be run in parallel.

In this case, again, the size of the bonding jumper for each raceway is based upon the size of the service-entrance conductor in the raceway by referring to Table 250.66.

Grounding and Bonding of Remote Metering

As mentioned before, 250.92(A) requires all equipment containing service conductors to be bonded together and to the grounded service conductor. This includes remote (from the service equipment) meter cabinets and meter sockets.

Grounding and bonding of equipment such as meters, current transformer cabinets, raceways, and auxiliary gutters to the grounded service conductor at locations on the line side of and remote from the service disconnecting means increases safety.

This equipment should never be grounded only to a grounding electrode such as a ground rod. Figures 5-14 and 5-15 show why. If a ground fault occurred at this line-side equipment, and it is not bonded as required, the only means for clearing a ground fault would be through the grounding electrodes and earth. Given the relatively high impedance and low current-carrying capacity of this path through the earth and high resistance of grounding electrodes such as rods, little current will flow in this path. This leaves the equipment enclosure(s) at a dangerous voltage-to-ground potential just waiting to shock or possibly electrocute a person or animal that may contact it.

The voltage drop across this portion of the circuit can easily be calculated by using Ohms Law. (Resistance times the current gives the voltage.) There are many records of livestock being electrocuted while contacting electrical equipment that was improperly grounded. Sections 250.4 and 250.54 require that the earth not be used as the sole equipment grounding conductor or fault-current path.

The most practical method for grounding and bonding this line-side equipment is to bond the grounded service conductor to it. As can also be seen in figures 5-14 and 5-15, a ground fault to the equipment will have a low-impedance path

Equipment bonding jumper must be installed in parallel where routed with phase conductors that are installed in parallel per 250.102(C)

Figure 5-13. Parallel Bonding Conductors

Bond remote CT enclosure to neutral

Supplemental grounding electrode permitted

Service equipment

Figure 5-14. Grounding and Bonding of Remote Metering

back to the source through the grounded service conductor. This will allow a large current to flow in the circuit to cause the overcurrent protection on the line side of the transformer to clear the fault.

Supplementary Grounding Electrodes

In accordance with 250.54, it is permissible to install a grounding electrode at the remote meter location shown in figures 5-14 and 5-15 to supplement the grounded service conductor. This section refers specifically to grounding electrodes supplementing the equipment grounding conductors. Some electric utilities require a grounding electrode at meter equipment installed remote from service equipment such as on poles. Section 230.66 makes it clear that individual meter socket enclosures are not to be considered service equipment. The same is true for metering equipment installed in remote current-transformer enclosures. As mentioned earlier, it is critically important that these meter enclosures be properly bonded as they contain service conductors.

This additional grounding electrode will attempt to keep the equipment at the earth potential that exists at the meter location. In addition, the electrodes at the remote meter and at the service location are bonded together by the grounded service conductor installed between the metering and service equipment.

This brings the installation into compliance with 250.58 which requires a common grounding electrode or where two or more electrodes are installed, they must be bonded together.

Grounding electrodes should never be used as the only means for grounding or bonding these enclosures or to carry fault current.

As previously stated, these grounding electrodes should never be used as the only means for grounding or bonding these enclosures or to carry fault current. An extensive discussion of this subject is found in chapter six.

Bonding of Multiple Service Disconnecting Means

Installation of multiple services as permitted by 230.2(A) through (D) and installations of services that have multiple disconnecting means can take several forms.

Additional services are permitted by 230.2 for:
(A) Fire pumps, emergency, legally required, standby, optional standby, or parallel power production systems.
(B) By special permission, for multiple occupancy buildings where there is no available space for service equipment that is accessible to all occupants or, for a single building or structure that is large enough to make two or more services necessary.
(C) Capacity requirement where the service capacity requirements exceed 2000 amperes at 600 volts or less, where load requirements of a single-phase installation are greater than the serving utility normally provides through a single service, or by special permission (related to capacity requirements).

Metal raceway and meter enclosure must be bonded to system neutral

Supplemental electrode permitted

Figure 5-15. Bonding Remote Meter Loop

(D) Different characteristics of the services such as different voltages, frequencies, or phases, or for different uses, such as for different rate schedules.[3]

The basic rule for sizing the equipment bonding jumper for bonding these various configurations is found in 250.102(C). This section requires that the bonding jumpers on the line side of each service and the main bonding jumper be sized from Table 250.66. In addition, the size of the bonding jumper for each raceway is based on the size of service-entrance conductors in each raceway. As discussed earlier, conductors larger than given in Table 250.66 are required for larger services. Since different sizes of service-entrance conductors may be installed at various locations, the minimum size of the equipment bonding conductor and main bonding jumper is based on the size of the service-entrance conductors at each location.

For example, the appropriate size of bonding jumper for the installation in figure 5-16 with the assumed size of conductors is as follows: (all sizes copper)

a. 500 kcmil in service mast and nipple has a bonding jumper of 1/0 AWG

b. 1000 kmil in wireway has a bonding jumper of 2/0 AWG

c. 300 kcmil to 300 ampere service has a bonding jumper of 2 AWG

d. 3/0 AWG to 200 ampere service has a bonding jumper of 4 AWG

e. 2 AWG to 125 ampere service has a bonding jumper of 8 AWG

A practical method for bonding the current transformer enclosure and wireway (sometimes referred to as a *hot gutter*) is to connect the grounded service conductor directly to the current transformer enclosure or wireway. This may be done by bolting a multi-barrel lug directly to the wireway and connecting the neutral or grounded service conductors to the lug. Be sure to remove any nonconductive paint or other coating that might insulate the connector from the enclosure. As previously discussed, the grounded service conductor must also be extended to each service disconnecting means and be bonded to the enclosure.

---

[1] NFPA 70, *National Electrical Code 2002*, Article 100, (Quincy, MA, National Fire Protection Association, 2002), p. 70-33.

[2] NFPA 70, 250.92(B), p. 70-109.

[3] NFPA 70, 230.2, p. 70-76.

(All conductors not shown)

Figure 5-16. Bonding of Multiple Service Disconnecting Means

Chapter Five: The questions included here were developed using material included in this chapter. The answers can be found by reviewing the text. It is also important that students make use of the 2002 *NEC*, where many of the answers can be found.

1. The permanent joining of metallic parts to form an electrically conductive path that will assure electrical continuity and the capacity to conduct safely any current likely to be imposed is defined as ____.
   a. grounding
   b. bonding
   c. welded
   d. grounded

2. The connection between the grounded circuit conductor and the equipment grounding conductor at the service is defined as the ____.
   a. main bonding jumper
   b. grounding electrode conductor
   c. equipment bonding jumper
   d. neutral conductor

3. The main bonding jumper is permitted to consist of a ____ or other suitable conductor.
   a. wire
   b. bus
   c. screw
   d. all of the above

4. Where the main bonding jumper consists of a ____ only, it is required to have a green finish that is visible with the ____ installed.
   a. bus
   b. screw
   c. wire
   d. jumper

5. Where ____ kcmil aluminum service-entrance conductors are installed, the main bonding jumper is required to be 4 AWG copper or 2 AWG aluminum.
   a. 1/0
   b. 2/0
   c. 3/0
   d. 250

6. Where the service-entrance conductors are larger than the maximum sizes given in Table 250.66, the main bonding jumper cannot be less than ____ percent of the area of the largest phase conductor.
   a. $9\text{-}^1/_2$
   b. $10\text{-}^1/_2$
   c. $11\text{-}^1/_2$
   d. $12\text{-}^1/_2$

7. Where conductors are used in parallel, and the sum of the circular mil area exceeds ____ kcmil copper or ____ kcmil aluminum, the equipment bonding conductor must be not less than $12^1/_2$ percent of the area of the ungrounded phase conductors.
   a. 1100 - 1750
   b. 1000 - 1650
   c. 1050 - 1500
   d. 1075 - 1400

8. The *Code* requires that electrical continuity of the grounding circuit of service equipment and enclosures that contain service conductors be obtained by ____.
   a. grounding
   b. welding
   c. approval
   d. bonding

9. The grounded circuit conductor (may be a neutral) is permitted to be used for grounding and bonding on the ____ side of the service disconnecting means.
   a. load
   b. line
   c. subpanel
   d. control center

10. Grounding and bonding of equipment such as meters, current transformer cabinets and raceways to the grounded service conductor at locations on the line side of and remote from the service disconnecting means increases ____.
   a. voltage
   b. current
   c. safety
   d. cost

11. The equipment bonding jumper for raceways containing service-entrance conductors must be sized according to:
   a. Section 250.102(C), Table 250.66
   b. Table 250.122
   c. Table 310.16.
   d. none of the above

12. Which of the following are NOT permitted to be the sole means for bonding enclosures for service-entrance conductors
   a. bonding bushings
   b. bonding locknuts
   c. threadless couplings
   d. standard locknuts

Key Notes

Grounding Electrodes

OBJECTIVES

After studying this chapter, the reader will be able to understand

- Definition and general requirements for grounding electrodes.
- Grounding electrode system to be used.
- Sizing bonding jumper for grounding electrode system.
- Description and installation of made electrodes.
- Common grounding electrode.
- Objectionable current flow and resistance of grounding electrodes.

Grounding electrodes provide the essential function of connecting the electrical system or electrical equipment to the earth. The earth is considered to be at zero potential. In some cases, the grounding electrode serves to ground the electrical system. In other instances, the electrode is used to connect non-current carrying metallic portions of electrical equipment to the earth. In both situations, the primary purpose of the grounding electrode is to maintain the electrical equipment at the earth potential present at the grounding electrode.

Another essential function of the grounding electrode is to dissipate overvoltages into the earth. These overvoltages can be caused by high-voltage conductors being accidentally connected to the lower-voltage system such as by a failure in a transformer or by an overhead conductor dropping on the lower-voltage conductor. Overvoltages can also be caused from lightning.

Section 250.24(C) requires the equipment grounding conductors, the service-equipment enclosures, and, where the system is grounded, the grounded service conductor to be connected to a grounding electrode. The conductor used to make this connection is the grounding electrode conductor.[1]

Definition
Grounding electrode. A conducting element used to connect electrical systems and/or equipment to the earth.

Grounding Electrode System
Section 250.50 requires that, if available on the premises at each building or structure served, all grounding electrodes be bonded together to form the grounding electrode system. This includes metal underground water pipes, metal frames of

1. Connects the electrical system to earth.

2. Connects electrical equipment to earth.

3. Attempts to maintain equipment at earth potential.

4. Little effect in clearing ground faults.

Figure 6-1. Functions of Grounding Electrode

Figure 6-2. Dissipation of Overvoltages

(A)(1) Metal Underground Water Pipe

(A)(2) Metal Frame of the Building or Structure

(A)(3) Concrete-Encased Electrode

(A)(4) Ground Ring

(A)(5) Rod and Pipe Electrodes

(A)(6) Plate Electrodes

(A)(7) Other Local Metal Underground Systems or Structures

If available on the premises at each building or structure served item (A)(1) through (A)(6) shall be bonded together to form the grounding electrode system. Where none of these are available, one or more of the electrodes in item (A)(4) through (A)(7) shall be installed and used.

Figure 6-3. 250.50 Grounding Electrode System

buildings, concrete-encased electrodes, and ground rings. It should be noted that the term *available* is vague and can lead to controversy. The general requirement is that a bonding jumper is required to be installed between the grounding electrodes to bond them together. A grounding electrode conductor is run from the service enclosure to one of the grounding electrodes that are bonded together to form the grounding electrode system. The *NEC* also provides for the option of running a grounding electrode conductor to each grounding electrode individually as provided in 250.64(F).

Where the interior metal water pipe is used as a part of the grounding electrode system or as a conductor to bond other electrodes together to create the *grounding electrode system,* 250.52(A)(1) requires that all connections to metal water piping electrodes take place within the first 5 ft. from the point the water pipe enters the building. It does not require that the interior water pipe be used for the purpose of interconnecting other electrodes to form the grounding electrode system.

Any of the other electrodes, such as the metal frame of the building, concrete-encased electrode or ground ring, can be used for the purpose of interconnecting the other grounding electrodes. Where these other electrodes are used for this purpose, no restrictions are placed on where the connections are permitted to be made or how far inside the building they are permitted. Section 250.68(A) requires grounding electrode conductor connections to grounding electrodes to be accessible except for connections to a buried, driven or concrete-encased electrode.

Grounding Electrodes Required To Be Used

Section 250.52(A) lists the grounding electrodes permitted to be used to form a grounding electrode system and 250.52(B) gives the items that are not permitted to be used as grounding electrodes. All of the identified grounding electrodes are required to be used where "available on the premises at each building or structure served." The *Code* does not define what is meant by *available* nor does it require that the electrodes be made available where they are not. For example, if the building has encased concrete reinforcing rods when the electrical system is installed, it is not required that the rods be exposed for connec-

Figure 6-4. 250.52 Grounding Electrodes — (A) Electrodes Permitted for Grounding

tion. On the other hand, the concrete reinforcing rods are required to always be used when available. Several electrical inspection agencies require that a concrete-encased grounding electrode be connected to the system before approval of the service for utility connection is granted. The grounding electrodes are not listed in an order of preference nor is it optional to choose which ones to use.

Electrodes that are required to be used, in addition to any electrodes that already exist or are field installed at the building or structure served, where available are as follows:

1. Metal Underground Water Pipe. Defined in 250.52(A)(1) as "A metal underground water pipe that is in direct contact with the earth for 3.0 m (10 feet) or more (including any metal well casing that is effectively bonded to the pipe)...." There is no minimum or maximum pipe size given. Types of metal, such as steel, iron, cast iron, stainless steel or even aluminum are not distinguished. Different types of water pipes such as for potable water, fire protection sprinkler systems, irrigation piping, and so forth, are also not defined. As a result, all of these metal underground water pipes are required to be used where "available at each building or structure served."

Continuity of the grounding path of the water pipe grounding electrode or the bonding of interior piping systems cannot depend on water meters or on filtering devices or similar equipment. Where a water meter or filtering equipment is in this metal water piping system, a bonding jumper is required to be installed around the equipment to maintain continuity even if the water meter or filter is removed. The bonding jumper is required to be the same size as the grounding electrode conductor and long enough to allow the meter, filter, or other equipment to be removed without disconnecting the bonding jumper.

2. Metal Frame of the Building or Structure. Section 250.52(A)(2) requires the metal frame of the building be used as a grounding electrode where it is effectively grounded. *Grounded effectively* is defined in Article 100 and means that the metal frame of the building is, "Intentionally connected to earth through a ground connection or connections of sufficiently low impedance and having sufficient current-carrying capacity to prevent the buildup of

voltages that may result in undue hazards to connected equipment or to persons."[2]

To be an effective grounding electrode, the metal frame of the building is required to have a sufficiently low-impedance contact with the earth to pass current when called upon to do so and to maintain the electrical system at or near the electrical potential of the surrounding earth. The

> Continuity of the grounding path of the water pipe grounding electrode or the bonding of interior piping systems cannot depend on water meters or on filtering devices or similar equipment.

building steel can be connected to the earth by bolted or welded connection to reinforcing steel in foundations or footings that are in turn encased in concrete. Also, building structural steel may itself be encased in concrete that is in contact with the earth. In both of these cases, the concrete that encases the building steel or reinforcing steel is required to be in direct contact with the earth.

Certain back-fills such as gravel or vapor barriers may render the building steel an ineffective electrode. Building steel that is connected to concrete footings or foundations by only "J" bolts is not considered effectively grounded unless these "J" bolts are in turn connected to structural members such as reinforcing steel. The reinforcing steel needs to be near the base of the footing or foundation.

Structural steel should be tested with an earth resistance tester if there is any doubt about its resistance to ground and adequacy as a grounding electrode.

3. Concrete-Encased Electrodes. Section 250.52(A)(3) defines this grounding electrode as one or more steel reinforcing bars or rods that are not less than 20 feet in length and $1/2$ inch in diameter or 20 feet or more of bare copper conductor not smaller than 4 AWG. These electrodes are required to be located within or near the bottom of the foundation or footing and be encased by at least 50 mm (2 in.) of concrete. A single 6.0 m (20 ft.) length of reinforcing bar is not required. Reinforcing bars are permitted to be bonded together by the usual steel tie wires or other effective means like welding. Where subjected to high currents such as lightning strikes, welding might be preferred.

Reinforcing rods are required to be of bare, zinc galvanized or other electrically conductive coated steel material. Obviously, insulated reinforcement rods would not perform properly as a grounding electrode. Some complaints have been made that lightning surges, that are dissipated through this electrode, break out chunks of concrete where the surge exits the footing.

This grounding electrode is commonly referred to as the *Ufer ground* after H.G. Ufer who spent many years documenting its effectiveness (see figure 6-5). See the appendix for additional information on the development of the concrete-encased electrode.

Several electrical inspection agencies require that a concrete-encased electrode be installed or connected to the service prior to authorizing electrical service due to its effectiveness in most any climatic and soil condition.

4. Ground Ring. Section 250.52(A)(4) recognizes a copper conductor, not smaller than 2 AWG and at least 6.0 m (20 ft.) long, as a ground ring grounding electrode. The conductor is required to encircle the building or structure and be buried not less than 750 mm (30 in.) deep. Ground rings often are installed at telecommunication central offices, radio and cellular telephone sites. Where available on the premises served, ground rings are required to be used as one or more of the grounding electrodes making up the grounding electrode system.

Supplemental Electrode
Section 250.53(D)(2) requires that where the only grounding electrode available and connected at the building or structure served is a metal underground water pipe, it has to be supplemented by another grounding electrode.

Electrodes suitable to supplement the metal underground water pipe include: the metal frame of the building, a concrete-encased electrode, ground ring, other local metal underground systems or structures, rod, pipe, or plate electrodes.

Figure 6-5. Concrete Encased Electrode

Connect to:

• Grounded metal service raceway
• Grounded service-entrance conductor
• Grounded service enclosure
• Grounding electrode conductor

Supplemental grounding
electrode if rod, pipe
or plate must meet
25-ohm rule of 250.56

Figure 6-6. Supplemental Electrode Connection

This supplemental grounding electrode is required since metal underground water pipes are often replaced by plastic water services or the system continuity is interrupted by nonmetallic couplings or repairs. The effectiveness of the water pipe grounding electrode would thus be lost.

Specific locations are provided where the supplemental grounding electrode is permitted to be connected (see figure 6-6). Where an underground metal water pipe is the only grounding electrode, the supplemental grounding electrode is permitted to be connected only to the grounding electrode conductor, the grounded service-entrance conductor, the grounded service raceway or to any grounded service enclosure. An exception to this requirement permits the bonding connection to the interior metal water piping in qualifying industrial or commercial installations to be made at any location if the entire length of interior metal water pipe that is being used as a conductor is exposed (see definition of *exposed* in Article 100). Locations where the exposed piping passes through walls or floors are required to be perpendicular to those penetrations.

Often, changes, repairs, or modifications are made to the metallic water piping systems with nonmetallic pipe or fittings or dielectric unions. In this case, it is possible to inadvertently isolate portions of the grounding system from the grounding electrode conductor. This is another in several steps that has been taken over recent years to reduce the emphasis and reliance on the metal water piping system for grounding of electrical systems.

Where the supplemental grounding electrode is of the rod, pipe, or plate type, it is now required to meet the 25-ohm-to-ground rule in 250.56. This means that the supplemental grounding electrode has a resistance of not more than 25 ohms or a second supplemental grounding electrode is required to be used. This has the effect of the system being served by only the supplemental grounding electrodes in case the underground metal water pipe grounding electrode is interrupted for any reason.

Size of Bonding Jumper for Grounding Electrode System

Section 250.53(C) requires the bonding jumper used to connect the grounding electrodes of the grounding electrode system together to be installed in accordance with the requirements of 250.64(A), (B), and (E). The bonding jumper used to bond the grounding electrodes together to form the grounding electrode system is required to be sized in accordance with 250.66 based on the size of the ungrounded service-entrance conductor and shall be connected in a manner specified in 250.70. The conductor that connects the grounding electrodes together is a bonding conductor and not a grounding electrode conductor. The bonding conductors are not required to be installed in one continuous length as grounding electrode conductors are. In addition, the exceptions for sizing the ground-

Grounding electrode
conductor 250.66

Water pipe 250.66

Not larger than
ground ring

Minimum 4 AWG CU

Minimum 6 AWG CU

Concrete-encased
electrode

Ground ring
Minimum 2 AWG CU

Ground rods

Figure 6-7. Bonding Jumper for Grounding Electrode System

ing electrode conductor in 250.66 apply for the sizing of the bonding jumpers.

For example, if the service-entrance conductor is 500 kcmil copper, the minimum size of bonding jumper between grounding electrodes is determined by reference to 250.66 and Table 250.66, and its installation falls under the requirements of the rules in 250.64(A), (B) and (E), which are as follows:

- To metal underground water pipe and metal frame of a building; 2 AWG copper or 1/0 AWG aluminum conductor (from Table 250.66).

- To installed electrodes as provided in 250.52(A) such as pipes, rods, or plates; that portion of the bonding conductor that is the sole connection to the rod, pipe, or plate electrode; 6 AWG copper or 4 AWG aluminum. The term *sole connection* means that the bonding conductor is not connected to the made electrode being considered and then another grounding electrode is connected to it [see 250.66(A)].

- To a concrete-encased electrode as in 250.52(A)(3); that portion of the bonding conductor that is the sole connection to the concrete-encased electrode; 4 AWG copper conductor [see 250.66(B)].

- To a ground ring as in 250.52(A)(4); that portion of the bonding conductor that is the sole connection to the ground ring is not required to be larger than the ground ring conductor but a minimum size 2 AWG is required [see 250.66(C)].

Note that bare aluminum is not permitted to be installed as a grounding electrode conductor where in direct contact with masonry or the earth or where subject to corrosive conditions. Where used outside, aluminum or copper-clad aluminum grounding conductors shall not be terminated within 450 mm (18 in.) of the earth [see 250.64(A)].

No sequence for installing the bonding jumper or jumpers is given. However, the minimum wire size required to the various grounding electrodes is to be observed. In addition, the point where the grounding electrode connects to the grounding electrode system is required to provide for the largest required grounding electrode conductor. For example, it would be a violation to connect a 4 AWG bonding conductor from a concrete-encased

Figure 6-8. Size of Individual Grounding Electrode Conductor

grounding electrode to a building steel grounding electrode which would require a 3/0 copper grounding electrode conductor. The installation would be acceptable if the 3/0 copper grounding electrode conductor connects to the building steel and a 4 AWG copper bonding jumper extends from the building steel to the concrete-encased electrode. In addition, the unspliced grounding electrode conductor is permitted to run from the service equipment to any convenient grounding electrode.

Alternately, individual grounding electrode conductors are permitted to be installed from the service equipment to one or more grounding electrodes rather than the electrodes being bonded together in a circular or *daisy-chain* manner [see 250.64(F)]. The minimum size of each grounding electrode conductor to the individual grounding electrode is based on the electrode each individual grounding electrode conductor is connected to (see figure 6-8). Note that a grounding electrode conductor is permitted to *supply* or *serve* any number of grounding electrodes but is sized for the largest grounding electrode conductor required. For example, a bonding conductor is permitted to be run to a concrete-encased electrode and then to the underground metal water pipe. The bonding jumper is required to be sized for the largest grounding electrode conductor required for the grounding electrode or electrodes served.

Installed Electrodes

Where the electrodes described in 250.52(A)(1) through (A)(6) are not available at the service location, a grounding electrode is required to be installed and used. The grounding electrodes as provided in 250.52(A)(4) through (A)(7) may be either or any combination of the following types:

(A)(4) Ground Ring. A ground ring encircling the building or structure, in direct contact with the earth, consisting of at least 6.0 m (20 ft) of bare copper conductor not smaller than 2 AWG.

(A)(5)(a) Rod and Pipe Electrodes. Rod and pipe electrodes shall not be less than 2.5 m (8 ft) in length and shall consist of the following:

(a) Electrodes of pipe or conduit shall not be smaller metric designator 21 (trade size $^3/_4$) and, where of iron or steel, shall have the outer surface galvanized or otherwise metal-coated for corrosion protection.

(b) Electrodes of rods of iron or steel shall be at least 15.87 mm ($^5/_8$ in.) in diameter. Stainess steel rods less than 16 mm ($^5/_8$ in.) in diameter, nonferrous rods, or their equivalent shall be listed and shall not be less than 13 mm ($^1/_2$ in.) in diameter.

(A)(7)Local underground systems or structures

(A)(5) Rod and Pipe Electrodes

Minimum length 2.5 m (8 ft)

(A)(6) Plate Electrodes

Figure 6-9. Made Electrodes, 250-52

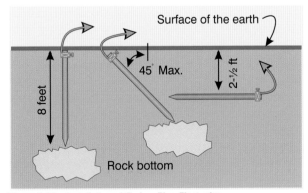

Surface of the earth

45° Max.

2-½ ft

8 feet

Rock bottom

Figure 6-10. Installation of Rod or Pipe Electrodes

(A)(6) Plate electrodes. Each plate electrode shall expose not less than 0.186 m² (2 ft²) of surface to exterior soil. Electrodes of iron or steel shall be at least 6.4 mm ($^1/_4$ in.) in thickness. Electrodes of nonferrous metal shall be at least 1.5 mm (0.06 in.) in thickness.

[*Editor's note:* underground metal gas piping systems are not permitted to be used as a grounding electrode. This does not eliminate the requirement that interior metal gas piping systems be bonded (see chapter eight for additional information on bonding of metal piping systems).]

(A)(7) Other Local Metal Underground Systems or Structures. Other local metal underground systems or structures such as piping systems and underground tanks.[3] [*Editor's note:* These objects are required to have the metal in direct contact with the earth. Protective coatings may render them ineffective as grounding electrodes.]

Installation of Grounding Electrodes

Installation requirements for grounding electrodes are covered in 250.53(A) through (H). Descriptions of the different types of grounding electrodes that are either permitted or not permitted are located in 250.52(A) and (B). Installation requirements for grounding electrodes such as depth, connections to, and electrode spacing between electrodes are found in 250.53.

Where practicable, field-installed electrodes such as rods, pipes, or plates are required to be installed below permanent moisture level. This is a key ingredient in establishing an effective electrode. They also are required to be free from nonconductive coatings such as paint and enamel.

Rod and pipe electrodes are required to be installed so at least 8 ft. is in contact with the soil. They are required to be driven vertically unless rock bottom is encountered. If rock bottom is encountered which prevents the rod from being driven 8 feet vertically, the rod is permitted to be installed at an oblique angle of not more than 45 degrees from vertical. Where rock bottom is encountered at an angle up to 45 degrees, the rod or pipe can be buried in a trench that is at least 750 mm (30 in.) deep [see 250.53(H)].

The upper end of the rod is required to be flush with or below ground level unless the aboveground end of the rod and the grounding

...

electrode attachment are protected from physical damage. This, of course, requires that a ground rod longer than 8 feet be used if any of the rod is exposed above ground level. For an eight-foot ground rod or pipe, the ground clamp is required to be listed for direct earth burial, as the electrode is required to be driven to its full length.

Installation requirements for plate electrodes are found in 250.53(H) and require the plate to be buried not less than 2½ feet below the surface of the earth.

Section 250.10 requires that ground clamps or other fittings be approved (acceptable to the authority having jurisdiction) for general use without protection or be protected from physical damage by metal, wood or equivalent protective covering.

Common Grounding Electrode

Often when more than one service supplies a building or structure, there is more than one utility transformer or source. Multiple services or sources can have differences of potential between them, and when installed on the same building or structure, use the same electrode. Section 250.58 requires that a common grounding electrode be used for all alternating-current system grounding in or at a building. In addition, where more than one service supplies a building, a common grounding electrode be used. This section recognizes that where two or more grounding electrodes are bonded together, they are considered to be a single grounding electrode or grounding electrode system.

Interestingly, no distance between electrodes is given beyond which the electrodes do not have to be bonded together. Buildings of *large area* are permitted by 230.2(B)(2) to have more than one service. However, nothing in the *Code* defines the dimensions of a *large building*. Some inspection authorities use voltage drop of major feeders for guidance in determining when a building is one of large area. Where feeder conductors would have to be increased in size unreasonably to maintain voltage regulation, one or more additional services are permitted.

Section 250.58 requires the grounding electrodes for the multiple services be bonded together no matter how far apart they are in the same building. This is important so there is not more than one earth potential impressed on equipment in or on the building. Section 250.53(C)

Same grounding electrode must be used for all services. Two or more electrodes, bonded together, are considered one.

Figure 6-11. Common Grounding Electrode 250.58

requires the bonding jumper(s) used for this purpose to be sized from Table 250.66 and be installed in accordance with 250.64(A), (B) and (E) (see chapter seven for additional information on installation of grounding electrode conductors). It is permitted to use the steel frame of a building that is effectively grounded or a concrete-encased grounding electrode to bond grounding electrodes from other services together.

Section 250.58 also requires that a common grounding electrode be used to ground conductor enclosures and equipment in or on the building and that the same grounding electrode be used to ground the system. This does not mean that one cannot use more than one grounding electrode. But, if more than one is used, then all the grounding electrodes are required to be bonded together to form a common grounding electrode. Where multiple grounding electrodes are bonded together as cited above, such multiple grounding electrodes become, in effect, a common grounding electrode system. Effectively grounded building steel grounding electrodes can be used to bond grounding electrodes together. For example, a large building is served by four services. Where each of the services is connected to the building steel for grounding electrode purposes and also has a ground rod installed at each service location, such action serves to bond the ground rods together (see figure 6-11).

Enhanced Grounding Electrodes

Grounding electrodes and grounding electrode systems are covered in Part III of Article 250. These are the minimum requirements for grounding electrodes for use in grounding services, systems, and equipment. With ever-increasing installations and use of information technology equipment and sensitive electronics, there is sometimes the need to exceed the minimum requirements which are established for safety of persons and property and install electrodes or electrode systems that are extensive in nature and designed to establish and maintain the lowest level of resistance in the connection to earth through the electrode. There are many products

Courtesy of Harger

Figure 6-12. Enhanced Grounding Electrodes

available to accomplish this additional grounding when desired for the electrical system.

Electrolytic Grounding Systems. Electrolytic grounding was invented by the XIT Rod Company in the late 1960s. It was specifically designed to eliminate the drawbacks found in other electrodes. This active electrode consists of a hollow copper pipe filled with natural earth salts. The salts extract moisture from the air which forms a highly conductive electrolytic solution. The solution continually weeps into the surrounding backfill material providing improved conductivity and seasonal stability. The electrode is installed in an augured hole or trench and backfilled with a specially processed bentonite clay. The clay is very conductive with nearly a neutral pH which helps protect the electrode from corrosion. Due to its high conductivity, the bentonite improves the ground system performance and also provides an excellent electrical bond between the electrode and the surrounding earth. These active electrodes are the only ones that improve with time; all others begin deterioration upon installation (see figure 6-13).[4]

Enhanced Grounding Electrode Types

These supplementary electrodes can be installed in a variety of methods. The nature of these elec-

Figure 6-13. Enhanced Grounding Electrode – Anatomy

Concept Courtesy of Lyncole XIT Grounding

trodes can also vary. Some are intended for a one-time install and no additional maintenance of the electrode is required. Others are installed and have to be maintained with chemicals or other effective means. The maintainable electrodes are more commonly installed under controlled conditions where there is qualified staff that will ensure proper supervision and maintenance of the electrode system. For additional information on these enhanced grounding electrodes and grounding electrode systems and their uses, see chapter seventeen.

Earth Return Prohibited

No mention is made to providing a low-resistance, low-impedance, common grounding electrode for clearing ground faults. The high impedance of the earth makes it an ineffective path for current flow. The earth should never be used as an equipment grounding conductor, as it is a very poor conductor. Figure 6-14 shows that the grounding electrode is in the earth return circuit. Even if the grounding electrode resistance to the earth was very low, it would have little effect on the clearing of a ground fault, because of the reactance of the earth and the soil in the ground-fault return circuit is very high. As discussed in chapter one, the greater the resistance or impedance, the less the amount of current flow. Where a parallel path exists through the earth and through the grounded service conductor, almost all of the ground-fault return current will flow to the source over the grounded service conductor. A low-resistance, common grounding electrode is valuable to the electrical installation in holding equipment and the grounded con-

ductor at or close to earth potential. It simply is not effective in clearing line-to-ground faults. Sections 250.4(A)(5) and 250.54 make it clear that the earth and grounding electrodes are not to be depended upon to function as equipment grounding conductors or used in the place of equipment grounding conductors. Both sections specifically state that the earth is not to be used as the sole or only equipment grounding conductor. However, grounding electrodes are permitted to supplement equipment grounding conductors and connected to the equipment grounding conductor(s).

If a ground fault should develop as shown in the upper drawing in figure 6-14 where two separate grounding electrodes are used, the fault-current flow will be through the service conductor then through the impedance of the ground fault,

> There is no requirement that additional electrodes such as ground rods or plates be installed until the 25 ohm-to-ground resistance is obtained.

the grounding electrode conductor, the grounding electrode, the path through the earth to the grounding electrode at the transformer and finally through the grounding conductor to complete the circuit to the transformer. It would be a rare case where that circuit resistance would add up to less than 12 ohms (while the impedance would be higher). At best, therefore, the fault current would not reach a value high enough to operate a 15-ampere overcurrent device on a 120 volt-to-ground circuit (120 ÷ 12 = 10 amperes).

Considering resistance only, the circuit shown has two grounding electrodes in series. Compared to the much lower resistance parallel path of the grounded circuit conductor, a resistance ratio between the two parallel paths is about 50 times for a 100-ampere service to well over 100 times for the larger services. When impedance of the two paths is considered, the ratio will be higher. Thus, almost all the current from a line-to-ground fault will return to the transformer over the grounded service conductor.

Under normal operating conditions some un-

Ground-fault at service attempts to return to source (transformer) through the high impedance path through the earth.

A low impedance path for clearing ground-faults is provided through the grounded service conductor.

Figure 6-14. Earth Return Prohibited

balanced return current will flow in the neutral back to its source. Some unbalanced neutral current will thus flow through the earth, but it will be small in comparison to that which will flow through the grounded service conductor.

Any belief that the circuit to the grounding electrode can be depended on to clear a ground fault is clearly erroneous no matter how large a grounding electrode conductor is used or how good a grounding electrode is. However, when the high-impedance earth path is short-circuited by installing the grounded circuit conductor as shown in the lower drawing in figure 6-14, a low-impedance ground-fault return path is established as required in 250.4(A)(5). This will allow a large current to flow over the equipment grounding and service grounded conductor to allow the branch-circuit, feeder, or service overcurrent device to clear the fault and thus provide the safety contemplated by the *Code*.

Resistance of Grounding Electrodes

There is no requirement in Article 250 that the grounding electrode system required by 250.50 (consisting of metal underground water pipe, metal frame of the building, concrete-encased electrode) meet any maximum resistance to ground. No doubt it is felt that the grounding electrode system will have a resistance to ground of 25 ohms or less.

• No maximum resistance value for grounding electrode system.

• Maximum 25 ohms for rod, pipe, or plate electrodes.

• Where rod, pipe, or plate eletrodes exceed

Minimum 6 feet apart or follow manufacturer's instructions.

Made electrode that exceeds 25 ohms must have additional supplemental electrode. See 250.56

Figure 6-15. Supplemental Electrode

25 ohms, they must be supplemented with one additional electrode.

The rules change for *man-made and installed* electrodes. The *Code* states, in 250.56, that where a single rod, pipe, or plate electrode does not achieve a resistance to ground of 25 ohms or less, it shall be supplemented by one additional electrode. This means that where driven ground rods are installed, two ground rods would be the maximum required under any condition. There is no requirement that additional electrodes such as ground rods or plates be installed until the 25 ohm-to-ground resistance is obtained.

In general, metallic underground water piping systems, metallic well piping systems, metal frame buildings and similar grounding electrodes may be expected to provide a ground resistance of not over 3 ohms and in some cases as low as 1 ohm.

However, from a practical standpoint, no grounding electrode, no matter how low its resistance, can ever be depended upon to clear a ground fault on any distribution system of less than 1000 volts.

If a system is effectively grounded as pointed out under 250.4(A)(5), a path of low impedance (not through the grounding electrode) is required to be provided to facilitate the operation of the overcurrent devices in the circuit (see chapter eleven).

The lowest practical resistance of a grounding electrode is desirable and will better limit the voltage-to-ground when a ground fault occurs. It is more important to provide a low-impedance path to clear a fault promptly, for a voltage-to-ground can only occur during the period of time that a fault exists. Clearing a ground fault promptly thus will enhance safety.

Even though the grounding electrode has low resistance, it is a part of a high-impedance circuit and plays virtually no part in the clearing of a fault on a low-voltage distribution system. This is due in part to being a higher resistance path through the earth than through the grounded service conductor. In addition, the remote path through the grounding electrode and earth is a high-impedance path compared to the circuit where the grounded service conductor is installed and routed with the ungrounded phase conductors.

Ground System Monitoring

Recently developed is a practical ground system monitoring capability. This innovative product

Figure 6-16. Ground Resistance Monitor.   Courtesy of Lyncole XIT

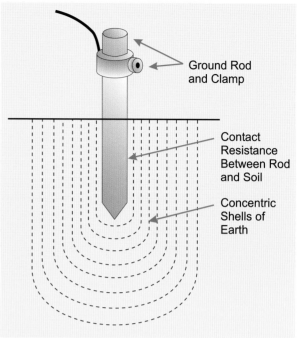

Ground Rod and Clamp

Contact Resistance Between Rod and Soil

Concentric Shells of Earth

Figure 6-17. Ground Rod Resistance.

is the solution to the difficult and often neglected testing of the grounding system. Grounding systems are the cornerstones of electrical protection of a site or facility. Grounding determines the effectiveness of the lightning protection and surge suppression systems. The GRM 2000 RS monitors ground system performance and provide an early warning of ground system degradation or loss of integrity.

The system consists of a permanent wall-mounted meter and a sensing head (attached to the grounding system conductor). The meter features both high and low level alarms for instantaneous notification when the pre-set resistance values are exceeded.

Features:

• Continuous monitoring of ground system resistance and current
• RS 232 output, with remote reading and control capability
• Local audible alarm
• High and low alarm values
• Adjustable sampling rate[5]

Ground Electrodes

The term *ground* is defined as a conducting connection by which a circuit or equipment is connected to the earth. The connection is used to establish and maintain as closely as possible the potential of the earth on the circuit or equipment connected to

it. A *ground* consists of a grounding conductor, a bonding connector, its grounding electrode(s), and the soil in contact with the electrode.

Grounds have several protection applications. For natural phenomena, such as lightning, grounds are used to discharge the system of current before personnel can be injured or system components damaged. For foreign potentials due to faults in electric power systems with ground returns, grounds help ensure rapid operation of the protection relays by providing low resistance fault-current paths. This provides for the removal of the foreign potential as quickly as possible. The ground should drain the foreign potential before personnel are injured and the power or communications system is damaged.

Ideally, to maintain a reference potential for instrument safety, protect against static electricity, and limit the system to frame voltage for operator safety, a ground resistance should be zero ohms. In reality, as described further in the text, this value cannot be obtained.

Last, but not least, low ground resistance is essential to meet *NEC*, OSHA, and other electrical safety standards.

Figure 6-17 illustrates a grounding rod. The resistance of the electrode has the following components:

(A) the resistance of the metal and that of the connection to it.

(B) the contact resistance of the surrounding earth to the electrode.

(C) the resistance in the surrounding earth to current flow or earth resistivity which is often the most significant factor.

More specifically:

(A) Grounding electrodes are usually made of a very conductive metal (copper or copper clad) with adequate cross sections so that the overall resistance is negligible.

(B) The National Institute of Standards and Technology has demonstrated that the resistance between the electrode and the surrounding earth is negligible if the electrode is free of paint, grease or other coating, and if the earth is firmly packed.

(C) The only component remaining is the resistance of the surrounding earth. The electrode can be thought of as being surrounded by concentric shells of earth or soil, all of the same thickness. The closer the shell to the electrode, the smaller its surface; hence, the greater its resistance. The farther away the shells are from the electrode, the greater the surface of the

shell; hence, the lower the resistance. Eventually, adding shells at a distance from the grounding electrode will no longer noticeably affect the overall earth resistance surrounding the electrode. The distance at which this effect occurs is referred to as the *effective resistance area* and is directly dependent on the depth of the grounding electrode.

Grounding Electrode Resistance Testing

Section 250.56 covers the resistance requirements of man-made grounding electrodes and reads as follows: "A single electrode consisting of a rod, pipe, or plate that does not have a resistance to ground of 25 ohms or less shall be augmented by one additional electrode of any of the types specified by 250.52(A)(2) through (A)(7). Where multiple rod, pipe or plate electrodes are installed to meet the requirements of this section, they shall be not less than 1.8 m (6 ft) apart."

The *Code* states that the resistance to ground for rod, pipe, or plate electrodes shall not exceed 25 ohms. This is an upper limit and guideline, since much lower is required in many instances.

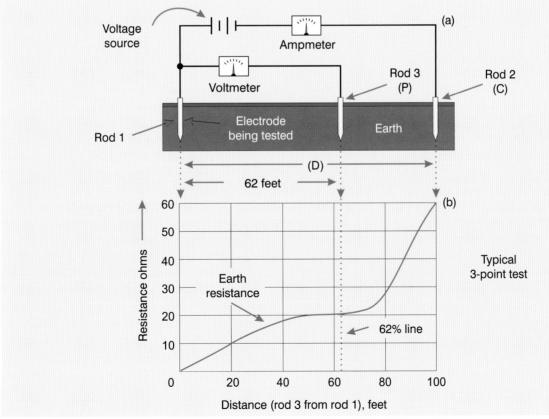

Figure 6-18. Principles of Earth Resistance Testing

"How low in resistance should a connection to ground be?" An arbitrary answer to this in ohms is difficult. The lower the ground resistance, the safer, and for positive protection of personnel and equipment, it is worth the effort to aim for less than one ohm. It is generally impractical to reach such a low resistance along a distribution system or a transmission line or in small substations. In some regions, resistances of 5 ohms or less may be obtained without much trouble.

> There will always exist circumstances that will make it difficult to obtain the ground resistance required by the *NEC* or other safety standards.

In other regions, it may be difficult to bring resistance of driven grounds below 100 ohms.

Accepted industry standards stipulate that transmission substations should be designed not to exceed one ohm resistance. In distribution substations, the maximum recommended resistance is for 5 ohms or even 1 ohm. In most cases, the buried grid system of any substation will provide the desired resistance.

In light industrial or in telecommunication central offices, 5 ohms resistance is often the accepted value. For lightning protection, the arrestors should be coupled with a maximum ground resistance of 1 ohm. These parameters can usually be met with the proper application of basic grounding theory. There will always exist circumstances that will make it difficult to obtain the ground resistance required by the *NEC* or other safety standards. When these situations develop, several methods of lowering the ground resistance can be employed. These include parallel rod systems, deep driven rod systems utilizing sectional rods and chemical treatment of the soil. Additional methods, discussed in other published data, are buried plates, buried conductors (counterpoise), electrically connected building steel, and electrically connected concrete reinforced steel.

Electrically connecting to existing water and gas distribution systems was often considered to yield low ground resistance; however, recent design changes utilizing non-metallic pipes and insulating joints have made this method of obtaining a low resistance to ground questionable and in many instances unreliable.

The measurement of ground resistances may only be accomplished with specially designed test equipment. Most instruments use the fall-of-potential principle of alternating current (ac) circulating between an auxiliary electrode and the ground electrode under test; the reading will be given in ohms, and represents the resistance of the ground electrode to the surrounding earth.

AEMC Instruments has also recently introduced clamp-on ground resistance testers. For additional details on grounding electrode testing methods and procedures, see appendix B.

Objectionable Currents

Section 250.6 recognizes that conditions may exist which may cause an objectionable flow of current over grounding conductors such as the grounding electrode conductor, other than temporary currents that may be set up under accidental conditions. We should recognize that grounding conductors are not intended to carry current under normal operating conditions. They are installed for and are intended to carry current to perform some safety function. (Additional information on ground resistance testing and studies can be found in Appendix C of this book.)

The *Code* does not define what is meant by "objectionable" currents. Clearly, any current over a grounding electrode conductor that would prevent it from maintaining the equipment at the earth potential would be objectionable. Since every conductor has resistance, current flow through the conductor will produce a voltage drop across it. Any voltage drop on a grounding conductor that would create a shock hazard certainly would also not be acceptable.

Section 250.6(B) permits the following corrective actions to be taken where there is an objectionable flow of current over grounding conductors:

1. If due to multiple grounds, one or more, but not all, of such grounds may be discontinued.
2. The location of the grounding connection may be changed.
3. The continuity of the grounding conductor or conductive path between grounding connections may be suitably interrupted.

4. Other means satisfactory to the authority enforcing the *Code* may be taken to limit the current over the grounding conductors.

The *Code* points out that temporary currents that result from accidental conditions such as ground-fault currents, that occur only while the grounding conductors are performing their intended protective functions, are not considered the "objectionable" currents covered in these sections.

Section 250.6(D) points out that currents that introduce noise or data errors in electronic equipment are not considered to be objectionable currents. Electronic data processing equipment is not permitted to be operated ungrounded or by connection to only its own grounding electrode.

Chapter seventeen includes special grounding provisions for installations, such as data processing rooms and the like, where EMI (electro-magnetic interference) or noise in the grounding circuits or systems can cause data errors and loss of data. In these types of installations, the minimum grounding and bonding requirements in Article 250 are required to be expanded upon to provide better grounding (earthing) electrodes and cleaner equipment grounding conductors.

Lightning Protection System

Any comprehensive discussion of lightning protection systems is beyond the scope of this text. Lightning protection systems should be installed in accordance with the NFPA-780, *Standard for the Installation of Lightning Protection Systems*.

The *Code* prohibits, in 250.60, the use of air terminal conductors and grounding terminals used for lightning protection such as driven pipes, rods, or other electrodes in place of the grounding electrodes required for a wiring system and for equipment. Note that where two grounding electrodes are installed, they are required to be bonded together. [See 250.106 for the requirement that lightning protection ground terminals (grounding electrodes of the counterpoise protection system) be bonded to the building or structure power grounding electrode system.]

The fine print notes following 250.106 provide valuable information regarding lightning protection systems. The *NEC* no longer has specific spacing requirements for separation of lightning protection system conductors from metal raceways and other metal enclosures of the building electrical system. The required spacing is typically 6 ft. through air and 3 ft. through dense construction materials such as concrete, brick, or wood. Specific requirements are given in the NFPA-780, *Standard for the Installation of Lightning Protection Systems*.

Conclusion

No potential exists between a conductor or equipment enclosure and earth if the system is properly and adequately grounded, except during a fault. By careful and intelligent design, one is required to reduce the time of clearing a fault to a minimum.

While it is desirable to obtain a grounding electrode having a resistance as low as practical, it is more important that to provide a path of low impedance for the return of ground-fault current which will clear a ground fault when it occurs.

No hazard can exist in the distribution system if there is no insulation failure to create a ground fault. Further, the hazard can exist only for the period of time it takes to clear the fault. If a ground fault clears promptly, it is extremely unlikely that any loss of life would occur and property damage would be kept to a minimum.

For maximum safety, one is required to follow one cardinal rule: only one grounding electrode system is required to be used in or on a building, with everything that should, or is required to, be grounded connected to that same grounding electrode system. If more than one grounding electrode is required, they may be used providing all the grounding electrodes used are bonded together to form a grounding electrode system which, in effect, becomes one common grounding electrode. Grounding electrodes should never be relied upon to provide the ground-fault return path for equipment. That is not their intended function in the electrical system. They cannot be relied upon to provide that effective ground-fault current path.

[1] NFPA 70, *National Electrical Code* 2002, 250.24(C), (Quincy, MA, National Fire Protection Association, 2002), p. 70-100.
[2] NFPA 70, Article 100, (Quincy, MA, National Fire Protection Association, 2002), p. 70-35.
[3] NFPA 70, 250.52(A), p. 70-104 & 70-105.
[4] This text provided by Lyncole XIT Grounding.
[5] This text provided by Lyncole XIT Grounding.

Chapter Six: The questions included here were developed using material included in this chapter. The answers can be found by reviewing the text. It is also important that students make use of the 2002 *NEC*, where many of the answers can be found.

1. A conducting element used to establish a ground for connecting electrical systems to the earth is defined as the ____.
   a. equipment grounding
   b. grounding electrode
   c. main bonding jumper
   d. earthing conductor

2. Intentionally connected to earth through a ground connection or connections of sufficiently low impedance and having sufficient current-carrying capacity that will prevent buildup of voltages that may result in undue hazard to connected equipment or to persons, defines ____.
   a. effectively grounded
   b. sufficiently grounded
   c. approved
   d. identified

3. One or more steel reinforcing bars or rods not less than 13 mm ($^{1}/_{2}$ in.) in diameter installed in concrete, or 6.0 m (20 ft) more of bare copper conductor not smaller than 4 AWG installed in concrete is defined as a ____.
   a. butt ground
   b. pole ground
   c. feeder electrode
   d. concrete-encased electrode

4. A concrete-encased electrode must be located within or near the bottom of a foundation or footing and is required to be encased by not less than ____ inches of concrete.
   a. 0
   b. 1
   c. 2
   d. 6

5. A copper conductor not smaller than 2 AWG at least 6.0 m (20 ft) encircling a building or structure, and buried at not less than 750 mm (2 1/2 ft) deep defines a ____.
   a. concrete-encased electrode
   b. ground ring
   c. Ufer ground
   d. common bonding grid

6. Where an underground metal water pipe is the only grounding electrode that is available, and is connected at the building or structure served, it must be supplemented by another ____.
   a. grounding electrode
   b. main bonding jumper
   c. equipment grounding conductor
   d. circuit bonding jumper

7. An electrode that is considered as being suitable for supplementing the metal underground water pipe, includes a concrete-encased electrode, ground ring, other local metal underground systems or structures, or ____ electrodes.
   a. rod
   b. pipe
   c. plate
   d. all of the above

8. If available on the premises at each building or structure served, all grounding electrodes, including ____ electrodes, are required to be bonded together to form the grounding electrode system.
   a. identified
   b. approved
   c. rod, pipe, or plate
   d. gas pipe

9. Where the electrodes described in 250.52(A)(1) through (A)(6) are not available at the premises or structure served, a grounding electrode must be installed. It may be ____.
   a. local metallic underground systems or structures, tanks, etc.
   b. pipe or conduit electrodes not less than 2.5 m (8 ft) in length not smaller than metric designator 21 (3/4 in.), and if of iron or steel, must be galvanized or metal-coated for corrosion protection.
   c. rod electrodes of steel or iron at least 15.87 mm (5/8 in.) diameter.
   d. any of the above

10. Where two or more grounding electrodes are bonded together, they are considered to be a _____ electrode.
- a. single
- b. an identified
- c. an approved
- d. a listed

11. The *Code* recognizes that conditions may exist that may cause an objectionable flow of current over the grounding electrode conductor, other than temporary currents that may be set up under accidental conditions. Permitted alterations include _____.
- a. if due to multiple grounds, one or more, but not all, of such grounds may be abandoned
- b. their location can be changed
- c. continuity of the grounding conductor may be suitably interrupted
- d. any of the above

12. Currents that introduce noise or data errors in electronic equipment are not considered to be _____.
- a. dangerous currents
- b. objectionable currents
- c. harmonic currents
- d. unsafe currents

13. Where a single rod, pipe, or plate electrode does not achieve a resistance to ground of _____ ohms or less, it is required to be supplemented by only one additional electrode.
- a. 50
- b. 40
- c. 30
- d. 25

14. The use of continuous metallic underground water and metal well casings, as well as the metal frames of buildings, should provide a ground resistance not exceeding _____ ohms.
- a. 3
- b. 6
- c. 12
- d. 25

15. In general, doubling the length of a rod type grounding electrode reduces the resistance by about _____ percent.
- a. 10
- b. 20
- c. 30
- d. 40

16. The use of air terminal (lightning rod) conductors and ground terminals of lightning protection systems are _____ for the grounding electrode(s) or grounding electrode system required by the *Code* for a wiring system and for equipment.
- a. permitted to be used
- b. prohibited from being used
- c. permitted by special permission
- d. identified when labeled to be used

17 Supplementary grounding electrodes are permitted by the *Code* and are required to be connected to the _____ conductor, and the _____ shall not be used as the sole equipment grounding conductor.
- a. grounding electrode, conduit
- b. equipment grounding, earth
- c. bonding, ground rods
- d. grounded, equipment grounding bus

CHAPTER SEVEN

Grounding Electrode Conductors

The grounding electrode conductor is used to connect the electrical system, or the equipment grounding conductor, or both, to a grounding electrode, hence the specific term, *grounding electrode conductor*. It is required to be properly sized as per Table 250.66 based on the size of the service-entrance conductors but is not required to exceed 3/0 AWG copper or 250 kcmil aluminum. Specific requirements are given regarding conductor material, how it is to be installed, protected from physical damage, and connected.

Definition:

"**Grounding electrode conductor:** The conductor used to connect the grounding electrode to the equipment grounding conductor, to the grounded conductor, or to both, at the service equipment, at each building or structure where supplied from a common service, or at the source of a separately derived system."[1]

Function of Grounding Electrode Conductor

The grounding electrode conductor is the sole connection between the grounding electrode and the grounded system conductor (may be a

> In many installations, the length of the grounding electrode conductors is short enough that their resistance can be ignored.

neutral) and the equipment grounding conductor for a grounded system; or the sole connection between the grounding electrode and the service equipment enclosure and equipment grounding conductor for an ungrounded system [see 250.24(C)].

A common grounding electrode conductor is required to ground both the system grounded con-

ductor and the equipment grounding conductor. In other words, one grounding electrode conductor cannot be used to ground the system conductor and another grounding electrode conductor be used to ground the equipment grounding conductor even though both grounding conductors are connected to the same grounding electrode. A single grounding electrode conductor is required to be used for both the circuit and the equipment.

Maximum Current on Grounding Electrode Conductors

In all grounded systems of 600 volts or less, the maximum current that will flow in the grounding electrode conductor is limited by the impedance of the circuit. The circuit consists of the resistance of the service conductor, the grounding electrode at the service, the grounding electrode at the transformer bank plus the resistance of the earth path between the two grounding electrodes. In addition, there will be the resistance of the grounding electrode conductors themselves. In many installations, the length of the grounding electrode conductors is short enough that their resistance can be ignored. In addition to these resistances, there will be an inductive reactance that will vary as the spacing between the supply path and return path increases.

As illustrated in figure 7-3, if we assume that the sum of these resistances is equal to 22 ohms (a figure that may be lower than is found in ac-

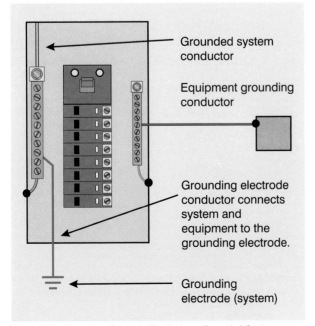

Figure 7-1. Grounding Electrode Conductor – Grounded System

Grounded system conductor

Equipment grounding conductor

Grounding electrode conductor connects system and equipment to the grounding electrode.

Grounding electrode (system)

Figure 7-2. Grounding Electrode Conductor – Ungrounded System

Equipment Grounding Conductor

Grounding electrode conductor connects equipment to the grounding electrode.

Grounding electrode (System)

Maximum current of ground fault is limited by the high impedance series circuit through grounding electrodes and the earth.

Figure 7-3. High Impedance Earth Return

tual practice), then, in a 120-volt-to-ground system, the maximum current through the grounding electrode conductor and grounding electrodes will be 5.5 amperes (120 volts ÷ 22 ohms = 5.5 amperes). As can be seen, this grounding connection is ineffective. Equipment grounded only in this manner is unsafe as a ground fault would not cause an upstream overcurrent protective device to clear the fault from the equipment. Obviously, the current through the high-impedance path provided by the grounding electrodes and the earth is not nearly enough to cause a 15-ampere circuit breaker or fuse to operate. Voltage levels that could be fatal could appear on equipment that is grounded in this manner. This maximum current value is arrived at by considering resistance only. If the reactance of the circuit is included, the net result will be an impedance value much higher than the resistance value. The actual current that will flow in the grounding electrode conductor will thus be considerably less than the figures as calculated based on resistance only. Ground-fault currents through the grounding electrode conductor are thus not likely to attain a value anywhere near the continuous rating of the conductor even under ground-fault conditions.

As covered in chapter six in detail, the main purpose of the grounding elec-

trode is to establish and maintain an earth reference for the system and noncurrent-carrying parts of the system and not to clear ground faults.

One of the reasons for requiring the grounding electrode conductors to be larger than for the normal system current they might carry is grounding electrode conductors also carry current from other events on the system, such as lightning strikes or from accidentally being energized from high-voltage sources such as transformers.

Sizing the Grounding Electrode Conductor at a Single Service

The grounding electrode conductor is required to be sized in accordance with Table 250.66. That conductor is required to be a minimum size of 8 AWG and need not be larger than 3/0 AWG copper. Where aluminum or copper-clad aluminum grounding electrode conductors are installed they are required to be not smaller than 6 AWG nor larger than 250 kcmil. The size of this grounding electrode conductor is based upon the size of the

Size of Largest Ungrounded Service-Entrance Conductor or Equivalent Area for Parallel Conductors[a] (AWG/kcmil)		Size of Grounding Electrode Conductor (AWG/kcmil)	
Copper	Aluminum or Copper-Clad Aluminum	Copper	Aluminum or Copper-Clad Aluminum[b]
2 or smaller	1/0 or smaller	8	6
1 or 1/0	2/0 or 3/0	6	4
2/0 or 3/0	4/0 or 250	4	2
Over 3/0 through 350	Over 250 through 500	1/0	3/0
Over 350 through 600	Over 500 through 900	2/0	4/0
Over 600 through 1100	Over 900 through 1750		
Over 1100	Over 1750	3/0	250

Notes:
1. Where multiple sets of service-entrance conductors are used as permitted in 230.40, Exception No. 2, the equivalent size of the largest service-entrance conductor shall be determined by the largest sum of the areas of the corresponding conductors of each set.
2. Where no service-entrance conductors, the grounding electrode conductor size shall be determined by the equivalent size of the largest service-entrance conductor required for the load to be served.

[a] This table also applies to the derived conductors of separately derived ac systems

[b] See installation restrictions in 250.64(A)

Table 250.66. Grounding Electrode Conductor for Alternating-Current Systems service entrance conductor by reference to Table 250.66 and not on the rating of the circuit breaker or fuse in the service equipment. Table 250.66 is reprinted in chapter nineteen.

For example, where 3/0 AWG copper service-entrance conductors are installed, the minimum size grounding electrode conductor is 4 AWG copper or 2 AWG aluminum. If 750-kcmil aluminum service-entrance conductors are installed, a 1/0 AWG copper or 3/0 AWG aluminum grounding electrode conductor is required to be used.

Sizing Grounding Electrode Conductor for Service with Parallel Conductors

Where service-entrance conductors are installed in parallel as allowed by 310.4, the circular mil area of one set of parallel conductors is added together and treated as a single conductor for purposes of sizing the grounding electrode conductor. For example, if four 250-kcmil copper conductors are installed in parallel, they are considered to be a single 1000-kcmil conductor. By reference to Table 250.66, we find the minimum grounding electrode conductor for this set to be a 2/0 copper or 4/0 AWG aluminum conductor.

If 1/0 AWG through 4/0 AWG conductors are installed in parallel, they are required to first be converted to circular mil area before applying Table 250.66. For example, if two 2/0 AWG alu-

minum service-entrance conductors are installed in parallel, first refer to *NEC* Table 8 of chapter nine for determining the conductor's circular mil area (this table is reprinted in chapter nineteen). There, we find the conductor to have an area of 133,100 circular mils. Then multiply the circular mil area of the conductors by the number of conductors connected in parallel.

133,100 circular mils;
133,100 cm x 2 = 266,200 cm.

By reference to Table 250.66 we find that a 2 AWG copper or 1/0 AWG aluminum grounding electrode conductor is required.

Sizing Grounding Electrode Conductors for Multiple Enclosure Services

Services are permitted to be installed in up to six enclosures installed at one location or at separate locations. The method of sizing grounding electrode conductors is a matter of choice. One basic rule must be followed: size the grounding electrode conductor for the circular mil area of service-entrance conductor(s) at the point of connection, or for the size of the main service-entrance conductor(s) [see 250.64(D)].

In figure 7-5 with assumed 500-kcmil copper service-entrance conductors, the service-entrance grounded conductor is shown grounded inside the wireway. By reference to Table 250.66, we find that the grounding electrode conductor to a water pipe or building steel grounding electrode is required to be 1/0 AWG copper or 3/0 AWG aluminum.

• Four 250 kcmil aluminum conductors per set.
• 4 x 250 = 1000 kcmil
• Refer to Table 250.66.
• Select 2/0 copper or 4/0 AWG aluminum GEC.
• See exceptions for GEC to made electrodes.

Figure 7-4. Grounding Electrode Conductor for Parallel Service-Entrance Conductors

Grounded service Conductor

500 kcmil ungrounded service entrance conductor

1/0 AWG copper or 3/0 AWG aluminum Grounding electrode conductor

Figure 7-5. GEC Connection in Wireway (Gutter)

Figure 7-6. GEC Connection to Electrode

Figure 7-7. Taps to Main Grounding Electrode Conductor

Figure 7-8. Tapping Grounding Electrode Conductor

If the installer chooses to, an individual grounding electrode conductor may be installed from each service disconnecting means enclosure to the grounding electrode. In this case, the grounding electrode conductor is sized for the service-entrance conductor serving that enclosure (see figure 7-6). In addition, a single grounding electrode conductor is permitted to serve several enclosures. The main or common grounding electrode conductor is sized for the main service-entrance conductors. The *NEC* does not use the term *main or common grounding electrode conductor* to identify the grounding electrode conductor that connects directly to the grounding electrode. We use it here to distinguish it from the grounding electrode tap conductors. Taps that are sized based on the individual service-entrance conductors supplying each service disconnect are connected from each service disconnecting means to the common grounding electrode. Grounding electrode conductor taps are covered in 250.64(D).

In figures 7-7 and 7-8, it is assumed that 2 AWG copper conductors serve each service disconnecting means from the wireway. Shown are taps to a main or common grounding electrode conductor. The tap conductors are required to be connected to the main or common grounding electrode conductor in a manner that it remains continuous without a splice or joint. This means that the tap conductor is required to be connected to the main grounding electrode conductor with a device that allows the main grounding electrode conductor to remain unbroken as the connection

is made. The concept of tapping the grounding electrode conductor applies whether the sets of service-entrance conductors are tapped from a wireway or are installed individually as overhead or underground systems.

For figure 7-8, the size of the common grounding electrode conductor is determined as follows. Assume that 500-kcmil copper service-entrance conductors supply the service and are connected in the wireway to the service-entrance conductors that serve each enclosure.

Refer to Table 250.66. The minimum size common grounding electrode conductor is 1/0 AWG copper or 3/0 AWG aluminum. This conductor is installed from the grounding electrode to the vicinity of the wireway. The grounding electrode tap conductors from the individual enclosures are sized from Table 250.66 and are 8 AWG copper or 6 AWG aluminum. Note that there is no minimum or maximum length for these grounding electrode tap conductors. If the installer chooses to, the grounding electrode conductors from the service disconnects may be connected individually to the grounding electrode rather than being tapped to the common grounding electrode conductor.

Exceptions to Size of Grounding Electrode Conductor

The grounding electrode conductor is generally required to be not smaller than the values in Table 250.66 based on the size of the largest ungrounded service-entrance conductor. What amounts to a three-part exception to the general rule for sizing the grounding electrode conductor is provided in 250.66(A) through (C).

Section 250.66(A) permits the grounding electrode conductor to be not larger than 6 AWG copper or 4 AWG aluminum wire where it is the sole connection to a rod, pipe or plate electrode. Section 250.66(B) provides that the grounding electrode conductor that is the sole connection to a concrete-encased grounding electrode need not be larger than 4 AWG copper wire. Note that aluminum wire is not permitted for this application. The use of this smaller 4 AWG conductor is based on the fact that it can never be called upon, even under ideal conditions, to carry a current beyond its safe short-time rated capacity.

Section 250.66(C) provides that, where connected to a ground ring, that portion of the grounding electrode conductor that is the sole connection to the ground ring need not be larger than the ground ring conductor. The minimum size conductor and material for a ground ring is 2 AWG copper. That requirement assumes, and correctly so, that the resistance of such electrodes will limit the maximum current that may be expected to flow through them. Design engineers will sometimes specify ground rings of 4/0 AWG or 250 kcmil. In this case, follow Table 250.66 to determine the minimum grounding electrode conductor size. The grounding electrode conductor is not required to be larger than 3/0 AWG copper or 250 kcmil aluminum or copper-clad aluminum.

The term *sole connection* means that the grounding electrode conductor does not serve or connect to more than one grounding electrode. For example, a grounding electrode conductor does not connect to a concrete-encased electrode then to building steel or an underground metal water pipe. In this example, a grounding electrode con-

ductor larger than 4 AWG may be required for the water pipe electrode based on the size of the service-entrance conductors.

Grounding Electrode Conductor Connections

The *Code* requires generally that the point of connection of the grounding conductor to grounding electrodes shall be accessible and made in a manner that will assure a permanent and effective ground. An exception provides that a connection at a concrete-encased, driven, or buried grounding electrode is not required to be accessible [see 250.68(A)].

Specific rules for connection of the grounding electrode conductor to grounding electrodes are found in 250.70. Grounding electrode conductor connections are required to be by exothermic weld-

Figure 7-9. Grounding Electrode Conductor Connections

Photo 7-1. Ground Clamp Connector

- Grounding conductor connected to electrode by exothermic welding, listed lugs, listed pressure connectors, listed clamps or other listed means.

- Ground clamps shall be listed for materials of grounding electrode and grounding electrode conductor.

- Shall be listed for direct soil burial where used on pipe, rod or other buried electrodes.

Figure 7-10. Ground Clamps Listed for Application

Ground clamp to be approved for use without protection or be protected from "ordinary" damage unless it is:
(a) Not likely to be damaged, or it is
(b) Protected in metal, wood or equal.

Figure 7-11. Protection of Ground Clamp Attachment

Photo 7-2. Pipe Clamp and Ground Rod Style Connection. Courtesy of Greaves.

ing, listed lugs, listed pressure connectors, listed clamps or other listed means. The only connection means not required to be listed are those made by exothermic welding, although listed connections are now available. Connections depending solely on solder shall not be used. Other requirements state that not more than one conductor is permitted to be connected to the grounding electrode by a single clamp or fitting, unless the clamp or fitting is listed as being suitable for connecting multiple conductors. Ground clamps are required to be listed for both the materials of the grounding electrode and the grounding electrode conductor. For example, ground clamps are required to be listed for aluminum conductors to be used for such connections. Clamps used on a pipe, rod or other electrode that is buried are required to be listed for direct earth burial. Typically, these clamps are identified by the manufacturer with *direct burial* or DB or similar. Direct burial or DB means the clamp or fitting is suitable for concrete encasement.

Sheet-metal-strap type ground clamps attached

to a rigid metal base that are listed are permitted for indoor telecommunication purposes only [see 250.70(3)]. Also, Underwriters Laboratories' Guide Card information (KDER) states, "Strap type ground clamps are not suitable for attachment of the grounding conductor of an interior wiring system to a grounding electrode." Use strap type ground clamps for only the type of conductor and grounding electrode, as well as the environment for which it is listed and labeled, such as for communication circuits (see chapter eighteen for low voltage and intersystem bonding and grounding requirements).

It is required that ground clamps be protected from physical damage unless the fittings are approved for use without protection or are installed in a location where they are not likely to be damaged. Protection of the ground clamp is permitted to consist of metal, wood, or equivalent materials (see 250.10).

Clean Surfaces

It is required in 250.12 that nonconductive coatings such as paint, lacquer, and enamel be removed from threads and other contact surfaces of equipment to be grounded or be connected by means of fittings designed to make such removal unnecessary. (Underwriters Laboratories reported in a June 27, 1995 letter and confirmed on June 7, 1998 that no fittings listed by them incorporate this feature.) This will ensure a good electrical connection (see chapter eight for additional discussion on the subject).

Here, again, it is important to consider that all grounding electrode conductors and their connections form a part of a circuit which is required under certain conditions to carry current. In some cases the current is several times the full-load current rating of the conductor involved if the conductor was being used for continuous duty in an electrical circuit. Such currents are usually of short duration so the withstand rating of the conductor is not exceeded. Removal of paint under connections of raceways and cable fittings is critical to make a reliable connection that can carry fault current when necessary.

Grounding Electrode Conductor Material

The grounding electrode conductor is required to be of copper, aluminum, or copper-clad aluminum and may be solid, stranded, insulated, covered, or bare. The material selected shall be resistant to any corrosive condition it may be exposed to. As an option, it may be suitably protected against corrosion (see 250.62). Note that there is no color code for the grounding electrode conductor such as the requirements for identification that exist for grounded or equipment grounding conductors.

The current the grounding electrode conductor will carry is limited by the resistance of the grounding electrode(s) and the earth return path. In a grounded system, it is further limited by the fact that it completes a parallel circuit from the grounded conductor through both grounding electrodes. The other side of the parallel circuit is the grounded service conductor which is a circuit of much lower impedance so it will carry almost all of the current.

Aluminum or Copper-clad Aluminum Grounding Electrode Conductors

Specific limitations are placed on aluminum and copper-clad aluminum grounding electrode conductors. Note that these rules apply regardless of whether the conductors are insulated or bare [see 250.64(A)].

Bare conductors are not permitted where they are in direct contact with masonry or the earth, or where subject to corrosive conditions. This rule would not prohibit installing conduit on masonry and pulling a bare aluminum grounding electrode conductor in it.

Where used outside, aluminum or copper-clad aluminum grounding electrode conductors are not permitted to be terminated within 18 in. of the earth.

Because of these restrictions on their installation, they cannot be used for connection to concrete-encased electrodes, ground rods or pipes where within 18 in. of the earth or for connection to plate electrodes. Aluminum conductors are otherwise permitted to be used as grounding electrode conductors where the clamp or connector is listed for both the conductor material and the electrode. Typical connectors or clamps would be marked *AL/CU*. The AL indicates the clamp is suitable for aluminum conductor connections.

Securing and Protection from Physical Damage

Section 250.64(B) requires that the grounding electrode conductor or its enclosure be securely fastened to the surface on which it is carried. Size 4 AWG or larger conductors require protection where exposed to *severe physical damage*, a term that is not defined. This leaves it to a judgment call on the part of the inspector as to what constitutes severe physical damage. Size 6 AWG conductors are required to be run along the surface of the building construction and securely fastened. Otherwise, they are required to be protected by installation in rigid or intermediate metal conduit, rigid nonmetallic conduit, electrical metallic tubing, or cable armor. Grounding electrode conductors smaller than 6 AWG are required to be protected in one of the methods allowed for 6 AWG conductors.

Splicing Grounding Electrode Conductor

The general rule in 250.64(C) is that grounding electrode conductors are to be installed in one continuous length without a splice or joint.

The exception permits splices in busbars as they are at times used as a grounding electrode conductor such as in open bottom switchboards where the main bonding jumper is at the service disconnect end of the equipment and the grounding electrode conductor connects to the grounding bus at the other end of the equip-

Figure 7-12. Splicing Grounding Electrode Conductor

Figure 7-13. Protecting Grounding Electrode Conductor From Magnetic Field

ment. The bolted connections between the sections are actually splices in busbars. Busbars are often installed in electrical rooms to provide a common connection point for individual grounding electrode conductors. Busbars often come in standard lengths such as 10 ft. and have to be joined together to achieve the required length. The exception recognizes these instances where the splices have to be made through bolted busbar connections.

This section permits the grounding electrode conductor to be spliced by two methods: irreversible compression-type connectors listed for the purpose or by the exothermic welding process. It is vital that the manufacturer's instructions be carefully followed where either of these splicing methods is chosen. Where compression-type connectors are used, the correct splicing sleeve for the conductor material and size is required to be selected. Then, the compression tool, and in some cases the proper die for the sleeve to be crimped, is required to be used. These compression-type connectors are required to be specifically listed by a qualified electrical testing laboratory for splicing grounding electrode conductors. Where exothermic welding of the grounding electrode conductor is performed, the correctly sized form or mold for the conductor to be spliced is required to be used. In addition, unless specifically permitted otherwise by the manufacturer, the conductors to be spliced by this method are required to be clean and dry. Inspect the resulting splice carefully to be certain that it has been made satisfactorily without causing damage to the conductor or otherwise effecting its integrity.

Protecting Grounding Electrode Conductor from Magnetic Field

Where metal enclosures are provided for protection of the grounding electrode conductor, one has to follow some special procedures. This is required by 250.64(E) and 250.92(A)(3).

1. Metal conduit must be electrically continuous from the point of attachment to cabinets or equipment to the grounding electrode.

2. The conduit must be securely fastened to the ground clamp or fitting.

3. Metal conduit that is not physically continuous from the enclosure to the ground clamp must be made continuous by bonding both ends of the conduit to the conductor.

4. Bonding of the metal raceway or armor that encloses a grounding electrode must be done at each end and to all intervening raceways, boxes, and enclosures between the service equipment and the grounding electrode [see 250.92(A)(3)].

It is common practice to use an 8 AWG grounding electrode conductor protected by metallic armored cable. The need for bonding the metallic armor of such cable is required in 250.64(E). Where that bonding procedure is not followed, the impedance of the grounding electrode conductor is approximately doubled with the result that its effectiveness is markedly reduced. Such bonding requirements can be avoided by protecting the grounding electrode conductor in rigid nonmetallic conduit. Schedule 80 PVC is listed as impact and crush resistant and would provide suitable protection.

Impedance of Conduit and Conductor

Table two of chapter twenty compares the continuous rating of copper grounding electrode conductors with the service conductors with which they are required to be used.

Table three of chapter nineteen compares the resistance and impedance of copper grounding electrode conductors where enclosed in a steel conduit for physical protection. The last two columns of that table show how the impedance of the conductor is approximately doubled where the conduit is not properly installed as required in 250.64(E). The *Code* requires that the conduit be bonded at both ends of the grounding electrode conductor to form a parallel circuit with the copper grounding electrode conductor. That important rule, if not observed, will result in doubling the impedance of the grounding conductor. Test data to confirm the above are available. Of course, where the impedance of the installation is increased, the effectiveness is reduced.

Where grounding electrode conductors are used on an alternating-current circuit and enclosed in conduit, it is necessary to compare the impedance values and not the resistance values.

The data in table three of chapter twenty shows that where an 8 AWG copper conductor is installed in steel conduit and properly bonded at each end, the impedance of the circuit is about the same as where the 8 AWG copper conductor was used alone.

For all other sizes of copper conductors installed in the proper sized conduit, the impedance of the circuit is greater where the conduit is used, as compared to using the copper conductor alone. The impedance values are from about 40 percent more for a 6 AWG copper conductor in a metric designator 21 (³/₄) trade size conduit to about 500 percent more for a 3/0 AWG copper conductor in a metric designator 35 (1 ¹/₄) trade size conduit, as compared to not using a metal conduit enclosure for physical metric designator 21 (³/₄) trade size protection.

Over 100-foot Grounding Electrode Conductor

The short-time rating of a copper conductor is related to the I^2t (current x current x time) rating of the conductor for a given temperature rise which will not damage adjacent insulated conductors or affect the continuity established by the bolted joints. For a period of five seconds the short-time rating may be taken as approximately 1 ampere for every 42.25 circular mils area. A 6 AWG copper conductor has an area of 26,240 circular mils and is thus capable of carrying about 621 amperes for five seconds safely.

Based on the safe I^2t values for the circuit comprising the various grounding electrode conductors, it will be seen that for a five-second flow of current, the IR drop in the differ-

Conductor	Conduit Size	Total Amperes	Current in Conductor	Current in Conduit
6	½	100	3	97
6	½	300	5	295
2	¾	90	7	83
2	¾	350	10	340
2/0	1	150	15	135
2/0	1	590	5	585
4/0	1¼	225	15	210
4/0	1¼	885	15	870

The above test data confirm that, for all practical purposes, the impedance of a conductor enclosed in steel conduit, when the conduit is bonded at both ends, is approximately equal to the impedance of the conduit.

Table 7-14. Division of Current in Conduit

Service equipment

Metal enclosures that are not electrically continuous from the cabinet or equipment to the grounding electrode shall be made electrically continuous by bonding each end to the grounding electrode conductor

Metal raceway

Grounding electrode conductor

Grounding electrode

Figure 7-14. Securing and Protecting from Physical Damage

ent sizes of grounding electrode conductors will be approximately 37 volts per 100 ft. Using that figure as a standard, it is recommended that where a grounding electrode conductor exceeds 100 ft. in length, the conductor cross section be increased to keep the IR drop to not over 40 volts when carrying the maximum short-time current for the size conductor specified for five seconds (I^2t value). The *National Electrical Code* does not, now, place a limit on the length of the grounding electrode conductor.

An example of selecting the proper size grounding electrode conductor for a run exceeding 100 ft. is as follows.

Given: a 1/0 AWG copper service-entrance conductor. The grounding electrode conductor specified in Table 250.66 is a 6 AWG copper and the length of the grounding electrode conductor is 150 ft.

If a 6 AWG conductor is used which has 26,240 circular mils, resulting in a short-time rating of 621 amperes, and a dc resistance of 0.0737 ohms for 150 ft. (0.491 ohms/M ft.), the voltage drop would be 621 x 0.0737 or 46 volts.

A larger grounding electrode conductor, whose resistance times the short-time rating of the 6 AWG conductor in amperes, is required to be selected and it should not exceed 40 volts.

The next larger-sized grounding conductor, a 4 AWG, has a resistance of 0.0462 ohms for 150 ft. (0.308 ohms/M ft.), so the voltage drop would be 621 x 0.0462 or 28.7 volts. That would make a 4 AWG copper grounding conductor the proper size to use for a service using a 1 AWG or 1/0 AWG copper service-entrance conductor and having a grounding electrode conductor run of 150 ft.

Direct Current Systems

Where used on a direct-current circuit, we do not destroy the value of the grounding conductor if the conduit is properly installed, that is, bonded at both ends to the grounding conductor. The resultant resistance is lower where the conduit is used for physical protection. However, that assumes a steady direct-current flow. Special considerations are required to be given if a direct-current circuit is to be properly protected with a grounding electrode conductor against transient currents such as are produced by lightning. It is necessary to treat the selection of the grounding electrode conductor as would be done for an alternating-current circuit.

Table four of chapter twenty shows that where an aluminum conduit is used to enclose an aluminum grounding electrode conductor, the required conduit size has a lower resistance than the aluminum grounding electrode conductor in every case. In the case of aluminum wire size 6 AWG, the conduit is about one-tenth the resistance of the conductor.

For the largest aluminum grounding electrode conductor size, 250 kcmil, the conduit is about half the resistance of the conductor. Thus, where an aluminum grounding electrode conductor is protected with an aluminum conduit and properly bonded at both ends, we will always have a much lower impedance than where the wire is not installed in an aluminum conduit. However, aluminum is subject to certain restrictions owing to chemical corrosion.

Since aluminum conduit is nonmagnetic and has lower resistance as well as impedance values compared to steel, the use of aluminum conduit for physical protection will provide lower impedance values.

Conclusion

All of the above means that we decrease the safety of an installation (which requires a grounding electrode conductor larger than 8 AWG) where we enclose the conductor in a steel conduit. Although we decrease physical damage to assure the integrity of the grounding electrode conductor, another hazard is introduced by decreasing the effectiveness of the grounding electrode conductor through increasing the voltage rise on it. No better case could thus be made for not permitting the use of a steel conduit on a grounding electrode conductor larger than an 8 AWG.

The obvious solution, where physical protection is necessary, is to use nonmetallic or aluminum conduit for enclosing the grounding electrode conductor. That is especially true in view of the improvement in the art of manufacturing nonmetallic conduit which now can be obtained in ample physical strength to meet the requirements of proper physical protection.

---

[1] *NEC 2002 National Electrical Code*, (Quincy, MA, National Fire Protection Association, 2002), p. 70-36.

Chapter Seven: The questions included here were developed using material included in this chapter. The answers can be found by reviewing the text. It is also important that students make use of the 2002 *NEC*, where many of the answers can be found.

1. A conductor used to connect the grounding electrode to the equipment grounding conductor and/or to the grounded conductor of the circuit at the service equipment, at a separate building or structure supplied from a common service, or at the source of a separately derived system is defined as a ____.
   a.   main bonding jumper
   b.   grounding electrode conductor
   c.   feeder bonding jumper
   d.   grounded

2. A grounding electrode conductor must be properly sized and is based on the size or rating of the:
   a.   service breaker or fuse
   b.   transformer
   c.   service-entrance conductors
   d.   grounding electrode

3. For service conductors sized at over 1100 kcmil copper or 1750 kcmil aluminum or copper-clad aluminum, the grounding electrode conductor must be a ____ copper or ____ aluminum.
   a.   2/0 AWG - 4/0 AWG
   b.   1/0 AWG - 3/0 AWG
   c.   3/0 AWG - 250 kcmil
   d.   3/0 AWG - 4/0 AWG

4. In all grounded systems of ____ volts or less, the maximum current that will flow in the grounding electrode conductor is dependent on the sum of the resistance of the grounding electrode at the service, the grounding electrode at the transformer bank, plus the resistance of the earth path between the two grounding electrodes.
   a.   600
   b.   1000
   c.   700
   d.   800

5. Where of copper, the grounding electrode conductor is required to be sized at not less than ____ AWG.
   a.   10
   b.   14
   c.   8
   d.   12

6. Where a 3/0 AWG copper service-entrance conductor is installed, the minimum size grounding electrode conductor is ____ AWG copper or ____ AWG aluminum, or copper-clad aluminum.
   a.   8 - 4
   b.   6 - 6
   c.   4 - 2
   d.   4 - 4

7. Where four 250 kcmil copper service entrance conductors are installed per ____, they are considered to be a single 1000-kcmil conductor.
   a.   service
   b.   feeder
   c.   cable
   d.   phase

8. The minimum size for a grounding electrode conductor for a 1000-kcmil service-entrance conductor is ____ AWG copper or ____ AWG aluminum, or copper-clad aluminum conductor.
   a.   1/0 - 2/0
   b.   3/0 - 4/0
   c.   1/0 - 3/0
   d.   2/0 - 4/0

9. Services are permitted to be installed in up to ____ enclosures where they are installed at one location or at separate locations.
   a.   one
   b.   two
   c.   six
   d.   eight

10. A grounding electrode conductor is required to be sized at not less than _____ AWG copper or _____ AWG aluminum wire where it is the sole connection to a rod, pipe, or plate electrode.
    a. 8 - 8
    b. 6 - 4
    c. 6 - 6
    d. 4 - 8

11. A grounding electrode conductor that is the sole connection to a concrete-encased grounding electrode is not required to be larger than _____ AWG copper conductor.
    a. 6
    b. 4
    c. 8
    d. 10

12. Grounding electrode conductor connections are required to be made by _____, listed clamps, or other listed means.
    a. exothermic welding
    b. listed lugs
    c. listed pressure connectors
    d. any of the above

13. Which one of the following statements is INCORRECT?
    a. Exothermic welding connections are not required to be listed.
    b. Soldered connections are permitted for connections to a grounding electrode.
    c. Unless listed, not more than one conductor can be connected to the grounding electrode by a single clamp or fitting.
    d. Clamps used on a pipe, rod, or other buried electrode must also be listed for direct earth burial.

14. Where exposed to severe physical damage, _____ AWG or larger grounding electrode conductors require protection.
    a. 4
    b. 8
    c. 6
    d. 3

15. _____ AWG grounding electrode conductors must be run along the surface of the building construction and be securely fastened, or they must be protected by installation in rigid or intermediate metal conduit, rigid nonmetallic conduit, electrical metallic tubing, or cable armor.
    a. 8
    b. 6
    c. 4
    d. 2

16. Bare aluminum or copper-clad aluminum conductors are not permitted to be installed where in direct contact with masonry or earth and are not permitted to be terminated within _____ inches of the earth.
    a. 12
    b. 18
    c. 14
    d. 16

Key Notes

Bonding Enclosures & Equipment

OBJECTIVES

After studying this chapter, the reader will be able to understand

- Requirements for maintaining continuity of ground-fault path.
- Bonding of systems over 250 volts to ground.
- Bonding multiple raceway systems.
- Bonding of receptacles.
- Bonding of metal piping systems.
- Bonding of exposed structural steel.

Bonding is an ongoing action in any electrical system from the point of service delivery to the final outlet on the system. The act of bonding metal parts or enclosures of electrical components and conductors connects them together electrically and mechanically. Essentially the desired outcome when bonding metal parts together is to make them become one. Bonding has a very important function electrically for both grounded and ungrounded systems. Bonding metallic parts together puts the parts at the same potential, and at the potential of the earth, through the bonding connection to the grounding electrode at the service or source of separately derived system. Section 250.90 generally requires bonding to ensure electrical continuity and the capacity to conduct safely any fault current likely to be imposed on the bonded metal parts. Section 250.100 gets more specific and restrictive in the requirements when bonding metal components in hazardous locations. There are specific and more restrictive bonding requirements in Chapter 5 of the *Code,* that amend the general requirements found in Article 250.

Maintaining Continuity

Section 250.96(A) requires that bonding be done around connections of metal raceways, cable trays, cable armor, cable sheath, enclosures, frames, fittings and other metal non-current-carrying parts used as equipment grounding conductors where necessary. This may be necessary to assure that these systems have electrical continuity and the current-carrying capacity to safely conduct the fault current likely to be imposed on them. Note that this bonding of raceways, and so forth, is required to be performed regardless of whether or not an equipment grounding conductor is run within the raceway, and so forth. This will ensure that the raceway will not become energized by a line-to-enclosure fault without having the capacity and capability of clearing the fault by allowing sufficient current to flow to operate the overcurrent protective device on the line side of the fault.

- Observe "weakest link" rule.
- Ground-fault return path must be:
- Permanent and electrically continuous,
- Have adequate capacity,
- Have sufficiently low impedance.

Figure 8-1. Bonding to Maintain Continuity

Keep in mind that the *weakest link rule* applies to the ground-fault return path. To provide adequate safety, the effective ground-fault current path is required to be (1) permanent and electrically continuous, (2) have the capacity to conduct safely any fault current likely to be imposed on it, and (3) have sufficiently low impedance to limit the voltage to ground and to facilitate the operation of the circuit-protective devices [see 250.4(A)(5)]. This ground-fault path is required to meet all three conditions from the farthest

enclosure or equipment all the way back to the service equipment and ultimately to the source. This path can be through many boxes, conduits or other raceways, pull boxes, wireways, auxiliary gutters, panelboards, motor control centers

Over 300 conduit fitting assemblies from ten different manufacturers were subjected to a current test to simulate performance under ground-fault conditions.

and switchboards. Every connection is important. It only takes one loose locknut or broken fitting to break a link in the fault-current chain.

Section 250.96(A) also refers to conditions where a nonconducting coating might interrupt the required continuity of the ground-fault path, and it points out that such coatings must be removed unless the fitting(s) is (are) designed as to make such removal unnecessary.

In some cases, the locknut may pierce painted enclosures to establish a good electrical connection. This applies to the use of heavy-type, formed-steel locknuts. General instructions are that the locknuts be tightened by hand, then be further tightened $1/4$ turn by means of a screwdriver and hammer. At that point, examine the connection to be sure any paint under the locknut has been adequately broken. Remove the locknut and scrape the paint off or install a bonding bushing if there is any question about the adequacy of the connection.

Testing of Conduit Fittings

The importance of removing paint from enclosures where the conduit or raceway is intended to serve as the fault-current path is further emphasized in a report on "Conduit Fitting Ground-Fault Current Withstand Capability" issued by Underwriters Laboratories on June 1, 1992. Over 300 conduit fitting assemblies from ten different manufacturers were subjected to a current test to simulate performance under ground-fault conditions.

A sample assembly consisted of a conduit fitting secured to one end of a two-foot length of conduit and attached to a metal enclosure.

After securing the conduit fitting to the conduit properly, the conduit fitting was secured to the enclosure using the locknut provided by the manufacturer. The locknut was first hand-tightened and then further tightened $1/4$ turn with a hammer and standard screwdriver. The fittings were installed through holes in the enclosures that were punched rather than being installed in pre-existing knockouts. A pipe clamp, wire connector, conductors and a power supply were assembled to complete the testing. Thermocouples were placed at strategic locations to record pertinent temperature data. Figure 8-2 is a drawing of the sample assembly.

Fittings for conduit in the $3/8$-inch through 6-inch trade sizes were tested. The appropriate current applied to the fittings in the test program is as shown in Table 1.

This test should not be confused with a short-circuit withstand test and is not intended to test the maximum short-circuit current these fittings can withstand. Due to the time and current in-

volved, a great deal of heat is generated in the test assembly.

Seven of the more than 300 assemblies tested sustained damage. A visual examination of sample assemblies that failed showed that melting of the die cast zinc locknuts occurred as a result of the fault current. Melting of the die-cast zinc body occurred on five sample assemblies. The painted enclosures on which the fittings were tested were also examined. The examination indicated that melting of the die-cast zinc was probably due to the inability of the locknut to penetrate through the enclosure paint and provide good electrical contact between the fitting and metal of the enclosure.

A visual examination of all the conduit fittings with die-cast zinc locknuts showed that there were three different constructions of the locknuts. The three constructions differed in that the surface of the locknut contacting the enclosure was either flat, ribbed, or serrated. The sample assemblies with die-cast zinc locknuts that did not complete the current test with acceptable results had locknuts with flat or ribbed surfaces. All fittings having die-cast zinc locknuts with serrations completed the current tests with acceptable results. It appeared as though locknuts with serrations consistently penetrated through the enclosure paint and provided better electrical contact between the fitting and the metal of the enclosure than did the locknuts with flat or ribbed surfaces.

The fittings investigated in this work were formed of die-cast zinc, steel and malleable iron. The melting point of zinc is 420°C while the melting point of steel and malleable iron is much

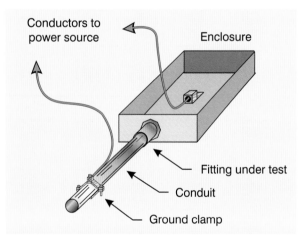

Figure 8-2. Testing of Conduit Fittings Sample Assembly

Conduit Size	Test Time	Current, Amperes
$3/8$	4 Seconds	470
$1/2$	4 Seconds	1,180
$3/4$,1	6 Seconds	1,530
$1\text{-}1/4$, $1\text{-}1/2$	6 Seconds	2,450
2	6 Seconds	3,900
$2\text{-}1/2$	6 Seconds	4,900
3, $3\text{-}1/2$, 4	9 Seconds	5,050
$4\text{-}1/2$	9 Seconds	6,400
5, 6	9 Seconds	8,030

Table 8-1. Test Current and Times

Photo 8-1. Testing of Conduit Fittings

higher, typically greater than 1400°C. Heat generated from the fault current in some sample assemblies was obviously greater than the melting point of the die-cast zinc fittings and locknuts, but not greater than the melting point of steel or malleable iron since no melting of the steel enclosure occurred. This was further evidenced by tests of two sample assemblies where the die cast zinc body of the fittings melted, but the steel locknuts did not.

All of the conduit fittings that were constructed of steel bodies and steel locknuts completed the test with acceptable results. A visual examination of the steel locknuts indicated that the nibs on these locknuts, which in most cases were sharp and well defined from the metal forming process, provided for better penetration through the enclosure paint than the nibs on the die-cast zinc locknuts.

For most of the sample assemblies which completed the current test with acceptable results, the maximum temperatures on the fitting bodies and locknuts were about the same as or less than the temperature of the conduit. In the case of the flexible metal conduit, the temperatures on the fittings were much less than on the conduit. This would seem to indicate that if the fitting can provide good electrical contact to the enclosure metal the fitting will provide for adequate equipment grounding.

Conclusions reached by Underwriters Laboratories as a result of the testing are as follows:

"1. Over 300 conduit fitting assemblies from ten different manufacturers were subjected to the Current Test to simulate performance under ground-fault conditions. As a result of the tests, only seven assemblies representing four different conduit fittings and three different manufacturers did not withstand the fault current without breaking or melting of the conduit fitting assembly. All seven of these sample assemblies were compression type connectors with die-cast zinc bodies, and all but one of these assemblies utilized a die-cast zinc locknut.

"2. An examination of the seven sample assemblies that did not complete the Current Test with acceptable results showed that the failures were probably due to high resistance from the inability of the fitting locknut to penetrate through the enclosure paint and provide good electrical continuity between the fitting and the metal enclosure. Heat generated by the high-resistance arcing was sufficient to melt the zinc, but not steel or iron.

"3. Some of the sample assemblies that did not exhibit breaking or melting did show signs of arcing and welding between the locknut and the enclosure and/or the fitting and the conduit. These sample assemblies usually had higher temperatures during the Current Test, however, the temperatures were not sufficient to cause melting of the zinc or steel parts nor loss of continuity between the conduit, fitting, and enclosure.

"4. Most of the sample assemblies that were subjected to the Current Test attained maximum temperatures on the fitting bodies and locknuts that were about the same as or less than the temperature of the conduit. For the tests with flexible metal conduit, the temperatures of the fittings were much less than the temperatures of the flexible conduit.

"5. As a result of the tests, it was observed that if the fitting provides good electrical contact to both the enclosure and the conduit, the fitting will provide a suitable equipment ground path for fault current."[1]

Bonding for Over 250 Volts

For systems having a voltage exceeding 250 volts to ground, the electrical continuity of metal race-

Listed fittings

- Threadless couplings and connectors for cables with metal sheaths
- Two locknuts, on rigid metal conduit or IMC, one inside and one outside
- Fittings with shoulders that seat firmly against the box or enclosure such as for EMT, flexible metal conduit and cable connectors with one locknut inside

Also permitted as well are:

- Threaded couplings and bosses
- Threadless couplings and connectors
- Bonding jumpers
- Other approved devices such as bonding-type locknuts.

Figure 8-3. Bonding for Over 250 Volt Circuits

ways or metal-sheathed cables that are not service-entrance cable must also be assured (see 250.97).

Acceptable methods include any of the methods approved for bonding at service equipment found in 250.92(B)(1) through (4). Note that standard locknuts and bushings without additional bonding means are not generally permitted for bonding equipment over 250 volts to ground.

The bonding methods permitted include those for services including:

(1) Bonding equipment to the grounded conductor as permitted in 250.142(A) in a manner specified in 250.8

(2) For rigid and intermediate metal conduit, connections made up wrenchtight with threaded couplings or threaded bosses on enclosures.

(3) Threadless couplings and connectors made up tight for rigid metal and intermediate metal conduit and electrical metallic tubing.

(4) Other approved devices like bonding type locknuts and bushings.[2]

In addition, bonding jumpers are permitted around concentric or eccentric knockouts that are punched or formed so as to impair the electrical connection to ground.

An exception to the requirement for bonding for these circuits provides that where oversized, concentric or eccentric knockouts are not encountered, or where concentric or eccentric knockouts have been tested and the box or enclosure is listed for the use, the following methods of ensuring continuity for these connections are permitted:

(a) threadless couplings and connectors for cables with metal sheaths,

Photo 8-2. Boxes Suitable for Bonding per 250.92(B)

(b) for rigid and intermediate metal conduit, two locknuts, one inside and the other outside the boxes and enclosures,

(c) fittings that seat firmly against the box or enclosure or cabinet such as for electrical metallic tubing, flexible metal conduit and cable connectors, with one locknut inside the enclosure, or

(d) listed fittings that are identified for the purpose (see Article 100 for definition of the word *identified*).

Outlet boxes are available that have specially designed and tested knockouts that perform satisfactorily for over 250-volt-to-ground applications. These boxes have only one eccentric knockout so when the solid knockout is removed, a conduit or fitting locknut makes contact with the base metal of the box to ensure good electrical and mechanical contact (see photo 8-2).

Oversized Knockouts

The installer needs to be cautious in the use of equipment that has concentric or eccentric knockouts, as their ability to carry fault current must be of concern. It is very common to find nibs of adjacent rings damaged during removal of the desired knockout. This leaves less material available for carrying fault current. The safest practice is to install bonding bushings around concentric and eccentric knockouts where there is any question about their integrity.

Concentric and eccentric knockouts in equipment such as panelboards, auxiliary gutters and wireways are not tested or certified by an electrical products testing laboratory for their current-carrying ability. As such, they should not be relied upon for such a purpose and should have bonding bushings and bonding jumpers installed around them.

In other areas, where oversized, concentric or eccentric knockouts are not present, threadless fittings which are made up tight with conduit or armored cable or the use of two locknuts, one inside and one outside of boxes and cabinets, are acceptable for bonding.

Where loosely-jointed metal raceways are used and especially where there may be expansion joints or telescoping sections of raceways, the *Code* requires that they be made electrically continuous by the use of equipment bonding jumpers or other means (see 250.98).

Photo 8-3. Knockout Reducing Washers

Reducing Washers
Reducing washers are commonly used in electrical installations where it is desirable or necessary to install conduit or fittings of a size that is smaller than the knockout available in the enclosure. These reducing washers, especially where they are installed on painted surfaces, are not suitable for bonding and should always be bonded around to provide an adequate fault-return path (see photo 8-3).

Bonding Jumper Sizes
Those equipment bonding jumpers form a part of the equipment grounding conductor and are, therefore, called upon to carry the same fault current that the equipment grounding conductor would carry and are required to be the same size.

The size of the bonding jumper will depend on its location and is always based on the size of the nearest overcurrent device in the circuit ahead of the equipment [see 250.102(D)]. Column 1 of Table 250.122 gives the size or setting of the overcurrent device in the circuit ahead of the equipment. Columns 2 and 3 give the minimum size of the equipment grounding conductor whether copper or aluminum or copper-clad aluminum.

Attaching Jumpers
Where bonding jumpers are used between grounding electrodes or around water meters and the like, the Code requires that good electrical contact be maintained and that the arrangement of conductors be such that the disconnection or removal of equipment will not interfere with or

Open-bottom switchboard or Motor Control Center

400 A, 300 A, 225 A and 125 A overcurrent devices. Size for 400 ampere = 3 AWG CU or 1 AWG AL.

Size bonding jumper for largest overcurrent device

Figure 8-4. Bonding Multiple Raceways

interrupt the grounding continuity of the jumper [see 250.68(B)].

Bonding jumpers are required to be attached to circuits and equipment by means of exothermic welding, listed pressure connectors, listed clamps or other suitable and listed means. A connection that depends solely on solder is not acceptable. Sheet metal screws are not permitted to connect grounding conductors to enclosures (see 250.8).

Bonding Multiple Raceway Systems
Where more than one raceway enters or leaves a switchboard, pull or junction box or other equipment, it is permissible to use a single conductor to bond these raceways to the equipment. The equipment grounding (bonding) conductor is sized for the largest overcurrent device ahead of conductors contained in the raceways [see 250.102(D)]. These are feeder or branch-circuit conductors and not service-entrance conductors.

For example, as shown in figure 8-4, four metallic raceways leave the bottom of an open switchboard or motor control center. The overcurrent protective devices ahead of the raceways are 400, 300, 225 and 125 amperes respectively. According to Table 250.122, the minimum size equipment bonding jumper for the raceway having conductors protected at 400 amperes is 3 AWG copper or 1 AWG aluminum. If this conductor were looped through a grounding bushing on each raceway, compliance with the Code would be obtained. Of course, the grounding bushings would need to be listed for both the size of conduit and conductor. In some cases, this method may require a larger conductor to bond some conduits than where individual bonding jumpers are installed.

In addition, as shown in figure 8-5, it is acceptable to install an equipment bonding jumper individually from each raceway to the equipment grounding terminals of the equipment. Each bonding conductor is sized for the overcurrent device ahead of the conductors in the raceway per Table 250.122.

It should be noted that the size of the bonding jumpers given in 250.122 is the minimum size.

Nonmetallic Boxes
Section 314.3 permits metal raceways or metal-armored cables to be used with nonmetallic boxes only where:
1. internal bonding means are provided between all raceways or metal-armored cables, or
2. internal bonding means with provision for attaching a grounding jumper inside the box are provided.

The bonding jumpers are required to, of course, be sized in accordance with Table 250.122. It should be noted that the size of the bonding jumpers given in 250.122 is the minimum size. Larger bonding jumpers may be required to comply with the available fault-current requirements in 110.10.

Bonding Receptacles
An equipment bonding jumper is required to connect the grounding terminal of a grounding type

Open-bottom switchboard or Motor Control Center

	CU	AL
400 A, =	3 AWG	1 AWG
300 A, =	4 AWG	2 AWG
225 A, =	4 AWG	2 AWG
125 A, =	6 AWG	4 AWG

Size bonding jumper for individual overcurrent device

Figure 8-5. Bonding Multiple Raceways

receptacle to a grounded box (see 250.146). Where more than one equipment grounding conductor enters a box, they must be spliced or joined inside the box with suitable devices. Four exceptions to the general rule requiring the bonding jumper are provided in 250.146(A) through (D) as paraphrased below.

(A) Where the box is mounted on the surface and direct metal-to-metal contact is made between the receptacle and the box. Note that the rule will now permit receptacles without an equipment bonding jumper only where the box is mounted on the surface. Cover mounted receptacles such as in a raised cover on 4-in. square boxes are not approved unless the box and cover combination are listed as providing adequate ground continuity.

(B) Contact devices or yokes designed and listed as providing bonding with the mounting screws to establish the grounding circuit between the device yoke and flush-type boxes. This is the device commonly referred to as a *self-grounding receptacle*. The device is designed and listed as maintaining good electrical contact between the yoke and box by means of a spring-type device

Splice or join equipment grounding conductors together inside box using suitable devices. Connect bonding jumper to grounding terminal of grounding receptacle unless:

(A) Boxes mounted on surface, metal-to-metal contact.
(B) Contact devices listed as providing grounding.
(C) Floor boxes listed for grounding.
(D) Isolated equipment grounding terminal.

Figure 8-6. Bonding of Grounding Receptacles

Photo 8-4. Self-Grounding Type Receptacles

Figure 8-7. Isolated Ground Receptacles

Isolated ground receptacles to be identified by orange triangle on face. To be used only with isolated grounding conductors per Section

Isolated receptacles installed in nonmetallic boxes shall be covered with a nonmetallic faceplate unless box has feature or accessory for grounding metal faceplate

Where installed outside the raceway or enclosure, the length is limited to not more than 1.8 m (6 ft).

The bonding jumper must be routed with the raceway.

See Section 250.102(E)

Figure 8-8. Installation of Bonding Jumper

that maintains good continuity between the device and the mounting screws. See Photo 8-4.

(C) Floor boxes that are listed as providing satisfactory grounding continuity.

(D) A receptacle having an isolated equipment grounding terminal that is required for reduction of electrical noise (electromagnetic interference). In this case, the equipment grounding terminal is required to be grounded by an insulated equipment grounding conductor that is run with the circuit conductors. This equipment grounding conductor is permitted to pass through one or more panelboards to terminate at an equipment grounding terminal of the separately derived system or service. However, this insulated equipment grounding conductor must be connected to the equipment grounding terminal bar at the building disconnecting means or the source of a separately derived system within the same building or structure.[3]

To provide effective grounding, the insulated equipment grounding conductor should never pass without being terminated at the separately derived system that is the source of power for the equipment being grounded. Also note that special rules are provided in 406.2(D) for grounding receptacle covers where isolated ground receptacles are installed in a nonmetallic outlet box.

Installation of Equipment Bonding Jumper

The equipment bonding jumper is permitted to be installed either inside or outside of a raceway or enclosure [see 250.102(E)].

Where the jumper is installed on the outside, the length is generally limited to not more than 1.8 m (6 ft.) In addition, the bonding jumper is required to be routed with the raceway. This is vital to keep the impedance of the equipment bonding jumper as low as possible (see figure 8-8). An equipment bonding jumper longer than 1.8 m (6 ft) is permitted at outside pole locations

Outside pole locations

Pole mounted transformer

Bonding bushing or clamp

Permitted to be longer than 1.8 m (6 ft)

Building or structure

Intermediate metal conduit or rigid metal conduit

Equipment location

Adaptor from PVC to Steel

ON OFF ON OFF ON OFF

ON OFF MAIN ON OFF ON OFF

Rigid nonmetallic conduit

PVC

Figure 8-9. Typical Bonding Jumper Installation at Pole Location.

for the purpose of bonding or grounding isolated sections of metal raceways or elbows that are installed in exposed risers of metal conduit or other metal raceway.

Bonding of Piping Systems

Section 250.104 requires that metal water piping and other metal piping systems installed within or attached to buildings or structures be bonded. This requirement for bonding is not to be confused with the requirement in 250.52(A)(1) that metal underground water piping is to be used as a grounding electrode. Some requirements change depending upon whether the piping is metal water piping or other metal piping systems.

Included among the items within a building that should be adequately grounded to the one common grounding electrode system are the water piping system (hot and cold) and gas and sewer piping, any metallic air ducts installed inside or on the exterior of the building, as well as such devices as TV towers, gutters provided with a deicing system, and so forth.

The bonding jumper is generally required to be sized in accordance with Table 250.66.

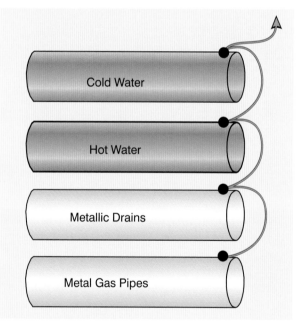

Cold Water

Hot Water

Metallic Drains

Metal Gas Pipes

Figure 8-10. Bonding Metal Piping Systems

Also, it is required to be installed in accordance with the general rules for installing grounding conductors in 250.64(A), (B) and (E). They shall be connected in a manner specified in 250.70. The point(s) of attachment of the bonding jumper(s) to the water piping system is required to be accessible.

Metal Water Piping

This requirement for bonding applies to all metal water piping system(s) installed within or on the exterior of the building [see 250.104(A)(1)]. The bonding jumper in the building where the service is located is generally required to be sized in accordance with Table 250.66 and, thus, is based on the size of the service-entrance conductor and not on the rating of the service overcurrent device. In addition, the points of attachment of the bonding jumper to the metal water piping system are required to be accessible.

The piping system is permitted to be bonded to the service equipment enclosure, the grounded conductor at the service, the grounding electrode

Points of attachment must be accessible per 250.104(A). Size conductor per Section 250.66.

Figure 8-11. Bonding Metal Water Pipes

It is permitted to bond metal interior water pipes to subpanel in multiple occupancy buildings under the conditions in Section 250.104(A)(2) and size per 250.122 based on feeder size.

Figure 8-12. Bonding Metal Water Piping systems

conductor where large enough, or to one or more grounding electrodes.

For example, if Table 250.66 requires a 2 AWG bonding jumper to the metal water piping, it cannot be connected to a 6 AWG or 4 AWG grounding electrode conductor.

Where a metallic underground water piping system exists and is connected to a metallic interior water-piping system and there is not an insulated coupling, the interior water piping system is automatically and adequately grounded (bonded) when the metallic underground water piping system is used as the grounding electrode. However, with the expanding use of nonmetallic piping and insulated couplings it becomes more important to be sure that the interior piping not only is electrically continuous, but that also it is adequately grounded by bonding it to the same grounding electrode used for the premises. That is a mandatory and essential requirement of the *Code*.

Multiple Occupancy Building

Section 250.104(A)(2) allows the metal water piping system(s) to be bonded to the panelboard or switchboard enclosure (other than service equipment) under specific conditions. The conditions are as follows:

(1) The building is multiple occupancy, and
(2) The metallic water piping is isolated from all other occupancies by nonmetallic water piping. In other words, the metallic piping system in each occupancy is isolated from all other piping systems of the other occupancies by nonmetallic means or individual isolation.

In this case, the bonding jumper to the water piping is sized in accordance with Table 250.122. Thus, the ampere rating of the overcurrent device supplying the feeder to the unit or occupancy determines the minimum size of the bonding conductor. The bonding jumper for the interior metal water piping runs from the equipment grounding terminal bar in the panelboard serving the unit to the piping. In this case, the bonding jumper does not connect to the neutral terminal bar in the panelboard.

Multiple Buildings or Structures Supplied from a Common Service

Where a building or structure is supplied from a common service by a feeder or branch circuit, bonding of metal water piping system(s) is required to be ensured by one of the following methods:

(1) Bonding to the building or structure disconnecting means where it is located at the building or structure.

(2) Bonding to the equipment grounding conductor that is run with the supply conductors to the building or structure. This connection would usually be made inside the building or

structure disconnecting means enclosure on the equipment grounding terminal bar. Note that the equipment grounding conductor is permitted to consist of the wiring method that supplies the building or structure if recognized in 250.118.

(3) Bonding to the one or more grounding electrodes (grounding electrode system) used.[4]

The bonding jumper to the interior water piping system(s) is required to be sized according

Figure 8-13. Multiple Buildings or Structures From a Common Service

Figure 8-14. Bonding of Metal Water Piping

Figure 8-15. Bonding Other Metal Piping Systems 250.104(B)

to Table 250.66 based on the size of the feeder or branch-circuit conductors that supply the building or structure.

Bonding Metal Water Piping to a Separately Derived System

The grounded conductor of a separately derived system is required to be bonded to the nearest available point on the interior metal water piping system in the area served by the separately derived system [see 250.104(A)(4)].

> The equipment grounding conductor for the circuit that may energize the piping is permitted to serve as the bonding means.

The bonding connection is required to be made to the same point on the separately derived system where the grounding electrode conductor is connected (see the rules on separately derived systems in chapter twelve).

This requirement assures that metal piping, such as process or manufacturing equipment piping, in the area served by the separately derived system is bonded. This ensures an adequate and low-impedance path for fault current to return to its source which is the separately derived system, not the service to the building or structure.

Bonding Metal Gas Piping

"Where installed in or attached to a building or structure, metal piping system(s), including metal gas piping, that may become energized are required to be bonded to the service equipment enclosure, the grounded conductor at the service, the grounding electrode conductor where of sufficient size, or to the one or more grounding electrodes used. The bonding jumper shall be sized in accordance

with Table 250.122 using the rating of the circuit that may energize the piping system(s)."[5] The equipment grounding conductor for the circuit that may energize the piping is permitted to serve as the bonding means.

This requirement applies to both indoor and outdoor metal gas piping systems. The *Code* includes requirements for sizing the bonding conductor at the minimum size of the equipment grounding conductor included with a circuit that may be likely to energize the gas piping.

Bonding Other Metal Piping

Other metal piping, including gas piping, which may become energized is required to be bonded. The *Code* does not give guidance on how to determine the conditions under which metal piping is likely to become energized. Since metal piping systems are conductive, from a safety standpoint, all metal piping systems should be bonded.

Common systems that have to be bonded include interior metal: pneumatic systems; waste, drain and vent lines; and oxygen, air, and vacuum systems.

The bonding conductor is sized from Table 250.122 using the rating of the overcurrent de-

Size bonding jumper per Section 250.122

Figure 8-16. Bonding Other Metal Piping

This is page 140.

vice in the circuit ahead of the equipment. The equipment grounding conductor for the circuit that may energize the equipment can be used as the bonding conductor. The point of connection

> The points of attachment of the bonding jumper to the structural steel are required to be accessible.

of these bonding conductors to the metal piping systems is not required to be accessible as the connections to metal water piping systems are.

The bonding conductor is permitted to be connected to the:
(1) Service equipment enclosure
(2) Grounded conductor at the service
(3) Grounding electrode conductor, where of adequate size
(4) One or more of the grounding electrodes used

Bonding Structural Steel

Section 250.104(C) requires exposed structural steel that is interconnected to form a steel building frame, not intentionally grounded, and may become energized to be bonded. This is requirement is applicable to interior or exterior structural framing members of buildings or structures. A bonding connection is required to be made to the service equipment enclosure, the grounded conductor at the service, the grounding electrode conductor where it is large enough, or to the one or more grounding electrodes used.

The bonding jumper is required to be sized in accordance with Table 250.66 and installed in accordance with the rules in 250.64(A), (B) and (E). The points of attachment of the bonding jumper to the structural steel are required to be accessible.

---

[1] "Conduit Fitting Ground-Fault Current Withstand Capability," Underwriters Laboratories, June 1, 1992.
[2] NFPA 70, *National Electrical Code* 2002, 250.92(B)(1), (National Fire Protection Association, Quincy, MA, 2002), p. 70-109.
[3] NFPA 70, 250.146(A)-(D), p. 70-117.
[4] NFPA 70, 250.104(A)(3), p. 70-110.
[5] NFPA 70, 250.104(B), p. 111.

- Exposed structural steel that is interconnected to form a building frame and is not intentionally grounded and may become energized shall be bonded.
- Bond to: service equipment enclosure, the grounded conductor at service, the grounding electrode conductor where of sufficient size, or to the one or more grounding electrodes used.

- Size bonding jumper per Table 250.66.
- Install per Sections 250.64(A), (B), and (E).
- Attachment point(s) of bonding jumper to be accessible.

Figure 8-17. Bonding Structural Steel

Chapter Eight: The questions included here were developed using material included in this chapter. The answers can be found by reviewing the text. It is also important that students make use of the 2002 *NEC*, where many of the answers can be found.

1. For systems having a voltage exceeding _____ volts to ground, the electrical continuity of metal raceways or metal-sheathed cables, that are not service-entrance cable, must also be assured.
   a. 125
   b. 150
   c. 100
   d. 250

2. Bonding jumpers are required to be attached to circuits and equipment by means of _____, or other listed means.
   a. exothermic welding
   b. listed pressure connectors
   c. listed clamps
   d. any of the above

3. Metal raceways or metal-jacketed cables are permitted to be used with nonmetallic enclosures only where _____.
   a. internal bonding means are provided between all raceways
   b. internal bonding means with provision for attaching a grounding jumper inside the box are provided
   c. internal bonding means are provided between all cables
   d. any of the above

4. Without exception, where circuit conductors are spliced within a box, or terminated to equipment or devices within or supported by the box, equipment grounding conductors associated with those circuit conductors must be spliced or joined with_____ devices.
   a. labeled
   b. listed
   c. suitable
   d. bonding

5. An equipment grounding conductor is permitted to pass through one or more panelboards within the same _____ to terminate at an equipment grounding terminal of the separately derived system or service.
   a. equipment
   b. building
   c. enclosure
   d. cabinet

6. Where the equipment bonding jumper is installed on the outside of a raceway, the length is limited to not more than _____ feet. In addition, the bonding jumper must be routed with the raceway.
   a. 7
   b. 8
   c. 6
   d. 10

7. The metal water piping system is permitted to be bonded to the service equipment enclosure, the grounded conductor at the service, the grounding electrode conductor where large enough, or to one or more grounding _____.
   a. clamps
   b. devices
   c. fittings
   d. electrodes

8. Common systems that must be bonded include interior metal _____.
   a. gas piping and pneumatic systems
   b. waste, drain and vent lines
   c. oxygen, air and vacuum systems
   d. all of the above

9. The equipment grounding conductor for the circuit that may energize any other metal piping systems can be used as the _____ conductor.
   a. equipment bonding
   b. grounded
   c. bonding
   d. identified

10. Expansion joints or telescoping sections of metal raceways must be made electrically continuous by the use of a ____.
   a. steel strap
   b. equipment bonding jumper
   c. welding cable
   d. equipment grounding conductor

11. The locknut/bushing and double-locknut types of installations are not acceptable for bonding in ____.
   a. hazardous (classified) locations
   b. commercial locations
   c. industrial location
   d. computer rooms

12. If four metal raceways have their conductors protected by overcurrent protective devices sized at 400, 300, 225 and 125 amperes and leave the bottom of an open switchboard or motor control center, the minimum size of a single equipment bonding jumper to be used to bond them together is ____.
   a. 4 AWG copper or 4 AWG aluminum
   b. 6 AWG copper-clad aluminum
   c. 3 AWG copper or 1 AWG aluminum
   d. 4 AWG copper or 3 AWG aluminum

13. The points of attachment of the bonding jumper to the metal water piping system are required to be ____.
   a. acceptable
   b. marked
   c. accessible
   d. soldered

14. Concentric and eccentric knockouts in enclosures like wireways and panelboards:
   a. are always suitable for grounding and bonding.
   b. are tested by a qualified electrical testing laboratory for their current-carrying ability.
   c. are not tested by a qualified electrical testing laboratory for their current-carrying ability.
   d. are capable of carrying large amounts of fault current.

15. Bonding connections to metal water piping systems:
   a. are permitted to be made with solder.
   b. must be accessible.
   c. are permitted to be connected to the neutral in a subpanel.
   d. are not required.

16. Structural building steel must be bonded where:
   a. it is exposed on the interior of the building.
   b. it is interconnected to form a steel building frame.
   c. it is not intentionally grounded.
   d. all of the above.

Key Notes

Equipment Grounding Conductors

OBJECTIVES

After studying this chapter, the reader will be able to understand

- General requirements for equipment grounding conductors on grounded and ungrounded systems.
- Rules applied to multiple raceways or cables.
- Rules for flexible cords.
- Use of building steel that is properly grounded by an equipment grounding conductor.
- Grounding of equipment by the grounded circuit conductor.

CHAPTER NINE

Equipment grounding conductors are intended to prevent an objectionable potential above ground on conductor and equipment enclosures and to provide a low-impedance path for fault currents by connecting the equipment grounding conductor to the grounded conductor if it is a grounded system. The equipment grounding conductor or path must also:

1. be permanent and electrically continuous;

2. have ample capacity to conduct safely any currents likely to be imposed on it; and

3. be of the lowest practical impedance [see 250.4(A)(5)]. The equipment grounding conductor or path is required to connect with the grounding electrode in a low-impedance path, and if the system is grounded, it is required to be connected through a low-impedance path to the grounded service conductor (often a neutral of a system or service).

Definition

"**Grounding conductor, equipment**. The conductor used to connect the non-current-carrying metal parts of equipment, raceways, and other enclosures to the system grounded conductor, the grounding electrode conductor, or both, at the service equipment or at the source of a separately derived system."[1]

Grounded Systems

The equipment grounding conductor or path is required to extend from the furthermost point on the circuit to the service equipment where it is connected to the grounded conductor and grounding electrode on a grounded system, or connected to the disconnecting means enclosure and the grounding electrode conductor for an ungrounded system. This connection is made through the main bonding jumper. Often, this equipment grounding conductor or path is the conductor enclosure (conduit, cable jacket, and so forth).

For enclosed panelboards typically installed at dwelling services, the grounded and equipment grounding conductors connect to the same terminal bar (see 408.20). Where this type of equipment is installed, the main bonding jumper connects the enclosure to the grounded service conductor as well as to a separate equipment grounding conductor terminal bar where one is installed.

Should insulation failure occur anywhere on a phase conductor and a ground fault develop between the energized conductor and the conductor enclosure, a ground-fault circuit will be established.

The ground-fault circuit will thus be from the source, through the supply conductors, through the overcurrent devices, to the point of fault on the phase conductor, usually through an arc, through the equipment grounding conductor, through the main bonding jumper, to the grounded conductor or neutral and back to the source.

If this circuit is complete, of adequate capacity and low impedance, the equipment and persons who may contact it are protected as an overcurrent device will open the faulted circuit. A break in this equipment grounding circuit or other grounding system failure will expose persons to possibly lethal shocks if a ground fault from a source having sufficient potential (usually more than 50 volts) energizes the enclosure and the person provides the path for current to flow through.

Ungrounded Systems

In an ungrounded system, the equipment grounding conductor or path is required to be permanent and continuous to ground to keep all equipment and conductor enclosures at or near ground potential. The equipment grounding conductor must be the same size as called for in a grounded system if we are to have maximum safety. The minimum sizes of equipment grounding conductors given in Table 250.122 apply equally to an ungrounded system and a grounded system.

In general, it may be said that any conductor or equipment enclosure, be it conduit, electrical metallic tubing, raceway or busway enclosure, provides a satisfactory equipment grounding conductor for an ungrounded system if all joints are made electrically continuous. It may be necessary to use bonding jumpers at certain points. Such bonding jumpers are also required to be sized per Table 250.122.

Figure 9-1. Ground Fault in Grounded System

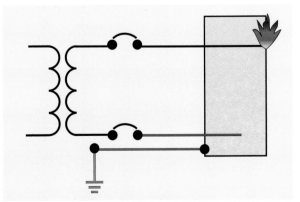

Figure 9-2. Ground Fault in Ungrounded System

Equipment Grounding Conductor Material

The *Code* specifies in 250.118 the conductors that are permitted to be used for equipment grounding conductors. They are as follows:

(1) A conductor of copper, aluminum, or copper-clad aluminum. The conductor shall be solid or stranded; insulated, covered, or bar; and in the form of a wire or busbar of any shape. [*Editor's note:* some sections of the *Code* may specifically require an equipment grounding conductor to be insulated or solid.]

(2) Rigid metal conduit.

(3) Intermediate metal conduit.

(4) Electrical metallic tubing.

(5) Flexible metal conduit where both the conduit and fittings are listed for grounding.

(6) Listed flexible metal conduit that is not listed for grounding, meeting all the following conditions:

a. The conduit is terminated in fittings listed for grounding

b. The circuit conductors contained in the conduit are protected by overcurrent devices rated at 20 amperes or less.

c. The combined length of flexible metal conduit and flexible metallic tubing and liquidtight flexible metal conduit in the same ground return path does not exceed 1.8 m (6 ft).

d. The conduit is not installed for flexibility." [*Editor's note:* Where installed for flexibility, 348.60 requires that an equipment grounding conductor be installed.][2]

Flexible metal conduit is commonly available as a listed product but has not been listed for grounding by a nationally recognized electrical testing laboratory. That is the reason for the phrase, "Listed flexible metal conduit that is not listed for grounding." The conduit has been recognized for several years for use as an equipment grounding conductor under the limitations indicated above. Note, however, the flexible metal conduit must be terminated in fittings that are listed for grounding.

(7) Listed liquidtight flexible metal conduit meeting all the following conditions:

a. The conduit is terminated in fittings that are listed for grounding.

Flexible Metal Conduit permitted as equipment grounding conductor as follows:

(a) Fittings listed for grounding.
(b) Maximum 20 ampere overcurrent protection of contained conductors.
(c) Combined length of flexible metal conduit, flexible metallic tubing and liquidtight flexible metal conduit in same ground return path not more than 1.8 m (6 ft)
(d) For connection of equipment where flexibility is required, additional equipment grounding conductor required.

Figure 9-3. Flexible Metal Conduit as Equipment Grounding Conductor

b. For metric designators 12 through 16 (trade sizes $^3/_8$ through $^1/_2$), the circuit conductors contained in the conduit are protected by overcurrent devices rated at 20 amperes or less.

c. For metric designators 21 through 35 (trade sizes $^3/_4$ through 1 $^1/_4$) the circuit conductors contained in the conduit are protected by overcurrent devices rated not more than 60 amperes and there is no flexible metal conduit, flexible metallic tubing, or liquidtight flexible metal conduit in trade sizes metric designators 12 through 16 (trade sizes $^3/_8$ or $^1/_2$) in the grounding path.

d. The combined length of flexible metal conduit and flexible metallic tubing and liquidtight flexible metal conduit in the same ground return path does not exceed 1.8 m (6 ft).

e. The conduit is not installed for flexibility." [*Editor's note:* Where installed for flexibility, 350.60 requires that an equipment grounding conductor be installed.]

(8) Flexible metallic tubing where the tubing is terminated in fittings listed for grounding and meeting all the following conditions:

a. The circuit conductors contained in the tubing are protected by overcurrent devices rated at 20 amperes or less.

b. The combined length of flexible metal conduit and flexible metallic tubing and liquidtight flexible metal conduit in the same ground return path does not exceed 1.8 m (6 ft).

(9) Armor of Type AC cable as provided in 320.108.

(10) The copper sheath of mineral-insulated, metal-sheathed cable.

(11) Type MC Cable where listed and identified for grounding in accordance with the following:

a. The combined metallic sheath and grounding conductor of interlocked metal tape-type MC Cable.

b. The metallic sheath or the combined metallic sheath and grounding conductors of the smooth or corrugated tube type MC Cable.

Liquidtight flexible metal conduit permitted as equipment grounding conductor as follows:

Fittings listed for grounding.

For sizes 3/8 and ½ maximum 20 ampere overcurrent device protecting contained conductors.

For sizes ¾ through 1-1/4 sizes maximum 60 ampere overcurrent device protecting contained conductors.

Maximum length in ground return path does not exceed 1.8 m (6 ft).

For connection of equipment where flexibility is required, additional equipment grounding conductor required.

Figure 9-4. Liquidtight Flexible Metal Conduit – Equipment Grounding Conductor

(12) Cable trays as permitted in 392.3(C) and 392.7.

(13) Cablebus framework as permitted in 370.3.

(14) Other electrically continuous metal raceways and auxiliary gutters listed for grounding.[3]

Included are auxiliary gutters (not specifically a raceway as defined in Article 100 but the same equipment as wireways), wireways with associated fittings; busway enclosures, and in some cases an additional ground bus, surface metal raceways, and pull and junction boxes that are installed in the ground-fault path.

It should be noted here that these requirements or provisions in 250.118 are general in nature. Many sections of the *Code* contain specific requirements that must be complied with. A few examples follow. Section 501.16 does not recognize the standard double locknut-type conduit connections for Class I hazardous (classified) locations.

Section 517.13(B) requires a supplementary insulated equipment grounding conductor in patient care areas of health care facilities. Section 550.33(A) generally requires the equipment grounding conductor for the feeder to a mobile home to be insulated. Several sections of Article 680 require an insulated equipment grounding conductor. It is always best to carefully examine the specific requirements for the equipment grounding conductor for the type of installation being made.

Conductor Enclosures

Conduit runs of rigid or intermediate metal, which are properly threaded and in which the couplings are set up tightly, preferably using a joint sealer that will not reduce continuity, may be expected to perform satisfactorily as an equipment grounding conductor for runs of limited length. A joint

Figure 9-5. Equipment Grounding Conductor Path

sealer aids in assuring an effective equipment grounding path because it acts as a lubricant and permits the joint to be screwed up tighter. Under poor conditions, the conduit impedance with couplings should not show an increase of over 50 percent when compared with a straight run of conduit. The use of this higher impedance figure would provide a factor of safety. In the case cited, there is no economic justification for using an additional equipment grounding conductor.

The *Code* further requires that where conduit is used as a grounding conductor, all joints and fittings shall be made up tight using suitable tools [see 250.120(A)]. This calls attention to the fact that conduit, where used as an equipment grounding means, is a current-carrying conductor under fault conditions and is required to be made electrically continuous by having joints made up tight.

Usually, large and often parallel conduits are installed from the transformer to the service equipment. Then, smaller and smaller conduits are installed for feeders and branch circuits. For instance, at one point the equipment grounding path may be three 4-inch conduits in parallel; at another point, two 4-inch conduits in parallel while down the line it may be only one 1¼-inch conduit, all being connected together to form a permanent and continuous path. As the circuit changes from large overcurrent protective means to smaller ones, the conductivity of the equipment grounding path becomes smaller and smaller. The conduit or tubing at the end of the circuit may be no larger than ½-inch electrical metallic tubing or ⅜-inch flexible metal conduit.

The *NEC* does not dictate any particular size of conduit or tubing to serve as the equipment grounding conductor for an upstream overcurrent device, other than as mentioned in the previous section. It is felt that the metallic raceway that is sized properly for the conductor fill will provide an adequate equipment ground-fault return path.

Cables as Equipment Grounding Conductors

Several cables used as wiring methods are suitable for use as an equipment grounding conductor or contain an equipment grounding conductor. These include:

Type AC cable. Armored cable (Article 320) is manufactured with conductors in sizes from 14 AWG through 1 AWG copper and from 12 AWG through 1 AWG aluminum. Type AC cable is required to have an armor of flexible metal tape. The insulated conductors are required to be in accordance with 320.104. Cables of the AC type are required to have an internal bonding strip of copper or aluminum in intimate contact with the armor for its entire length. It is suitable as an equipment grounding conductor in accordance with 250.118(9). In addition to the internal bonding strip, another means of identification of Type AC cable is the conductors are required to have an overall moisture-resistant and fire-retardant fibrous (paper) covering. Another type of AC cable construction includes an insulated equipment grounding conductor and is acceptable for use as the branch circuits serving patient care areas as provided in 517.13 and for use in branch circuits for isolated grounding receptacles as permitted in 250.146(D) and 408.20 Exception

Type MC cable. Type MC cable is covered in Article 330. Type MC cable is produced in three configurations: spiral interlocking metal tape, corrugated metal tube, and a smooth metal tube.

1. The spiral-interlocking-metal-tape Type MC cable must always have an equipment grounding conductor which may be insulated or bare. The jacket itself is not suitable as an equipment grounding conductor. The principal equipment grounding conductor may be divided (sectioned) into more than one conductor, often to facilitate spacing in the cable construction. Additional equipment grounding conductors have green insulation and either a yellow stripe or other identification.

2. The sheath of the smooth or corrugated tube Type MC cable or a combination of the sheath and a supplemental bare or green insulated conductor is suitable for the required equipment grounding conductor. The principal equipment grounding conductor may be divided (sectioned) into more than one conductor, often to facilitate spacing in the cable construction. Additional equipment grounding conductors have green insulation and either a yellow stripe or other identification.

Nonmetallic-sheathed cable. NM cable is covered in Article 334. This cable is permitted to be produced in three styles: Type NM, Type NMC and Type NMS. The power conductors are permitted to be in sizes 14 AWG through 2 AWG copper and 12 AWG through 2 AWG aluminum. Type NMS is permitted to contain signaling conductors in compliance with 780.5. These cables typically contain an equipment grounding conductor sized in compliance with Table 250.122.

Service-entrance cable. Type SE cable is produced in a variety of configurations. The type most commonly used for internal wiring is Type

Photo 9-1. AC Cable

Photo 9-2. MC cable.

SE style U and Type SE style R. Specific rules for Type SE cables are contained in 338.10.

Type SE service-entrance cables are permitted in interior wiring systems where all of the circuit conductors of the cable are of the rubber-covered or thermoplastic type.

Type SE service-entrance cables without individual insulation on the grounded circuit conductor are not to be used as a branch circuit or as a feeder within a building, except a cable that has a final nonmetallic outer covering and is supplied by alternating current at not over 150 volts to ground is permitted as a feeder to supply only other buildings on the same premises. Type SE service-entrance cables are permitted for use where the fully insulated conductors are used for circuit wiring and the uninsulated conductor is used for equipment grounding purposes.

Underground feeder and branch-circuit cable. Type UF cable is permitted to be produced in sizes 14 AWG copper or 12 AWG aluminum through 4/0 AWG. Multiconductor cables are permitted to be installed in accordance with Article 340. In addition to the insulated conductors, the cable is permitted to have an insulated or bare conductor for equipment grounding purposes only. As such, it is required to comply with Table 250.122.

Size of Equipment Grounding Conductor

The entire equipment grounding conductor or path of any raceway system will be as shown in figure 9-6. Starting at the service, we have a large overcurrent protective device which is in series with other and usually smaller feeder or branch overcurrent protection devices. The ungrounded (hot) conductor usually decreases in size as it progresses through smaller and smaller overcurrent devices.

Thus, a ground fault may be cleared through a 15-ampere overcurrent device so that the equipment ground path below the 15-ampere overcurrent device level only needs to be a 18 AWG copper wire (subject to *Code* limitations) as given in 250.122(E). That conductor has a short-time rating of about 38 amperes and it should thus be adequate to operate the 15-ampere overcurrent device if it operates in a reasonable time.

Section 250.122(A) provides the general rules for sizing the equipment grounding conductor. It refers to Table 250.122 for determining the minimum size of conductor that is permitted to be used as an equipment grounding conductor. The size is based on the ampere rating of the overcurrent protective device ahead of the conductor (Table 250.122 is reprinted as Table 6 in chapter nineteen).

For example, if the overcurrent protection ahead of the circuit or feeder is 225 amperes, the minimum size equipment grounding conductor is found as follows:

In Table 250.122, follow the first column, which gives the rating of the overcurrent device, down to find the rating which equals or exceeds 225 amperes. Since 225 amperes is not found, go to the next larger size which is 300 amperes. Follow that line across to find the minimum size copper wire to be 4 AWG and for aluminum, a 2 AWG minimum size conductor.

Follow a similar process to determine the minimum size conductor for any installation. In addition, the note below Table 250.122 requires that, "Where necessary to comply with 250.4(A)(5) or 250.4(B)(4), the equipment grounding conductor shall be sized larger than given in this table."[4] A comprehensive analysis of the withstand rating of these equipment grounding conductors can be found in chapter eleven.

Specific requirements are provided for: Equipment grounding conductors that are increased in size for any reason; for example, for voltage drop as provided in 250.122(B), for multiple circuits in 250.122(C), for motor circuits in 250.122(D), for flexible cord and fixture wire in 250.122(E), and for conductors in parallel in 250.122(F).

Size of equipment grounding conductor based on rating or setting of overcurrent device ahead of circuit. See Table 250.122.

Figure 9-6. Minimum Size of Equipment Grounding Conductor

Increasing the Size of Equipment Grounding Conductor

Section 250.122(B) requires that, "Where conductors are increased in size, [for example to compensate for voltage drop], equipment grounding conductors, where installed, shall be increased in size proportionately according to circular mil area of the ungrounded conductors."[5] This means that where a feeder or branch-circuit conductor is increased in size to compensate for voltage drop, the equipment grounding conductor, where run, is required to be increased at not less than the same ratio the feeder or circuit conductors are increased. For example, a 200-ampere feeder is to be installed. It is determined that the voltage drop would be excessive. A 250-kcmil conductor is selected for the feeder rather than installing the 3/0 copper conductor as is permitted by Table 310.16. Table 250.122 requires a 6 AWG equipment grounding conductor for the 200-ampere overcurrent device.

Determine the minimum size equipment grounding conductor required for the feeder by the following formula: (Use Table 8 of *NEC* Chapter 9 to determine the area in circular mils where the conductor size is given by a non-circular mil designation).

Selected Feeder Conductor Area ÷ Required Feeder Conductor Area = Ratio
Table 250.122 Equipment Grounding Conductor X Ratio = Required EGC
250,000 kcmil ÷ 167, 800 kcmil = 1.49
26240 (Circular mil area of 6 AWG) x 1.49 = 39098 circular mils
Next larger size = 4 AWG copper required equipment grounding conductor

Equipment Grounding Conductors for Multiple Circuits

The *Code* permits a single equipment grounding conductor to serve several circuits that are in the same raceway or cable. To use this concept, the equipment grounding conductor is required to be sized for the rating of the largest overcurrent device of the group.

For example, a conduit contains branch-circuit conductors that have conductors having overcurrent protection rated: 20-amperes, 30-amperes, 50-amperes and 60-amperes. A single 10 AWG equipment grounding conductor is permitted to serve all the branch circuits in the raceway. The minimum size is determined from Table 250.122 based on the rating of the 60-ampere overcurrent device.

Equipment Grounding Conductors for Motor Circuits

The general rule for sizing the equipment grounding conductor for motor circuits is contained in 250.122. Determine the minimum size conductor from Table 250.122 based on the rating of the overcurrent protective device. In some cases, this could result in an equipment grounding conductor that is the same size as the branch-circuit conductors. This is illustrated as follows: a 30-hp, 460-volt motor is being installed. From Table 430.150, the full-load amperes of the motor is 40 amperes. The minimum size branch-circuit conductors is 40 amperes x 1.25 = 50 amperes, from Table 310.16 = 8 AWG copper conductors (75°C insulation and terminations). Maximum rating of

> The *Code* permits a single equipment grounding conductor to serve several circuits that are in the same raceway or cable.

overcurrent device of a circuit breaker type is 250 percent of the motor full load amperes = 250 x 40 = 100 amperes (Table 430.152), unless one of the exceptions to 430.52(C) applies. From Table 250.122, the minimum size of equipment grounding conductor based on a 100-ampere overcurrent device is 8 AWG copper which is the same size as the branch-circuit conductors. Note that 250.122(A) provides that the size of the equipment grounding conductor is not required to be larger than the branch-circuit conductors.

Where the overcurrent device for the motor consists of an instantaneous-trip circuit breaker (rather than a more standard inverse-time circuit breaker) or a motor short-circuit protector, the equipment grounding conductor size is permitted to be based on the rating of the motor overload protective device but not less than Table 250.122. Note that the instantaneous-trip circuit breaker is permitted to be used only if it is a part of a listed combination motor controller having coordinated motor overload protection.

50 Hp 208 volt 3∅ motor

50 Hp motor draws 143 amps

D.E.T.D. fuse max = 250 amps

Equipment grounding conductor = 4 AWG CU

Branch circuit conductors = 3/0 THW

Size EGC per 250.122

Figure 9-7. Sizing Equipment Grounding Conductors

Figure 9-8. Sizing Equipment Grounding Conductors

Using the above example, the instantaneous-trip circuit breaker that serves as the branch-circuit, short-circuit and ground-fault protective device is permitted to be up to 1300 percent of the motor full-load current (up to 1700 percent for Design E and Design B energy efficient motors). The minimum size of branch-circuit conductors is determined as 40 amperes x 1.25 = 50 amperes. The minimum conductor from Table 310.16 is an 8 AWG copper conductors with 75°C insulation and terminations. The maximum rating of an overcurrent device of an instantaneous circuit breaker type is 1300 percent of the motor full load amperes = 40 x 13 = 520 amperes, unless one of the exceptions following 430.52(C)(3) applies. However, assume the motor FLA on the nameplate is 38.2 amperes and the running overload protection for the motor is set at 115 percent [see 430.32(A)]. The equipment grounding conductor is permitted to be based on 43.93 amperes (38.2 x 1.15). From Table 250.122, the minimum size equipment grounding conductor based on a 50-ampere overcurrent device is 10 AWG copper.

Equipment Grounding Conductors for Flexible Cord and Fixture Wire

The use of an equipment grounding conductor in a supply cord is permitted providing the cord is connected as specified in 400.7. The method of grounding non-current-carrying metal parts of portable equipment may be by means of the equipment grounding conductors in the flexible cord feeding such equipment. The proper type attachment plug is required to be used to terminate the conductors, and the attachment plug must have provision to make contact with a grounding terminal in the receptacle.

For the grounding of portable or pendant equipment, the conductors of which are protected by fuses or circuit breakers rated or set at not exceeding 20 amperes, 240.5 permits the use of an 18 AWG copper wire as an equipment grounding circuit conductor, provided the 18 AWG grounding conductor is a part of a listed flexible cord assembly [see 250.122(A)].

Equipment Grounding Conductors in Parallel

Special rules apply where more than one raceway or cable is installed with parallel conductors and an equipment grounding conductor is in-

stalled in the raceway. (Parallel conductors consist of two or more conductors that comply with 310.4 and are connected together at each end to form a single conductor.) In this case, 250.122(F) requires that an equipment grounding conductor be installed in each raceway or cable. Generally, each equipment grounding conductor is required to be sized in compliance with the ampere rating of the overcurrent device protecting the conductors in the raceway or cable.

Section 310.4 permits equipment grounding conductors, smaller than 1/0 AWG, to be sized in compliance with Table 250.122. However, all other requirements for installing conductors in parallel must be met. These rules require that each set: (1) be the same length; (2) be of the same conductor material (all copper or all aluminum); (3) be the same size in circular mil area; (4) have the same

Figure 9-9. Size of Equipment Grounding Conductors

insulation type; (5) be terminated in the same manner; and (6) the raceways or cables must have the same physical properties. However, the sets of conductors are not required to be identical

One reason for this requirement for installing equipment grounding conductors in parallel is shown in figure 9-9. In the event of a line-to-ground fault in the equipment supplied by the circuit, the fault current will divide equally between the equipment grounding conductors. However, should a line-to-ground fault occur in the raceway or cable, current will be fed to the fault from both directions. The equipment grounding conductor will thus be called upon to carry the entire amount of fault current until the overcurrent protective device ahead of the fault opens.

Equipment Grounding Conductors in Cables in Parallel

In some cases, where cables are installed in parallel, special constructions will be required to comply with *Code* rules on the equipment grounding conductor in the cable. Listed cables are generally produced with the equipment grounding conductors sized in compliance with a construction standard that complies with Table 250.122. For example, a copper cable construction suitable for a 300-ampere overcurrent device will have a 4 AWG copper equipment grounding conductor placed within the cable by the manufacturer. If two of these cables are installed in parallel and connected to a 600-ampere overcurrent protective device, a 1 AWG copper equipment grounding conductor would be required in each cable to comply with Table 250.122. These "special" cables can be ordered from the manufacturer although conditions such as minimum length requirements may apply. In addition, a significant amount of time may be required to produce these special cables.

Rules are now in place that permit an equipment ground-fault protection system for conductors in cables where the equipment grounding conductors are too small for the overcurrent device on the line side of the conductors. Conditions that must be met to use this concept include:

1. Conditions of maintenance and supervision ensure that only qualified persons will service the installation.
2. The ground-fault protection equipment is set to trip at not more than the ampacity of a single ungrounded conductor of one of the cables in parallel.
3. The ground-fault protection is listed for the purpose.

This is an effort to mitigate the impact of requiring, where multiconductor cables were installed in parallel, an equipment grounding conductor that is based on the rating of the overcurrent device. This requirement resulted in an equipment grounding conductor larger than is normally available in listed cables.

For example, a cable rated for 200 amperes will typically have a 6 AWG copper or 4 AWG aluminum equipment grounding conductor in it. If three of these cables are

installed in parallel and have overcurrent protection rated at 600 amperes, Table 250.122 requires a 1 AWG copper or 2/0 AWG aluminum equipment grounding conductor in each cable.

This provision allows the standard size equipment grounding conductor in multiconductor cables that are connected in parallel to be protected by ground-fault protection of equipment (not GFCI protection for personnel). The ground-fault protection is required to be set at the ampacity of a single ungrounded circuit conductor, not on the ampacity of the equipment grounding conductor.

For example, a 1600-ampere feeder is installed with four cables each having an ampacity of 400 amperes. The equipment ground-fault protection for these cables would have to be set at not more than 400 amperes. These cables would typically have a 3 AWG copper or 1 AWG aluminum equipment grounding conductor according to Table 250.122. The overcurrent protection for the feeder would be 1600 amperes.

In a circuit without equipment ground-fault protection, the equipment grounding conductor would typically be called upon to carry several times the rating of the overcurrent device on the line side of the circuit until the overcurrent device clears the fault. As a result, this new provision is more conservative than circuits installed without equipment ground-fault protection.

Supplementary Grounding Electrode

Engineers often specify that ground rods be installed to ground metal lighting standards or poles and at metal poles for electric signs. These rods are permitted to be used but are required to be

Size of parallel equipment grounding conductors in cables permitted to be sized from Table 250.122 on rating of ground-fault protection of equipment under specific conditions.

Figure 9-10. Equipment Grounding Conductors in Cables in Parallel

Equipment grounding conductor and supplementary electrode bonded together.

Grounding electrode conductor

Lugs

Pole

To Rod

EGC

Conduit

Wiring method containing required equipment grounding conductor size according to Section 250.122

Supplementary grounding electrode

Concrete pole base

Figure 9-11. 250.54 Supplementary Electrodes

considered supplementary ground rods. They can supplement the equipment grounding conductor that is run with the branch circuit but cannot be the only means of grounding this or similar equipment (see 250.54). To use these ground rods as the only means of grounding would constitute an earth return which is unsafe and prohibited by *Code*. The concept of the earth being used for a circuit conductor should never be considered. The *NEC* strictly prohibits this in multiple sections. The earth is a poor conductor.

Equipment Grounding Conductor with Circuit Conductors

A very important requirement for installing equipment grounding conductors is contained in 250.134(B). This requirement is that the equipment grounding conductor is generally required to be installed in the same raceway, cable or cord, or otherwise be run with the circuit conductors. This requirement is repeated in 300.3(B) where, in addition to the requirement for raceways, equipment grounding conductors are required to be contained in the same trench with other circuit conductors. This requirement is critical for the installation of alternating-current systems.

It has been proven that separating the equip-

ment grounding conductor from the circuit conductors greatly increases the impedance of the circuit. Separation of these conductors will increase the inductive reactance of an ac circuit which in turn increases grounding circuit conductor impedance values. The impedance of the equipment grounding conductor of a circuit should be kept as low as practicable.

This excessive separation can render an adequately-sized equipment grounding conductor ineffective in carrying enough current to operate the circuit protective device and clear the faulted equipment. In this case, providing properly sized equipment grounding conductor, but installing it improperly, results in an ineffective and possibly unsafe installation (see chapter eleven for additional information on this subject).

Nonmetallic Raceway

Where the wiring method or means is nonmetallic, it is necessary to run an equipment grounding conductor along with the circuit conductors. Do not separate them at any point in the circuit by any metallic material regardless of whether the metallic material is magnetic or not. It is true that if the material is nonmagnetic, the increase in impedance of the circuit will not be as great as if the material was magnetic. In any case, such separation is to be avoided.

Use of Building Steel for Grounding

Section 250.136(A) permits a metal rack or structure to ground electric equipment that is secured to it and in electrical contact, provided the support means is grounded by an equipment grounding conductor as specified by 250.134(B). However, building steel is not permitted to serve as an equipment grounding conductor to ground equipment. That is due to the uncertain path that ground-fault current must take in an effort to clear a fault (see figure 9-12).

This section emphasizes the requirement in 250.134(B) and 300.3(B) that the equipment grounding conductor must be in the same raceway, cable or cord, or otherwise be run with the circuit conductors. Again, this is so the grounding circuit impedance will be as low as possible to allow adequate ground-fault current to flow so the circuit protective device will clear the fault. Separating the equipment grounding conductors from the circuit conductors increases the inductive reactance of the circuit in ground-fault conditions and thus increases the impedance on the grounding circuit, which is required to be kept at a minimum.

In the same manner, and on the same basis, metal car frames, supported by metal hoisting cables attached to or running over sheaves or drums of elevator machines, are considered grounded when the machine is grounded as required by the *Code*.

Grounding for Direct-Current Circuits

All of the above applies to the grounding of alternating-current systems where reactance of the circuit plays a large part in the impedance of the ground-return path. In the case of direct-current circuits the concern is with ohmic resistance only. Owing to that fact, the current-carrying capacity of the grounding conductor for a direct-current supply system is required to be equal to that of the largest conductor of the system. However, if the grounded circuit conductor is a neutral derived from a balancer winding or a balancer set which has overcurrent protection as required un-

Structural steel is not permitted as the required equipment grounding conductor for ac equipment per Section 250.136(A)

Figure 9-12. Structural Metal Frame of a Building

der 445.12(D), then the grounding conductor size shall be not less than the size of the neutral.

The requirement for overcurrent protective devices here states that the two-wire direct-current generators used in conjunction with balancer sets shall be equipped with overcurrent protective devices which will disconnect the 3-wire system in the case of excessive unbalancing of voltages or currents.

Grounding Conductors for Direct-Current Circuits

For direct-current circuits, the size of the grounding conductor is specified in 250.166. The size may be larger than would be required for the same size alternating-current circuit. That is because resistance is the only factor in determining current flow in a direct-current circuit. Therefore, fault-currents are larger in a dc circuit. In an alternating-current circuit, impedance becomes the important factor for not only resistance, but reactance also must be taken into account.

Long Term Reliability of Metal Raceways

In the above discussions, it is assumed that a conductor enclosure (conduit or other raceway) has been properly installed with good tight joints that will provide a permanent and continuous electrical circuit when it is first installed. However, time and corrosion will take its toll and will tend to destroy the continuity of the conduit.

The safety of an electrical system will thus depend on how long we can expect the equipment grounding conductor of the conduit to remain permanent and continuous. The answer will vary depending on the type of metal raceway, the environment it is installed in, and the quality of the installation.

For design purposes, two categories can be created:

1. Where little corrosion will exist and where it can be reasonably expected that the equipment grounding conductor, in the form of a metal raceway, will remain permanent and continuous for a period of fifty years or more.

2. Where corrosion in varying degrees will exist and where the permanency of the equipment grounding conductor provided by the metal raceway can be questioned.

Most commercial and residential buildings are in the first category. That being the case, conductor enclosures, which are approved for the purpose can be used as part of the equipment grounding conductor (with the use of bonding jumpers where required).

Some industrial and most areas of petrochemi-

Care must be taken when installing metallic conduit and electrical metallic tubing in the earth, in concrete on or below grade, or where exposed to moisture.

cal plants are in the second category where a metallic equipment grounding conductor, sized per Table 250.122, is required to be run in parallel with and within the conductor enclosure so as to ensure continuity if the conduit circuit is broken owing to corrosion sooner or later.

Some electrical design engineers and local electrical inspection agencies require that an equipment grounding conductor be installed in each metal conduit and tubing to help ensure the reliability of the equipment grounding conductor path.

Metal Conduit Underground

Care must be taken when installing metallic conduit and electrical metallic tubing in the earth, in

In Concrete:	Required	Optional
Rigid Steel		X
Intermediate Steel		X
Aluminum Rigid	X	
Steel EMT	Below grade may need	On or above grade
Aluminum EMT	X	
In Soil:		
Rigid Steel		X
Intermediate Steel		X
Aluminum Rigid	X	
Steel EMT	Generally Required	
Aluminum EMT	X	

Table 9-1.

concrete on or below grade, or where exposed to moisture (344.10 and 300.6). The Underwriters Laboratories' 2001 guide card information for *Rigid Metal Conduit (DYIX)* and *Intermediate Metal Conduit (DYBY)* contains the following information regarding corrosion protection:

Galvanized rigid (and intermediate) steel conduit installed in concrete does not require supplementary corrosion protection. Galvanized rigid (and intermediate) steel conduit installed in contact with soil does not generally require supplementary corrosion protection.

In the absence of specific local experience, soils producing severe corrosive effects are generally characterized by low resistivity (less than 2000 ohm-centimeters).

Wherever ferrous metal conduit runs directly from concrete encasement to soil burial, severe corrosive effects are likely to occur on the metal in contact with the soil.

Conduit that is provided with a metallic or nonmetallic coating, or a combination of both, has been evaluated for resistance to atmospheric corrosion. Nonmetallic outer coatings that are part of the required resistance to corrosion have been additionally evaluated for resistance to the effects of sunlight.

Rigid metal conduit with or without a nonmetallic coating has not been evaluated for severely corrosive conditions.[6]

In addition, experience has shown that steel conduit fails rapidly where exposed to corrosive environments found at some seacoast marinas, boatyards and plants as well as at some chemical plants. In addition, experience has shown that metal conduit systems are particularly vulnerable to failure from corrosion where they pass from concrete that is on or below grade to exposure to an atmosphere containing corrosive elements, particularly in combination with atmospheres containing oxygen.

For electrical metallic tubing [see 358.10(B)], the following instructions are given in the UL 2001 guide card (FJMX):

Galvanized steel electrical metallic tubing installed in concrete, on grade or above, generally requires no supplementary corrosion protection. Galvanized steel electrical metallic tubing in concrete slab below grade level may require supplementary corrosion protection.

In general, galvanized steel EMT in contact with soil requires supplementary corrosion protection. Where galvanized steel electrical metallic tubing without supplementary corrosion protection extends directly from concrete encasement to soil burial, severe corrosive effects are likely to occur on the metal in contact with the soil.

Aluminum electrical metallic tubing used in concrete or in contact with soil requires supplementary corrosion protection. Supplementary nonmetallic coatings presently used have not been investigated for resistance to corrosion. [7]

As a result, the authority having jurisdiction is required to make a decision regarding the suitability of these raceways for these applications. This, of course, affects the reliability of the raceway serving as an equipment grounding conductor. Several reports have been made where electrical metallic tubing installed to provide an equipment grounding means has failed due to corrosion.

To maintain the integrity of the equipment grounding means, some inspection agencies require that a copper equipment grounding conductor be installed in parallel with the electrical metallic tubing.

In addition, the authority having jurisdiction must make a decision regarding the suitability of supplementary nonmetallic coatings intended for resistance to corrosion.

Grounding of Equipment by Using the Grounded Circuit Conductor

The *Code* does not generally permit the grounded circuit conductor (often a neutral) to be grounded again on the load side of the service disconnecting means [see 250.24(A)(5) and 250.142(B)]. Four exceptions to this rule exist:

1. Grounding separately derived systems.
2. Grounding the grounded circuit conductor at a remote building or structure.
3. For existing branch circuits only, grounding the frame of an electric range or electric clothes dryer.
4. Grounding meter enclosures that are near the service disconnecting means.

Where the electrical system produced by a separately derived system meets the conditions of 250.20(A) or (B), the system is required to be grounded according to 250.30. Systems that fall within the parameters that require it to be grounded

Separately derived systems. Section 250.142(A)(3)

Each separate building where required by Section 250.32(B)and 250.142(A)(2)

Figure 9-13. Grounding Equipment by Grounded Circuit Conductor

Ranges, counter-mounted, cooking units, wall-mounted ovens, clothes dryers. Section 250.142(B) Ex. No. 1

Load-side meter enclosures that are near the service disconnecting means where service ground-fault protection is not provided. Section 250.142(B) Ex. No. 2

Figure 9-14. Grounding Equipment by Grounded Circuit Conductor

is required to have a grounding electrode conductor connected to the grounded conductor of the separately derived system (see chapter twelve for additional information on this subject).

Under the specific conditions given in 250.32, a grounded circuit conductor is permitted to be grounded again at a separate building or structure. Where so installed, the grounded conductor serves as both a grounded conductor and an equipment grounding conductor between the buildings or structures (see chapter thirteen for additional information on this subject).

Section 250.142(B) covers rules on the use of the grounded circuit conductor for grounding equipment on the load side of the service equipment. As stated previously, such practice is generally prohibited. Four exceptions to the general rule are provided.

Exception No. 1: The frames of ranges, wall-mounted ovens, counter-mounted cooking units, and clothes dryers under the conditions permitted for existing installations by 250.140 shall be permitted to be grounded by a grounded circuit conductor. (See chapter ten for additional information on this subject.)

Exception No. 2: It shall be permissible to ground meter enclosures by connection to the grounded circuit conductor on the load side of the service disconnect if

(a) No service ground-fault protection is installed, and [Editors note: This condition is important as grounding the grounded circuit conductor downstream from the service will de-sensitize the equipment ground-fault protection system.]

(b) All meter enclosures are located near the service disconnecting means, and [Editor's note: No distance is given to explain what is meant by the term near.]

(c) The size of the grounded circuit conductor is not smaller than the size specified in Table 250.122 for equipment grounding conductors.[8]

Exception No. 3: Direct-current systems shall be permitted to be grounded on the load side of the disconnecting means or overcurrent device in accordance with 250.164.[9] [Editor's note: Rules are different depending on whether the direct-current supply is from an off-premises or on-premises source.]

Exception No. 4: Electrode-type boilers operating at over 600 volts shall be grounded as required in 490.72(E)(1) and 490.74.

Underground Metal-Sheathed Cable System
The *Code* recognizes that if an underground service originates from a continuous underground metal sheath cable system and the sheath or armor is metallically connected to the underground system, or if an underground service conduit is used which contains a metal-sheathed cable which is itself bonded to the underground system, then the service run need not be grounded at the building and may be insulated from the interior conduit or piping (see 250.84).

The *Code* here recognizes that since the service armor is already bonded to a continuous underground metal sheath cable system, a path for fault-currents in the service cable is provided so there is no necessity to bond back to the service equipment in order to provide a return path to the neutral.

[1] NFPA 70, *National Electrical Code* 2002, Article 100, (National Fire Protection Association, Quincy, MA, 2002), p. 70-35.
[2] NFPA 70, 250.118, p. 70-112 & 70.113.
[3] NFPA 70, 250.118, p. 70-113.
[4] NFPA 70, Table 250.122, p. 70-114.
[5] NFPA 70, 250.122(B), p. 70-114.
[6] *General Information for Electrical Equipment 2001*, (Underwriters Laboratories, Northbrook, IL, 2001), p. 20.
[7] *General Information for Electrical Equipment 2001*, (Underwriters Laboratories, Northbrook, IL, 2001), p. 28.
[8] NFPA 70, 250.142(B), p. 70-116 & 70.117.
[9] NFPA 70, 250.142(B), p. 70-116.

Chapter Nine: The questions included here were developed using material included in this chapter. The answers can be found by reviewing the text. It is also important that students make use of the 2002 *NEC*, where many of the answers can be found.

1. The conductor used to connect the non-current-carrying metal parts of equipment, raceways, and other enclosures to the system grounded conductor and/or the grounding electrode conductor at the service equipment, or at the source of a separately derived system is defined as ____.
    a. equipment grounding conductor
    b. main bonding jumper
    c. grounding systems conductor
    d. circuit bonding jumper

2. An equipment grounding conductor is intended to prevent an objectionable voltage above ground on conductor and equipment enclosures and to provide a low-impedance path for fault-currents. This path must also ____.
    a. be permanent and electrically continuous
    b. have ample capacity to conduct safely any currents likely to be imposed on it
    c. be of the lowest practical impedance
    d. all of the above

3. The equipment grounding conductor or path must extend from the ____ point on the circuit to the service equipment where it is connected to the grounded conductor.
    a. closest
    b. service
    c. furthermost
    d. bonding

4. Where the overcurrent protection ahead of a copper branch circuit or feeder is sized at 225 amperes, the minimum size of the equipment grounding conductor is to be a ____ copper conductor.
    a. 4 AWG
    b. 6 AWG
    c. 8 AWG
    d. 10 AWG

5. Which of the following is NOT recognized as a conductor or raceway for use as an equipment grounding conductor?
    a. a conductor of copper or other corrosion-resistant material such as aluminum
    b. rigid or intermediate metal conduit
    c. electrical metallic tubing
    d. surface metal raceways

6. Listed flexible metal conduit and listed flexible metallic tubing are permitted to be used for equipment grounding purposes. Which of the following statements is NOT true?
    a. The total combined ground return in the same path cannot exceed 1.8 m (6 ft)
    b. They must be terminated with fittings listed for grounding
    c. They cannot be used on a circuit exceeding 15 amperes
    d. They cannot be used on a circuit exceeding 20 amperes

7. Listed liquidtight flexible metal conduit in sizes metric designator 21 through 35 ($^3/_4$-in. through $1^1/_4$-in.) trade size is to be used as an equipment grounding conductor. Which of the following statements is NOT true? ____
    a. The total length of the combined ground return in the same path cannot exceed 6 feet.
    b. Listed fittings must be used for grounding at connections.
    c. The circuit is permitted to be protected by a 100 ampere or less overcurrent device.
    d. The circuit is permitted to be protected by a 60 ampere or less overcurrent device.

8. Where rigid metal conduit is used as an equipment grounding conductor, all joints and fittings are required to be ____.
    a. approved
    b. tested
    c. made up wrenchtight
    d. sealed

9. For the grounding of portable or pendant equipment and where protected by fuses or circuit breakers rated or set at not over ____ amperes, the *Code* permits the use of a 18 AWG copper wire as an equipment grounding conductor provided it is a part of a listed flexible cord assembly.
   a.   25
   b.   20
   c.   30
   d.   35

10. Under what conditions can building steel serve as an equipment grounding conductor to ground equipment? ____
   a.   Where it is effectively grounded.
   b.   Never.
   c.   Where approved.
   d.   By special permission.

11. Where equipment grounding conductors are installed in parallel in separate nonmetallic raceways, which of the following statements is true?
   a.   A full-size equipment grounding conductor is required in only one of the conduits.
   b.   A smaller equipment grounding conductor than required by Table 250.122 is permitted if the total area is not less than given in the table.
   c.   A full-size equipment grounding conductor is required in each of the conduits.
   d.   Various sized copper and aluminum conductors can be used together so long as they are not smaller than given in Table 250.122.

12. Equipment grounding conductors in parallel listed cables are permitted to be smaller than given in Table 250.122 if
   a.   they are protected by an equipment ground fault protection device
   b.   the total area of the conductors is not less than the area required divided by the number for conductors
   c.   they do not leave the building or structure they originate in
   d.   not more than four cables are installed in parallel.

Enclosure & Equipment Grounding

OBJECTIVES

After studying this chapter, the reader will be able to understand

- General requirements and definitions for enclosure and equipment grounding.
- Grounding of fixed and specific equipment.
- Grounding of nonelectrical equipment.
- Methods of effective grounding and grounding of equipment.
- Grounding of metal enclosures and panelboards.
- Installation of grounding-type receptacles.
- Grounding of cord- and plug-connected equipment.
- Equipment spacing from air terminals.

Both enclosures for service conductor and other conductor enclosures, where of metal, are required to be grounded (see 250.80 and 250.86). That does not mean connecting equipment to a grounding electrode is acceptable or permitted. The installation must comply with the requirements of 250.4(A)(5) where the concept of the effective path for fault current is carefully outlined. It is important to realize that wherever the *Code* states, *shall be grounded*, it means effectively grounded as spelled out in 250.4. Note that nothing in the *Code* permits equipment that is supplied by a grounded system to be grounded to only a grounding electrode. *Effective grounding* always includes providing a low-impedance path consisting of an equipment grounding conductor that has adequate capacity to conduct the maximum fault current it is likely to carry. It also must be permanent and electrically continuous.

While the grounding methods are different, electrical equipment associated with both grounded and ungrounded systems must be effectively grounded. Where grounding is not effectively accomplished, the situation, while bad in an ungrounded system, becomes worse in a grounded system.

Figure 10-1. Grounded Improperly

Definitions

"**Bonding Jumper:** A reliable conductor to ensure the required electrical conductivity between metal parts required to be electrically connected."[1]

"**Bonding Jumper, Equipment:** The connection between two or more portions of the equipment grounding conductor."[2]

Equipment Grounding Conductor

It is important to recall that an equipment grounding conductor is required to be used for grounding equipment. This equipment grounding conductor is permitted to consist of any of the conductors or wiring methods identified in 250.118.

Installing a grounding electrode to ground equipment without having an equipment grounding conductor connected to the grounded service conductor is unsafe and not permitted (see 250.4 and 250.54).

By referring to figure 10-1, it can be seen that the grounding connections as shown comply literally with the wording of the *Code* in that the enclosures are *grounded* which means "connected to earth ...." But the wiring does not comply with 250.4(A)(5) as an effective ground-fault current return path has not been provided. In addition, the grounding shown in this figure violates 250.4(A)(5) and 250.54 as an earth return grounding circuit is indicated.

The wiring in figure 10-2 complies literally with 250.80 and 250.86 and also complies with 250.4 as a grounded system conductor is installed.

It is obvious that only a high-impedance fault-current return path is indicated in figure 10-1, while in figure 10-2 there is a path having sufficiently low impedance to limit the voltage to ground and to facilitate the operation of the circuit protective devices in the circuit.

Note that in figure 10-2 there are two paths for current to return to the source. The primary and low-impedance path is over the grounded system conductor (often a neutral) while a second, high-impedance path in parallel with the grounded service conductor is through the grounding electrodes and the earth.

Figure 10-2. Grounded Properly

Service Raceways and Enclosures

We have dealt in earlier chapters with the requirement for bonding service raceways and equipment. By bonding the service equipment to the grounded service and grounding electrode, we have complied with the requirements of 250.80.

An exception to 250.80 exempts a metal elbow from the requirement that it be grounded where it is installed in an underground run of rigid nonmetallic conduit and is isolated from possible contact by a minimum cover of 18 inches to any part of the elbow. These metal elbows are often referred to as *pulling elbows* and are commonly installed in duct banks or other underground runs of rigid nonmetallic conduit as they are more durable than PVC elbows during the cable-pulling process.

A similar exception regarding metal elbows used in underground runs of rigid nonmetallic conduit

Figure 10-3. Metal Elbows Used Underground

(Exception No. 3) for other than service raceways has been added to 250.86. This exception exempts a metal elbow from the requirement that it be grounded where it is installed in nonmetallic raceway and is isolated from possible contact by a minimum cover of 450 mm (18 in.) to any part of the elbow or is encased in a minimum 50 mm (2 in.) of concrete such as in decks above finish grade where these isolated metal elbows might be installed. There are additional requirements for bonding of isolated sections of metallic raceways or enclosures installed at pole locations. Section 250.102(E) permits a bonding jumper to be installed either inside or outside the raceway or enclosure. Where installed on the outside of the raceway or enclosure, it is required to be routed with the raceway and not exceed 6 feet in length. The exception to this rule allows a bonding jumper to exceed the length of 1.8 m (6 ft) at pole locations for the purposes of bonding isolated portions of metallic raceways or enclosures or elbows in a run of rigid nonmetallic conduit.[3]

Exceptions from Grounding Requirements

For other than service conductor enclosures, three exceptions are provided from the requirement that metallic conductor enclosures be grounded (see 250.86).

Exception No. 1 covers "metal enclosures and raceways for conductors added to existing installations of open wire, knob-and-tube wiring, and nonmetallic-sheathed cable." Conditions that must be met are as follows:

1. An equipment grounding conductor is not provided by the wiring method.

2. The metal enclosure or raceway must be less than 7.5 m (25 ft) long.

3. The metal enclosure or raceway must be free of probable contact with ground or a grounded object.

4. The metal enclosure or raceway is guarded against contact by persons.

Exception No. 2 exempts short sections of metal enclosures from the requirement to be grounded where used to protect cable assemblies from physical damage. No explanation is given for the meaning of "short sections" of metal enclosures. Since the standard length is 10 feet, "short sections" probably are less than 10 feet long.

Exception No. 3 permits a metal elbow to not be grounded where it is installed in an underground run of rigid nonmetallic conduit and is isolated from possible contact by a minimum cover of 450 mm (18 in.) to any part of the elbow or is encased in not less than 50 mm (2 in.) of concrete. These provisions are similar to that for services as provided in 250.80 Exception but include the isolated metal elbows that may be installed in runs of nonmetallic conduit above ground such as in deck slabs (see figure 10-3).[4]

Grounding of Fixed Equipment

It is mandatory that non-current-carrying metal parts of fixed equipment, which are likely to become energized, be grounded under the six conditions cited in 250.110. Here, again, the term *shall be grounded* must be interpreted literally to mean "connect to a grounding electrode" but must be interpreted in the light of all of Article 250 where providing an effective fault-current path is outlined in 250.4(A)(5). The *Code* does not define what is meant by *likely to become energized*. Generally, if the equipment has exposed non-current-carrying metal parts and is supplied by electric current it should be grounded where any of the following six conditions exist at the equipment.

The six conditions under which exposed non-current-carrying metal parts of equipment are required to be grounded are:

(1) Where within 2.5 m (8 ft) vertically or 1.5 m (5 ft) horizontally of ground or grounded metal objects and subject to contact by persons

(2) Where located in a wet or damp location and not isolated

(3) Where in electrical contact with metal

(4) Where in hazardous (classified) locations as covered by Articles 500 through 517

(5) Where supplied by a metal-clad, metal-sheathed, metal raceway, or other wiring method which provides an equipment ground, except as permitted by 250.86, Exception No. 2, for short sections of metal enclosures.

(6) Where equipment operates with any terminal at more than 150 volts to ground.[5]

The three exceptions from the requirement that this equipment must be grounded are as follows:

Exception No. 1. Metal frames of electrically heated devices, exempted by special permission [Editor's note: the written approval of the authority having jurisdiction], *in which case the frames are required to be permanently and effectively insulated from ground.*

Exception No. 2. Distribution apparatus, such as transformers and capacitor cases, mounted on wooden poles, at a height exceeding 2.5 m (8 ft) above ground or grade level.

Exception No. 3. Listed equipment that is protected by a system of double insulation or its equivalent, shall not be required to be grounded. Where such a system is employed, the equipment shall be distinctively marked.[6]

Grounding Specific Equipment

The *Code* requires that exposed, non-current-carrying metal parts of certain specific equipment, regardless of voltage, shall be grounded. Those items are spelled out in 250.112 and include:

(A) **Switchboard Frames and Structures.** Switchboard frames and structures supporting switching equipment, except frames of 2-wire dc switchboards where effectively insulated from ground.

(B) **Pipe Organs.** Generator and motor frames in an electrically operated pipe organ, unless effectively insulated from ground and the motor driving it.

(C) **Motor Frames.** Motor frames, as provided by 430.142.
(1) Where supplied by metal-enclosed wiring
(2) Where in a wet location and not isolated or guarded
(3) If in a hazardous (classified) location as covered in Articles 500 through 517
(4) If the motor operates with any terminal at over 150 volts to ground. [7]

(D) **Enclosures for Motor Controllers.** Enclosures for motor controllers unless attached to ungrounded portable equipment.

(E) **Elevators and Cranes.** Electrical equipment for elevators and cranes.

> The *Code* requires that exposed, non-current-carrying metal parts of certain specific equipment, regardless of voltage, shall be grounded.

(F) **Garages, Theaters, and Motion Picture Studios.** Electric equipment in commercial garages, theaters, and motion picture studios, except pendant lampholders supplied by circuits not over 150 volts to ground.

(G) **Electric Signs.** Electric signs, outline lighting, and associated equipment as provided in Article 600 [Editor's note: see chapter sixteen for additional information on this subject].

(H) **Motion Picture Projection Equipment.** Motion picture projection equipment.

(I) **Power-Limited Remote-Control, Signaling, and Fire Alarm Circuits.** Equipment supplied by Class 1 power-limited circuits and by Class 1, Class 2, and Class 3 remote-control and signaling circuits, and by fire alarm circuits, shall be grounded where system is required by Part II or Part VIII of this article.[8]

(J) **Luminaries (Lighting Fixtures).** Luminaries (lighting fixtures) as required by Part V of Article 410.

(A) **Exposed Conductive Parts.** Exposed metal parts shall be grounded or insulated from ground and other conductive surfaces or be inaccessible to unqualified personnel. Lamp tie wires, mounting screws, clips, and decorative bands on glass spaced at least 38 mm (1 ½ in.) from lamp terminals shall not be required to be grounded [see 410.18(A)].

(B) **Made of Insulating Material.** Luminaires (fixtures) directly wired or attached to outlets supplied by a wiring method that does not provide a ready means for grounding shall be made of insulating material and shall have no exposed conductive parts [see 410.18(B)].

Luminaires (fixtures) with exposed metal parts shall be provided with a means for connecting an equipment grounding conductor for such luminaries (see 410.20).[9]

(K) Skid Mounted Equipment. Permanently mounted electrical equipment and skids shall be grounded with an equipment bonding jumper sized as required in 250.122.

(L) Motor-Operated Water Pumps. Motor-operated water pumps, including the submersible type. [*Editor's note:* See 547.5(F) and 547.9 for specific requirements for grounding in agricultural buildings.]

> For equipment that is fastened in place or connected by permanent wiring methods, the equipment grounding conductor must be connected to the enclosure in a proper manner.

(M) Metal Well Casings. Where a submersible pump is used in a metal well casing, the well casing shall be bonded to the pump circuit equipment grounding conductor.[10]

Grounding of Cord- and Plug-Connected Equipment

Section 250.114 covers grounding of equipment connected by cord and plug. It is mandatory under certain conditions that non-current-carrying metal parts of cord- and plug-connected equipment which are liable to become energized be grounded. Listed tools, listed appliances, and listed equipment that is protected by a system of double-insulation are not required to be grounded. This double-insulated equipment is required to be distinctively marked.

Specifically, cord- and plug-connected equipment is required to be grounded where located in:

(1) hazardous (classified) locations,

(2) if the equipment operates at more than 150 volts to ground. Exempted from this requirement are:

(a) motors, where guarded, and

(b) metal frames of electrically-heated appli-

ances exempted by special permission, in which case the frames shall be permanently and effectively insulated from ground.[11]

The *Code* cites in 250.114(3) specific equipment of the cord- and plug-connected type which must be grounded in residential occupancies. In addition, 250.114(4) lists specific cord- and plug-connected equipment in other than residential occupancies that must be grounded. In this section, recognition is given to the use of a listed system of double insulation which would provide the safety required without the use of grounding.

For tools and portable handlamps in other than residential occupancies, an exception from the requirement for grounding is provided for cord- and plug-connected equipment that is supplied through an isolating transformer with an ungrounded secondary of not over 50 volts.

Nonelectric Equipment

The grounding of nonelectric equipment is covered in 250.116. The equipment mentioned is considered as being likely to become energized and is thus required to be grounded as a safety measure. Included are:

(1) Frames and tracks of electrically operated cranes and hoists

(2) Frames of nonelectrically driven elevator cars to which electric conductors are attached

(3) Hand-operated metal shifting ropes or cables of electric elevators

The fine print note following this section recommends that "where extensive metal in or on buildings may become energized and is subject to personal contact, adequate bonding and grounding will provide additional safety."[12]

Methods of Grounding

To provide the reliable and effective ground-fault return path required, it is important to connect the equipment grounding conductor recognized in 250.118 to the equipment in such a manner that the requirements for effective grounding are met.

For equipment that is fastened in place or connected by permanent wiring methods, the equipment grounding conductor must be connected to the enclosure in a proper manner.

Grounding conductors and bonding jumpers

must be connected to equipment that is required to be grounded by exothermic welding, listed pressure connectors, listed clamps or other listed means (see 250.8).

The only connecting means not required to be listed is the exothermic welding method. It is most important that equipment to be welded be clean and dry and that manufacturer's instructions be followed to ensure a satisfactory connection.

These welds must be examined and tested after completion to be certain that a reliable connection has been made. It is common to test the welds in the field by x-ray where required by the job specifications or by striking the weld with a hammer after it has cooled for less demanding installations.

In the case of a metallic raceway being used as the equipment grounding conductor, the raceway must be connected to the enclosure by using fittings designed for the purpose. All connections must be made up tight using proper tools. This includes locknuts, bushings, conduit, and electrical metallic tubing couplings and connectors [see 250.120(A)].

Solder Connections

Connections that depend solely upon solder cannot be used for grounding connections (see 250.8 and 250.148).

The reason for this prohibition is that when equipment grounding conductors carry fault-current, they can get very hot. This elevated temperature may exceed the melting point of the solder and weaken or destroy the connection. This may create a hazard by leaving equipment at a dangerous potential above ground, creating a shock hazard to those who may contact the equipment.

It is permissible to secure the connections mechanically and then apply solder to the joint to make the electrical connection.

Special Provisions for Grounding Certain Appliances

Special requirements are set forth in 250.140 for the grounding of frames of electric ranges, electric clothes dryers and similar appliances. These provisions apply only to existing branch-circuit installations. New installations must comply with the requirements for an insulated neutral as well as an equipment grounding conductor given in 250.134 and 250.138.

Figure 10-4. Wiring Methods for Appliances Permitted To Be Grounded by the Grounded Circuit Conductor

Figure 10-5. Wiring Methods for Appliances Permitted To Be Grounded by the Grounded Circuit Conductor

Appliances or equipment to which 250.140 applies include:

1. electric ranges;
2. wall-mounted ovens;
3. counter-mounted cooking units;
4. clothes dryers, and
5. outlet or junction boxes that are part of the circuit for these appliances.

These appliances and equipment, such as the junction boxes in the circuit, are permitted to be grounded in two ways: either by use of an equipment grounding conductor; or, except for mobile homes and recreational vehicles, by the use of the grounded circuit (neutral) conductor, provided all the conditions of this section are complied with. If a 3-wire with ground circuit is installed, simply use the equipment grounding conductor for grounding metal equipment and install a grounding type receptacle.

The conditions that must be met for grounding the above equipment to the grounded circuit conductor in existing installations are as follows:

(1) The appliances are supplied by a 120/240-volt, single-phase, 3-wire circuit or by a 3-wire circuit derived from a 208Y/120-volt, 3-phase, 4-wire, wye-connected system.

(2) The grounded (neutral) conductor is 10 AWG copper or 8 AWG aluminum or larger.

(3) The grounded (neutral) conductor is insulated. The grounded (neutral) conductor is permitted to be uninsulated if it is part of a type SE service-entrance cable and it originates at the service equipment, and

(4) Grounding contacts of receptacles that are a part of the appliance are bonded to the equipment.[13]

It is important to note that all of these special conditions must be met before the appliances, and outlet and junction boxes that are a part of the circuit to the appliances, are permitted to be grounded by use of the circuit neutral. Note that the supply cable must have an insulated neutral conductor where supplied from a sub-panel.

Use caution when applying the provisions of 250.140(3) which permits the use of a Type SE cable having a bare neutral for wiring these appliances when the circuit originates at service equipment. Section 338.10(B) no longer permits Type SE cable without an insulated neutral conductor to be used for the purposes of 250.140.

Wiring for New Ranges and Dryers

New installations for ranges, dryers, and similar appliances require a three-wire with ground circuit having an insulated neutral conductor and an equipment grounding conductor. The equipment grounding conductor is permitted to be any of those included in 250.118 including conduit and cables. Receptacles, where installed, must be of the 3-pole, 4-wire grounding type. Supply cords, where used, must be of the 3-wire with ground type.

Neither the frame of the appliances nor the outlets or junction boxes that are a part of the supply to the appliances are permitted to be grounded to the circuit neutral. The frame of the appliances and junction boxes must be grounded by means of an equipment grounding conductor run with the branch circuit.

Care must be taken when these appliances are moved from a location employing one grounding scheme to a location having a different one. In the case where a 3-wire cable is used, the bonding jumper in the appliance junction box must be connected between the frame of the appliance and the neutral. Where a four-wire supply is used, the bonding jumper must be removed or disconnected, the neutral conductor is isolated from the appliance frame and the equipment grounding conductor connected to the frame.

Outlet, Device, Pull and Junction Boxes

Metal outlet, device, pull and junction boxes are required to be grounded in accordance with Article 250 (see 314.4). It is important that the grounding of these enclosures be accomplished by the wiring method that supplies the enclosure (see 250.134).

Under no circumstances should a metal enclosure be connected to a local grounding electrode in lieu of grounding them by means of the suitable wiring method or equipment grounding conductor unless specifically permitted by *Code* rules. The metal raceway containing circuit conductors is permitted to be used for grounding these enclosures if installed properly with all connections made up wrenchtight. In addition, where fittings are used for connecting cables or raceways being used for grounding, the fittings used for connecting the wiring methods must be suitable for grounding. Some wiring methods have restrictions on their use as an equipment grounding conductor. These include flexible metal conduit, liquidtight flexible metal conduit and flexible metallic tubing.

Figure 10-6. Grounding New Range and Dryer Circuits

Equipment grounding conductors that are supplied for grounding these enclosures must be sized in accordance with Table 250.122 based upon the rating of the overcurrent device protecting the circuit conductors in the raceway.

Section 314.40(D) requires that "a means be provided [by the manufacturer] in each metal box for the connection of the equipment grounding conductor. This means shall be permitted to be a tapped or equivalent."[14]

Grounding of Panelboards

Section 215.6 requires that "where a feeder supplies branch circuits in which equipment grounding conductors are required, the feeder shall include or provide a grounding means, in accordance with the provisions of 250.134 to which the equipment grounding conductors of the branch circuit shall be connected."[15]

Specific requirements for grounding panelboards

Figure 10-7. Neutrals and Grounds Separated in Subpanel

Figure 10-8. Neutrals and Grounds Separated in Subpanel

are contained in 408.20. All panelboard cabinets and panelboard frames, if of metal, are required to be in physical contact with each other and must be grounded. Section 408.3(C) requires that where used as service equipment, the panelboard must be provided with a main bonding jumper located inside the cabinet for the purpose of bonding the enclosure to the grounded service conductor.

"Where the panelboard is used with nonmetallic raceway or cable or where separate grounding conductors are provided, a terminal bar for the grounding conductors shall be secured inside the cabinet. The terminal bar shall be bonded to the cabinet and panelboard frame if of metal; otherwise it shall be connected to the grounding conductor that is run with the conductors feeding the panelboard."[16] Usually, the manufacturer provides matching and tapped holes along with appropriate screws or bolts for attaching the bar to the enclosure.

Grounding and equipment grounding conductors are not permitted to be connected to the terminal bar for grounded (may be a neutral) conductors unless the terminal bar is identified for the purpose. This identification will usually be on the manufacturer's label that is located within the cabinet. Another requirement is that the panelboard is used at a location where the grounded conductor terminal (neutral) bar is connected to a grounding electrode, as permitted or required by Article 250.

These locations include at services, at the building disconnecting means (see 225.32) and for separately derived systems (see 250.30). An exception permits an insulated equipment grounding conductor for an isolated grounding scheme to pass through the panelboard without being connected to the equipment grounding terminal bar in compliance with 250.146(D). Generally, the grounded (may be a neutral) conductor is isolated from the enclosure downstream from the service equipment. Often, the term *floating neutral* is used to describe the grounded conductor's relationship to the enclosure as it is insulated electrically from the enclosure. Section 250.24(A)(5) generally prohibits a grounding connection to a grounded conductor (may be a neutral) on the load side of the service disconnecting means. Exceptions to the general rule are provided for separately derived systems, at separate buildings, and for certain appliances such as electric ranges and clothes dryers on existing branch circuits.

See chapter twelve for additional information on separately derived systems and chapter thirteen for grounding at more than one building on the premises.

Grounding-Type Receptacles

Where grounding-type receptacles are installed, the grounding terminal of the receptacle must be connected to an equipment grounding conductor of the circuit supplying the receptacle [see

> The grounding terminal of the receptacle must be grounded by means of an insulated equipment grounding conductor that is run with the circuit conductors.

406.3(C)]. Where more than one equipment grounding conductor enters a box, they must be connected together using a suitable and listed connector. In addition, it is permitted to connect each of the equipment grounding conductors to the metal box individually using a listed clip or screw. A jumper is used to the receptacle so grounding continuity is not disturbed if the device is removed.

A grounding jumper from the receptacle to the box is not required where a device with listed grounding means is installed in a metal box that is

properly grounded. These devices are often referred to as *self-grounding receptacles* and are specially designed so one or more of the mounting screws are maintained in contact with the device's metal yoke (see the requirements in chapter eight for additional information on this subject).

Isolated Grounding Receptacles

Receptacles that have the equipment grounding terminal isolated from the mounting strap, and thus from the box, are commonly installed at computer terminals and cash registers. This is permitted by 250.146(D) for the purpose of reducing electrical noise (electromagnetic interference). The grounding terminal of the receptacle must be grounded by means of an insulated equipment grounding conductor that is run with the circuit conductors.

This insulated equipment grounding conductor is permitted to pass through one or more panelboards without connection to the terminal bar within the panelboard on its way back to, usually, the service disconnecting means [see 250.146(D) and 408.20, Exception]. However, the insulated equipment grounding conductor must terminate within the same building at the building or structure disconnecting means or source of a separately derived system (see chapter twelve for additional information on the subject of grounding separately derived systems).

Isolated Equipment Grounding

Section 250.96(B) permits an equipment enclosure supplied by a branch circuit to be grounded by

Figure 10-9. Isolated Grounding Receptacles 250.146(D)

Figure 10-10. Isolated Equipment Grounding

means of an insulated equipment grounding conductor contained within the raceway with the branch circuit that supplies the equipment. A listed isolating raceway fitting at the point of connection to the equipment must be installed. This provision typically applies to listed data processing (information technology) equipment.

Underwriters Laboratories performed tests of a similar grounding scheme to determine whether isolating the metal conduit from the equipment had an adverse effect on the grounding circuit impedance. They found no appreciable increase in impedance with the metal conduit isolated from the equipment and being grounded by means of the insulated equipment grounding conductor.

Short Sections of Raceway

Where isolated sections of metal raceway or cable armor are required to be grounded, the *Code* requires in 250.132 that grounding of such sections be performed in accordance with the requirements of fixed equipment found in 250.134. While the *Code* does not identify what is meant by a "short section," perhaps this is a length less than the standard length of 10 ft. These short sections of raceways are often installed as physical protection of cables.

As mentioned previously, short sections of metal enclosures or raceways used to provide support or protection of cable assemblies from physical damage are not required to be grounded due to Exception No. 2 to 250.86.

Where these short sections of raceways or cable

armor are required to be grounded, 250.134 generally requires that they be grounded by one of the equipment grounding conductors recognized by 250.118.

Section 250.106 now simply requires that the ground terminals for the lightning protection system be bonded to the building or structure grounding electrode system.

Spacing from Air Terminals

Section 250.106 no longer requires that metal raceways, enclosures, frames, and other non-current-carrying metal parts of electric equipment be kept at least 6 feet away from lightning rod conductors. It is recognized that the subject of lightning protection systems is covered extensively in the NFPA 780, *Standard for the Installation of Lightning Protection Systems*.

Fine print note (FPN) No. 2 to 250.106 points out that the *Standard for the Installation of Lightning Protection Systems* may require bonding of adjacent portions of the electrical wiring system to the lightning protection system. The spacing differs depending on whether the separation is through air or through differing construction materials.

Section 250.106 now simply requires that the ground terminals for the lightning protection system be bonded to the building or structure grounding electrode system.

Air terminal(s)

Ground terminal(s) for lightning protection system must be bonded to the building or structure grounding electrode system.
See 250.106 and NFPA 780

Ground terminal(s)

Figure 10-11. Bonding Ground Terminals to Service Electrode

[1] NFPA 70, *National Electrical Code* 2002, Article 100, (National Fire Protection Association, Quincy, MA, 2002), p. 70-33.
[2] NFPA 70, Article 100, p. 70-33.
[3] NFPA 70, 250.102(E), p. 70-110.
[4] NFPA 70, 250.86, p. 70-108.
[5] NFPA 70, 250.110, 70-111.
[6] NFPA 70, 250.110, p. 70-111.
[7] NFPA 70, 430.142, p. 70-308.
[8] NFPA 70, 250.112, p. 70-111 & 70-112.
[9] NFPA 70, 410.20, p. 257.
[10] NFPA 70, 250.112, p. 70-111 & 70-112.
[11] NFPA 70, 250.114, p. 70-112.
[12] NFPA 70, 250.116, p. 70-112.
[13] NFPA 70, 250.140, p. 70-116.
[14] NFPA 70, 314.40(D), p. 70-140.
[15] NFPA 70, 215.6, p. 70-61.
[16] NFPA 70, 408.20, p. 70-253.

Chapter Ten: The questions included here were developed using material included in this chapter. The answers can be found by reviewing the text. It is also important that students make use of the 2002 *NEC*, where many of the answers can be found.

1. A reliable conductor to assure the required electrical conductivity between metal parts required to be electrically connected is defined as a ____.
   a. grounding electrode conductor
   b. grounded conductor
   c. bonding jumper
   d. identified conductor

2. The connection between two or more portions of the equipment grounding conductor is defined as ____.
   a. equipment bonding jumper
   b. grounded
   c. neutral
   d. main bonding jumper

3. Exposed non-current-carrying metal parts of equipment must be grounded where within____ feet vertically or ____ feet horizontally of ground or grounded metal objects, and subject to contact by persons.
   a. 9 - 6
   b. 8 - 5
   c. 8 - 8
   d. 9 - 9

4. Exposed non-current-carrying metal parts of equipment are required to be grounded where located in wet or damp locations and are not ____.
   a. guarded
   b. identified
   c. shielded
   d. isolated

5. Where other than short sections of metal enclosures are installed as per Section 250.86, exposed non-current-carrying metal parts of equipment are required to be grounded where supplied by any of the following wiring methods EXCEPT ____.
   a. metal-clad, metal-sheathed cables
   b. knob-and-tube wiring
   c. other wiring methods that provide an equipment ground
   d. approved metal wireways

6. Exposed non-current-carrying metal parts of equipment not required to be grounded include metal frames of electrically heated devices unless exempted by ____ in which case the frames must be permanently and effectively insulated from ground.
   a. the product instructions
   b. the local government
   c. special permission
   d. the code

7. Exposed non-current-carrying metal parts of equipment not required to be grounded include distribution apparatus, such as transformers and capacitor cases, that are mounted on wooden poles at a height of more than ____ feet from the ground or grade level.
   a. 6
   b. 8
   c. 5
   d. 7

8. Exposed non-current-carrying metal parts of equipment not required to be grounded include ____ equipment that is distinctively marked.
   a. hospital
   b. triple insulated
   c. single insulated
   d. double insulated

9. Where added to existing installations of open wire, knob-and-tube wiring, and nonmetallic-sheathed cable without an equipment grounding conductor, a metal enclosure for conductors run in lengths not to exceed 7.5 m (25 ft), free from probable contact with ground, grounded metal, metal lath, or other conductive material, and if guarded against contact by persons are ____.
   a. required to be grounded
   b. not required to be grounded
   c. required to be protected by a GFCI
   d. not permitted unless approved

10. Motor-operated water pumps including the submersible type are required to be grounded when they operate at _____.
    a.   any voltage
    b.   120 volts
    c.   240 volts
    d.   208 volts

11. All of the connecting means included below are required to be listed by a qualified electrical testing lab EXCEPT _____.
    a.   pressure connectors
    b.   lugs
    c.   clamps
    d.   exothermic welding

12. Connections that depend solely upon solder for grounding connections _____.
    a.   are only permitted when approved
    b.   cannot be used
    c.   are permitted on the load side of the service
    d.   are permitted on the line side of the service

13. Metal raceways that enclose service conductors are required to be grounded unless it is:
    a.   installed on a pole
    b.   more than 2.5 m (8 ft) above the ground
    c.   an elbow in PVC run covered by 450 mm (18 in.) of earth
    d.   buried at a depth given in Table 300.5

14. All the following statements about grounding the frame of ranges and dryers are true EXCEPT:
    a.   A three-wire with ground circuit is permitted for existing installations.
    b.   The frame of the appliances is permitted to be grounded to the neutral in new installations.
    c.   A two-wire with ground circuit is required for new installations.
    d.   New installations must be grounded to comply with Sections 250.57 and 250.59.

15. All the following statements about grounding of panelboards supplied by a feeder in the same building as the service are true EXCEPT:
    a.   The feeder must supply or provide the grounding means.
    b.   The neutral conductor is permitted to ground the enclosure.
    c.   The neutral conductor must connect to a neutral bar that is isolated from the enclosure.
    d.   An equipment grounding conductor must be connected to an equipment grounding terminal bar.

Clearing Ground Faults & Short Circuits

OBJECTIVES

After studying this chapter, the reader will be able to understand

- Ground faults and circuit impedance.
- Fundamentals of equipment grounding, circuit design, and test procedures.
- Common elements in clearing ground faults and short circuits.
- Sizing of equipment grounding conductors.
- Purposes served by neutral on grounding systems.
- Conductor withstand ratings.

Section 310.2(A) requires, generally, that conductors be insulated. The exception allows covered or bare conductors to be used where specifically permitted. Ungrounded or hot conductors must be insulated for the applied voltage. Typical voltage rating of conductors is 300, 600, 2000, 5000 and 15,000. Conductors are available with much higher rated insulations, although these are primarily used for distribution of electrical energy rather than for premises wiring.

Table 310.13 gives conductor application and insulations. It gives the trade name, type letter such as MI, maximum operating temperature and application provisions. It also gives the conductor insulation, size in American Wire Gage (AWG), thickness in mils and outer covering, if any.

Bare or covered conductors are permitted to be used in certain cases. Grounded service-lateral conductors are permitted to be uninsulated under the conditions given in 230.30 Exception. Uninsulated service-entrance conductors are permitted to be used as provided in 230.41. Of course, insulated, covered or bare equipment grounding conductors are permitted to be used by 250.118(1) and 250.119. Some sections require insulated conductors for specific applications such as for certain electrical equipment associated with swimming pools and in patient care areas of certain health care facilities.

If no hazard can exist on a distribution system unless there is an insulation failure of the ungrounded or hot conductor, it follows that every precaution should be taken to furnish the best possible insulation consistent with an economical installation. Good installation practices should be followed carefully to ensure that cable insulation is not damaged during the installation process. Cable insulation can be easily damaged by pulling operations if raceways are not cleaned prior to the installation of conductors, if excessive bends are in the run, or where long or heavy pulls require the use of heavy-duty pulling equipment.

Further, if insulation breaks down, the hazard can exist only as long as the circuit remains energized. Every effort should be made to de-energize, as quickly as is practical, any circuit on which a fault to ground has developed.

Figure 11-1. All Conductors To Be Together

When first considered, the use of a copper or aluminum equipment grounding conductor may seem to offer an ideal method of getting a low-impedance path. It will in most cases, but that is not always true. Each individual circuit must be studied in relation to the size and length of the circuit, the size of its overcurrent device, and the size of the conductor enclosure before any conclusion can be reached.

Fundamentals of Equipment Grounding Circuit Design

Briefly, it can be stated that to get low impedance of the grounding system in an ac system, the circuit conductors and the equipment grounding conductor must be kept intimately together at all times [see 250.134(B) and 300.3(B)]. The equipment grounding conductor may, of course, be a copper or aluminum conductor or the metal enclosure of the conductors such as conduit or wireways, where acceptable by *Code* for such use.

In an ac electrical distribution system, whether grounded or ungrounded, inductive reactance exerts a powerful influence by directing the return current to flow in a path closely paralleling the outgoing power conductor. In addition to resistance, an inductive reactance value is associated with every conductor in an ac system. The inductive reactance (expressed in ohms) increases as the spacing between the conductors or the circuit is increased. This indicates that the inductive reactance of a current path closely paralleling the phase conductors offers lower impedance to the ground-fault current than any other current path regardless of a lower resistance of the other current paths. Inductive reactance will be the predominate factor in determining current division in parallel ground return paths in heavy or larger circuit constructions. This has been proven in actual tests where it is shown that there is a strong tendency for ground return currents to seek a path physically close to the power conductor over which the outgoing current flows.[1]

Usually, the conduit or metallic raceway that encloses the conductors provides an excellent fault return path. The presence of magnetic material in the power conductor enclosure (conduit or raceway) introduces additional inductive effects tending to confine the return ground currents within the magnetic enclosure. Installation of an external equipment grounding conductor is quite ineffective. Connections to nearby structural building steel members, in an effort to serve as an equipment grounding means, are equally ineffective as well and are not permitted.

The intentional or accidental omission of using the metallic conductor enclosure as an equipment grounding conductor may lead to large induced voltages in nearby metallic structures, which may appear as a dangerous shock hazard or unwanted circulating currents. Only by the installation of an internal equipment grounding conductor, in parallel with the raceway, can the current carried by the raceway be reduced. Joints in conduit and raceways must be connected wrenchtight, using proper tools, for the raceway to function effectively as an equipment grounding conductor.

Fault-Current Test Procedure

As illustrated in figure 11-2, a special installation of 2 1/2-inch rigid steel conduit and 4/0 AWG copper cable was made for this investigation. It was installed in a building previously used for short-circuit testing because the building had heavy steel column construction, and all columns were tied to an extensive grounding mat composed of 250-kcmil AWG copper cables. The conduit was supported on insulators throughout the 100-foot length. The conduit was about 5 feet from a line of building columns. The external 4/0 AWG cable was spaced about 1 foot from the conduit on the side opposite the building columns.

The setup was intended to simulate an electrical feeder circuit as may be found in many commercial and industrial plants. It allowed a study of a wide variety of equipment grounding arrangements. In every case the supply current was

Figure 11-2. Grounding Current Test Procedure

through the "A" cable run through the conduit. The "B" conductor served as an internal equipment grounding conductor. The "C" conductor was connected to the run of steel metal conduit, the "G" conductor was connected to the steel building columns, and the "H" conductor was the external equipment grounding conductor spaced about one foot away from and parallel to the conduit. A copper shorting bar was used at the far end of the conduit. With this arrangement it was possible to simulate various fault conditions at the right end and to verify various fault-current return paths.

Tests Performed, Low-Current

One series of tests was made at low current of 200 and 350 amperes using an ac welding transformer as a source of 60-cycle power. At these low-current magnitudes, the current could be maintained for extended periods. Voltage measurements were made with high quality indicating meters. Current measurements were made with a clip-on ammeter.

Tests Performed, High Current

A second series of tests was made at high current, around 10,000 amperes using a 450 kVA, 3-phase, 60-cycle transformer with a 600-volt secondary as a source of power. Switching was done at the 13,800 primary voltage. An induction relay was used to control the duration of current flow to about 1/4 second. An oscilloscope was used for all measurements of current and voltage.

Fault-Current Test Results

The results of the tests performed are shown in Table 11-1. The test number (for example, A2) is for reference purposes. The next two columns show the cable connections used to determine current flow (for example, out on A and return on C). Next, the current values are shown, first the total input current through cable A (350), and next the return current through the conduit (350) and its percentage of the total (100). The other columns with an "I" heading, along with a subscript indicating the circuit, show the amount of current returning over other possible paths.

"No further analysis is needed to show conclusively that only by the use of an internal grounding conductor can any sizable fraction of the return current be diverted from the raceway.

Test No.	Current Flow Out On	Return On	IA Total	Ic Amperes	Per Cent of Total	I_B	I_H	I_G	E_{AC}	E_{CG}	E_{AB}	E_{AH}	E_{AG}	E_{GB}
Low-Current Tests														
A1	A	B	350	0		350	0	0	2.47		4.85			
A2	A	C	350	350	100	0	0	0	15.9	0.45	2.5			
A3	A	C	200	200	100	0	0	0	9.05	0.15	1.51			
A4	A	CH	350	340	97	0	12	0	16.0	0.05	2.55			
A5	A	CH	200	190	95	0	8	0	9.13	nil	1.55			
A6	A	CG	350	340	97	0	0	12	14.6		2.54	14.4		
A7	A	CG	200	180	90	0	0	8	9.5		1.50	9.4		
A8	A	CB	350	62	18	290	0	0	4.55	nil	4.55			
A9	A	CB	200	40	20	150	0	0	2.68	nil	2.68			
A10	A	GH	350	0	0	0	160	160	14.0	12.5	2.5	26.4	26.4	24.6
A11	A	GH	200	0	0	0	98	98	9.2	8.1	1.5	17.1	17.1	15.1
High-Current Tests														
B2	A	C	11,200	11,200	100	0	0	0	168	36*				
B3	A	C	11,070	11,070	100	0	0	0	173	38*				
B4	A	CH	11,070	11,200	101	0	1,140	0	173	18*				
B5	A	CH	11,080	11,090	100	0	1,220	0	173	17*				
B6	A	CG	10,830	10,770	99	0	0	1,080	168		71			
B7	A	CG	10,910	10,780	99	0	0	1,145	173	9*				
B8	A	CB	11,620	5,810	50	5,660	0	0	0	27*			155	
B9	A	CB	11,380	6,070	53	5,620	0	0	146	25*				
B10	A	CH	8,710	0	0	0	4,300	4,500	146				268	

*Distorted wave shape. Tabulated values are crest $/ \sqrt{2}$

Table 11-1. Measured Electrical Current Quantites

In spite of the extremely low resistance of the building structural frame, it was ineffective in reducing the magnitude of the return current in the conduit (see tests A6, A7, B6 and B7).

"Some interesting secondary effects were observed in the course of the tests. The first high-current test produced a shower of sparks from about half of the couplings in the conduit run. From one came a blowtorch stream of sparks which burned out many of the threads. Several small fires set in nearby combustible material would have been serious if not promptly extinguished.

"The conduit run had been installed by a crew regularly engaged in such work and they gave assurance that the joints had been pulled up to normal tightness and perhaps a little more. A short 4/0 copper (wire) jumper was bridged around this joint but, even so, some sparks continued to be expelled from this coupling on subsequent tests. Other couplings threw no more sparks during subsequent tests. Apparently, small tack welds had occurred on the first test.

"In one high-current test, the conduit termination was altered to simulate a connection to a steel cabinet or junction box (see figure 11-3). The bushing was applied finger-tight. In one test, with about 11,000 amperes flowing for about $^1/_4$ second, a fan-shaped shower of sparks occurred parallel to the plate. In the process, a weld resulted and the parts were separated only with considerable difficulty, with the use of wrenches and a hammer. This suggested that a repeat shot (of current) would have produced no disturbance (shower of sparks).

"During high-current test B10 (conduit circuit open), a shower of sparks was observed at an intermediate building column. Careful inspection

Figure 11-3. Alternate Termination of "C" Cable

disclosed that the origin was at a spot at which a water pipe passed through an opening cut in the web of the steel beam involved. Here is evidence of the objectionable effects of forcing the short-circuit current to seek return paths remote from the outgoing conductor. The large spacing between outgoing and returning current creates a powerful magnetic field which extends far out in space around the current-carrying conductors."[2]

Fault-Current Study Circuit Analysis

"The reactance of the circuit including the 'B' conductor will be the lowest. Next will be the innermost tube of the conduit, followed by others in successive order until the outer tube is reached. The inductance of these tubular elements of the steel conduit assumes unusual importance because of the high magnetic permeability. Next in spacing (impedance) is the external grounding conductor (H cable) and last, the structural members of the building frame and their interconnecting grounding cables buried below floor level."[3]

"In test B10, both the exterior grounding conductor (H cable) and the building frame (G terminal) were connected to provide parallel paths for the return current, but the conduit circuit was left open (C terminal not connected). The test results clearly evidence the powerful forces tending to maintain current flow in the conduit circuit. Note that across the open connections at the 'C' terminal, a voltage of 146 volts (or, more significantly, over 50 percent of the impressed driving voltage), is required to force the current to return via the 'H' cable and the building frame in parallel. Such a voltage could be a serious shock hazard.

"Furthermore, unless the conduit was well insulated throughout its entire length (which is usually impossible or impractical in typical commercial or industrial installations), there would be a significant number of sparks at stray points to constitute a serious fire hazard. It was during this test that a shower of sparks occurred between magnetic members in the building system that was caused simply by the

strong magnetic field extending far out from the power conductors." [4]

Conclusions on Fault-Current Path Study

"The significance of this investigation points unmistakably to the conclusions presented earlier in this paper. Effective use of the conduit or raceway in the equipment-grounding system is paramount. Additional work is needed to develop joints which will not throw fire during faults. To improve effectiveness requires greater conductivity in the conductor enclosure or the use of an

Figure 11-4. Square D Trip Curves Graph   Courtesy of Square D

internal (equipment) grounding conductor. Grounding cables (grounding electrode conductors) connecting building structure with grounding electrodes (connection to earth) are needed to convey lightning currents or similar currents seeking a path to earth, but these conductors will play a negligible part in the performance of the equipment grounding system. Of course, the importance of proper equipment grounding becomes greater with the larger size feeder circuits and the availability of higher short-circuit currents." [5]

Common Elements of Fault-Current Path

Clearing faults involves one or two parts, depending on whether the service is grounded or ungrounded. In a grounded distribution system, there are two parts: (a) the grounded system conductor and (b) the equipment and conductor enclosure grounding system. In an ungrounded system, the grounding system covers only (b) the equipment and conductor enclosure grounding system.

Effective Grounding

To have an effective path for fault current for both grounded systems and ungrounded systems, the fault-current path must: (1) be installed as permanent and continuous; (2) must have ample capacity to conduct safely any fault currents likely to be imposed on it; and (3) have lowest possible impedance to limit the voltage to ground and to facilitate the operation of overcurrent protective devices in the circuit [see 250.4(A)(5) and 250.4(B)(4)].

OVERCURRENT DEVICE (AMPS)	MAXIMUM Z OF GROUND-FAULT CIRCUIT IN OHMS INCLUDES FAULT IMPEDANCE
60	0.4
100	0.24
200	0.12
400	0.06
600	0.04
800	0.03
1200	0.02
1600	0.015
2000	0.012
2500	0.0096
3000	0.008
4000	0.006

Table 11-1.

"Impedance sufficiently low" means, for all practical purposes, that the grounding path and the circuit conductors for an ac system must always be within the same metallic enclosure such as a conduit, wireway or cable. The metallic enclosure (such as a conduit or cable) under certain conditions may be used to provide the equipment grounding path for the circuit conductors within it. In addition, where a nonmetallic wiring method is used for an ac system, all the circuit conductors, including the equipment grounding conductor, must be in the same raceway.

The *Code* gives no maximum impedance of the ground-fault circuit for grounded or ungrounded systems but states in 250.4(A)(5) that it shall be sufficiently low to facilitate the operation of the circuit-protective devices. This path should have an impedance no greater than will allow the circuit breaker or fuse to reach its instantaneous pickup operating range. This will cause the overcurrent device to operate quickly to remove the fault from the system. Any fault current less than the instantaneous operating value will extend or delay, by some value, the opening time of the overcurrent device.

Every manufacturer of circuit breakers and fuses publishes operating or trip-curve charts for their products (for example, see figure 11-4). These charts should be carefully reviewed to be certain the ground-fault path has an impedance low enough to allow the overcurrent device to operate quickly to reduce thermal damage to the circuit and equipment. As can be seen by reviewing these charts, circuit breakers have a different trip curve depending on whether they are single-, double- or three-pole configurations. In addition, the same family of circuit breakers will have different trip curves for different ampere-rated circuit breakers. Look for the multiple of the circuit breaker rating at which it reaches its instantaneous pickup rating which is no time delay in the operation of the breaker.

A similar review of the fuse operating characteristics chart should be performed to select a ground-fault circuit that will result in instantaneous operation of the fuse.

Some engineering designs may incorporate the concept of the fault current being not less than five times the rating of the overcurrent device. For example, if the overcurrent device is 400 amperes, not less than 2000 amperes must flow in the circuit for it to open the faulted circuit in a reasonable time. For some overcurrent devices, current of five times the device rating may not reach the instantaneous trip range of the device. It can generally be said that the longer the excessive current flows in the circuit, the greater the thermal and mechanical stress to the conductor insulation.

Clearing Short Circuits

The method employed for clearing a short circuit is the same regardless of whether the system is grounded or ungrounded. Essentially, it involves placing an overcurrent device in series with each ungrounded system conductor.

In the event of a *short circuit*, which is a fault from conductor to conductor, the words "as quickly as is practical" means a very short period of time down to as low as a fraction of a cycle depending on the amount of short-circuit current that will flow and the characteristics of the overcurrent device. The high-interrupting capacity current-limiting fuse does that automatically within virtually all ranges of short-circuit current values. In addition current-limiting circuit breakers and circuit breaker current limiters are available from a wide variety of manufacturers. Where properly applied, these current-limiting devices significantly reduce the thermal and mechanical damage to electrical equipment.

Current-limiting fuses and their correct application are covered in Underwriters Laboratories' *Electrical Construction Materials Directory Guide Card* (JCQR) information. This section reads in part: "The term 'Current Limiting' indicates that a fuse, when tested on a circuit capable of delivering a specific short-circuit current (RMS amperes symmetrical) at rated voltage, will start to melt within 90 electrical degrees and will clear the circuit within 180 electrical degrees ($\frac{1}{2}$ cycle).

"Because the time required for a fuse to melt is dependent on the available current of the circuit, a fuse which may be current limiting when subjected to a specific short-circuit current (RMS amperes symmetrical) may not be current limiting on a circuit of lower maximum available current."[6]

A current-limiting circuit breaker is defined in the Underwriters Laboratories' *Electrical Construction Materials Directory Guide Card* (DIVQ) information as, "A current-limiting circuit breaker is one that does not employ a fusible element and

Note the current flow in the circuit.

Short circuits are cleared by the overcurrent device in series with the fault.

Figure 11-5. Clearing Short Circuits

that when operating within its current-limiting range, limits the let-through I^2t (current squared time) to a value less than the I^2t of a $\frac{1}{2}$ cycle wave of the symmetrical prospective current. Current-limiting circuit breakers are marked 'current-limiting' and are marked either to indicate the let-through characteristics or to indicate where such information may be obtained."[7]

Circuit breaker current limiters are covered in Underwriters Laboratories' *Electrical Construction Materials Directory in Guide Card* (DIRW). They are described as "Circuit breaker current limiters are designed to be used in conjunction with specific circuit breakers and to be directly connected to the load terminals of the circuit breakers. They contain fusible elements which function only to increase the fault current interrupting ability of the combination which is intended for use in the same manner as circuit breakers when installed at the service and as branch circuit protection. The limiters are rated 600 v or less."[8]

It is vital to carefully follow manufacturer's instructions when applying current-limiting fuses or circuit breakers.

In some cases, depending on the available fault current, this equipment may not be current limiting at all. This is due to the *inverse time* nature of these overcurrent devices. Inverse time means that as the amount of current flowing through the device increases, the operating time of the device reduces. Overcurrent devices see fault current at lower operating ranges as a load and react much slower than at their current-limiting range.

Many installations of fuses or circuit breakers

are installed in a series configuration, that is, two or more fuses or circuit breakers are in series for the circuit. These devices that have been tested for their operating compatibility may be marked by the manufacturers with a *series-combination rating*. In this case, usually the downstream overcurrent device is rated below the fault current that is available at its line terminals, while the overcurrent device closest to the source is rated at or above the fault current that is available at its line terminals.

Where a series-rated system is installed, only the equipment that has been tested to determine its suitability can be installed in this manner. Suitability is determined by reference to manufacturer's identification of components on decals located on the equipment. For additional information, see "Interplay of Energies in Circuit Breaker and Fuse Combinations."[9]

The time-current characteristic curve of such fuses starting from a value of from five to six times the rating of the fuse indicates a clearing time of about one second and a time of well below one cycle for values of 50 times fuse rating. A further study indicates that such high-interrupting capacity current-limiting fuses have a pronounced current-limiting effect at values as high as fifty times the fuse rating.

Whether a fault current is a short circuit or a ground fault, the overcurrent devices ahead of the point of fault will see all the fault current. Placing an overcurrent protective device at the point the conductor receives its supply, as required by 240.21, will provide the short-circuit protection that is necessary. The rating of the overcurrent device is based upon 240.4 and the ratings of standard overcurrent protective devices given in 240.6. The basic requirement is that conductors be protected in accordance with their rated ampacity in accordance with 310.15.

Clearing Ground Faults

A ground-fault circuit is different in ways than a short circuit. In a ground fault, there can be such a high impedance of the fault circuit that the controlling factor of current flow is the impedance of the ground-fault circuit. In this case, the only part the available short-circuit capacity plays is in its ability to maintain voltage. In some circuits, the amount of current that will flow in a ground-fault circuit is not dependent on the available ca-

Clearing ground faults involves two major factors:
1. Impedance of circuit
2. Available short-circuit current of system.

Consider impedance of total circuit to allow enough current to flow to operate overcurrent device.

Review operating or trip curve of overcurrent device.

Figure 11-6. Clearing Ground Faults

pacity of the system, other than its ability to maintain full voltage during a ground fault.

When it comes to clearing a ground fault, two things control the flow of current, these are:
1. the impedance of the circuit, and
2. the available short-circuit amperes of the system.

A relatively large voltage drop can be tolerated in this grounding conductor. The impedance of the complete ground-fault circuit, from the point of fault back to the transformer, should never be higher than what would permit a minimum current flow for the overcurrent device to operate within its instantaneous range (review the operating characteristics chart for the overcurrent device for the correct value).

This will provide a factor of safety to allow for the variable impedance at the point of the fault. Some overcurrent devices operate at their instantaneous current range at about five times the rating of the overcurrent device. For example, some 50-ampere overcurrent devices have an instantaneous trip rating of about 250 amperes. In other cases, the instantaneous rating could be from six to ten times the rating of the overcurrent device.

Clearing Faults in Service Equipment

The service equipment enclosure is probably the most vulnerable for ground faults. There is no overcurrent protection on the line side of the service conductors, only short-circuit protection provided by the serving utility's transformer primary overcurrent protection. Overload protection is provided by the service overcurrent device in series with the service-entrance conductors. This is one of the main reasons the *Code* requires the service disconnecting means to be located outside of the building, or, if installed inside, nearest the point of entrance of the conductors into the building. This limits the risk of having conductors with no short-circuit protection inside the building.

If, on the other hand, a ground fault does not develop on the load side of the service, but at a point on the line side of the service, then that fault can only be cleared by the primary overcurrent device on the supply side of the utility transformer. This protection is the primary fuses or cutouts for the transformer which can be many times the rating of the secondary overcurrent device. In many cases, a ground fault on the line side of the service will not clear through the primary overcurrent devices and can only clear by developing into a short circuit or by burning itself clear.

If the electrical equipment on the line side of the service is not properly bonded and a properly-sized main bonding jumper installed, it is highly unlikely that enough current will flow in the path to clear the ground fault through the overcurrent devices on the line side of the utility transformer.

Analysis of Clearing Ground-Faults

The following examples illustrate a simple electrical system and analyze the conditions existing in the event of a ground fault. In the first case,

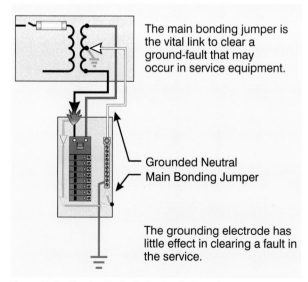

The main bonding jumper is the vital link to clear a ground-fault that may occur in service equipment.

Grounded Neutral Main Bonding Jumper

The grounding electrode has little effect in clearing a fault in the service.

Figure 11-7. Clearing Faults in Service Equipment

the neutral is run from a grounded system where it is connected to the service equipment. In the second example, the grounded system (service) conductor has not been run from the system to the service disconnecting means.

Purposes Served by the Grounded Conductor (Neutral) On a Grounded System

For maximum safety and to comply with 250.24(B), the neutral or grounded conductor must be run from a grounded system transformer to all services and be bonded to each service disconnecting means enclosure even though all power may be utilized at line voltage only. This is required even though the service supplies only line-to-line loads [see 250.24(B) and 250.28].

The grounded conductor (usually a neutral) of any grounded system serves two main purposes. First, it permits utilization of power at line-to-neutral voltage, and thus will serve as a current-carrying conductor to carry any unbalanced current back to the source. Second, it plays a vital part in providing a low-impedance path for the flow of fault currents to facilitate the operation of the overcurrent devices in the circuit, as required by 250.4(A)(5).

The grounded conductor (neutral) provides the lowest impedance return path for fault currents to the transformer neutral point as can be traced in figure 11-8. If the neutral is not needed for voltage requirements, it still must be run to the service and connected to the equipment grounding conductor at the service. In that event, the neutral conductor no longer serves as a neutral but as a grounded conductor. If that is not done, it is difficult, if not impossible, to clear a fault on the system.

Grounded (Neutral) Conductor Installed

In figure 11-8, the neutral of the system serving as the grounded conductor is carried to the service equipment and bonded to the equipment grounding conductor to provide a low-impedance path directly to the transformer. Here it can be seen that a fault current is not required to go through the earth to complete the circuit, but will go through the grounded service conductor, which is a low impedance path. This will very likely permit sufficient current to flow to operate the overcurrent device. The parallel path through the grounding electrode conductor and the ground-

High resistance and impedance path

Figure 11-8. Grounded Service Conductor Run to Service

ing electrode and the earth still exists. The grounded service conductor, forming a relatively low impedance path, will carry most of the fault current, probably 90 percent or more.

It is obvious, then, that to get all the protection afforded by a grounded system, the grounded conductor must be run to the service and must be bonded to the equipment grounding conductor even though the neutral is not required to serve any load whatever. For the same reason, where the neutral is needed for voltage requirements, the neutral size should be based not only on the neutral load demands but also on the basis of the service overcurrent device and the amount of fault current that must flow to operate the overcurrent devices.

Neutral Not Installed

Figure 11-9 shows a 120/240-volt single-phase grounded system where all the load is supplied at 240 volts. A grounding electrode conductor is properly connected to a low-resistance grounding electrode, the water supply system, and bonded to the equipment grounding conductor, all in accordance with the requirements of the *Code* (prior to the 1962 *Code*). However, the neutral is *not* carried into the service equipment since all power utilization is at 240 volts only. For serving the load, the neutral has no useful purpose, and it would at first appear as if it could be omitted.

From the standpoint of limiting the voltage between equipment and ground under normal conditions and from the standpoint of utilization of power at 240 volts, the circuit, as shown in

figure 11-9, will work perfectly, so long as the insulation remains perfect and no ground faults occur on the system.

If, however, a ground fault should occur on the system in the equipment, as illustrated in figure 11-9, owing to insulation failure, the voltage between equipment and ground will rise considerably. Starting at the point of fault, first is the impedance of the fault, then the conduit itself becomes a conductor in the fault-current circuit, then the grounding electrode conductor and the grounding electrode. However, the fault current must travel from the grounding electrode through the earth itself to the point where the neutral is grounded at the transformer, then through the transformer and back to the service and, finally, through the overcurrent device to the point of fault.

A study of the fault circuit as diagrammatically represented in figure 11-9 shows that it would be unlikely to have an impedance of less than 22 ohms as the sum of all the impedances shown.

That is an optimistically low figure at best. The maximum current that can flow in this circuit is 5.5 amperes for the 120/240-volt circuit shown (120 volts ÷ 22 ohms = 5.5 amperes). With a 100-ampere service and 20-ampere overcurrent device, it is obvious that the overcurrent device would not operate. A serious shock hazard, as well as fire hazard, will exist until the circuit is manually opened. If the circuit was opened as a result of being grounded properly, then the only period of time a shock hazard would exist would be for the duration of the fault. The fire hazard, as well as shock hazard, would be reduced to a relatively short period of time.

High resistance and impedance path

Figure 11-9. Grounded Service Conductor Not Run to Service

When a fault occurs on a system, it is only during the period while the fault exists that a potential hazard is present. It is of the utmost importance to safety that the fault clearing time be held to the shortest practical period of time.

With the wiring method as shown in figure 11-9, the fault will not clear and may exist for minutes, hours or even days before it is recognized. Further, that recognition may come from observing a fire, or from noting that a victim has

The probabilities are that the fault current would not reach a high enough value to operate the overcurrent device.

received a shock, which sometimes may even be fatal. It is a matter of record that many serious, and sometimes fatal, accidents have resulted from faults which were not cleared promptly because the system was not properly grounded.

If the neutral at the transformer was grounded to the same water pipe system as the service (may not be too likely), then the resistance of the fault path would be appreciably decreased. Because of the wide separation between the service conductor and the water pipe, the reactance and, thus, the impedance of the fault circuit would remain high. The probabilities are that the fault current would not reach a high enough value to operate the overcurrent device. Again, fire and damage to equipment would continue until the circuit was manually opened. To improve the safety of such a system, a low-impedance path must be provided to pass enough current to clear the circuit through the overcurrent devices.

Three-Phase Services

Identical reasoning to that for single-phase systems may be applied to any multi-wire, multi-phase grounded system. Any 3-phase power supply taken from such a grounded transformer bank must have the neutral or grounded conductor brought into the service of the power supply to satisfy the requirements of 250.24(B). This is true regardless of whether or not there is neutral load at the service. The grounded conductor provides a low-impedance path for fault current to return to the source.

Open System

The above statements applying to maximum ground-fault currents would not apply to an open system. If there are no metallic enclosures, the impedance of the ground-fault circuit would be much lower. Accordingly, greater ground-fault currents may be expected in actual practice in an open or nonmetallic installation. However, since most systems of 600 volts or less are metal enclosed, there may not be such high ground-fault currents in such systems. For open systems, it is vital that the equipment grounding conductor be run with the circuit conductors to maintain a low impedance ground-fault path.

Recommended Length of Conduit for Use as Equipment Grounding Means

The *National Electrical Code* places no restriction on the size or length of rigid metal, intermediate metal conduit or electrical metallic tubing where used as an equipment grounding conductor. Independent tests have shown that consideration must be given to both size and length of conduit.

Extensive work to determine the maximum safe length of conduit to serve as an equipment grounding conductor has been done by the School of Electrical and Computer Engineering at the Georgia Institute of Technology in Atlanta, Georgia.[10] Computer software has been developed which will allow the calculation of almost an unlimited combination of conductors and conduit for use in the ground-return path. This software can also be used for calculating the maximum length of equipment grounding conductors that are safe to use where not installed in metallic conduits, such as PVC (rigid nonmetallic conduit).

Where a metallic conductor enclosure is used as an equipment grounding conductor, it must have continuity as well as the required conductivity to pass enough current to facilitate the operation of the overcurrent devices. Some electrical inspection authorities require that an equipment grounding conductor be installed inside the conduit to account for poor workmanship of the raceway installation or to maintain continuity where fittings may be broken during use.

The engineer should examine the equipment grounding conductor (the metal enclosure) to assure it will function properly in the event of a ground fault. Where wireways, auxiliary gutters and busways have steel enclosures, it will be found that there may be enough cross section to serve as an equipment grounding conductor. It is questioned if electrical connections between lengths are adequate for carrying enough fault current to clear the fault. These raceways may require a supplemental equipment grounding conductor.

The conductor enclosure, conduit, raceway, and so forth, also may be acceptable as the equipment grounding conductor in lieu of the copper conductors given in Table 250.122. Generally, if a conduit, or electrical metallic tubing meets the *Code* requirements for conductor fill, the enclosure will provide an acceptable equipment grounding conductor.

In an average busway up to 1500 amperes rating, there is enough steel in the enclosure to provide an acceptable equipment grounding conductor if proper conductivity is assured at the joints. For busways of higher rating, it is doubtful if the steel enclosure is heavy enough. For all sizes of busways, the matter of good electrical connection at the joints must be checked carefully. Many busways have aluminum enclosures. In such busways, the enclosure has sufficient conductivity. Good electrical connections at the joints also must be checked.

Long Conduit Runs

Consider an installation where a 3-inch conduit run is 1000 feet long and has a 400-ampere overcurrent protective device protecting the contained conductors. Assume that the minimum fault current would be 600 percent of the overcurrent device rating, or, in this case, 2400 amperes. The impedance of 3-inch conduit at 2400 amperes is approximately 0.0875 ohms/1000 feet. For a 208Y/120-volt system with zero impedance at the point of fault, the maximum current that will flow will be about 1400 amperes. That value of current would operate a 400-ampere high-interrupting capacity current-limiting fuse in about two seconds. However, such ground faults are nearly always arcing faults. This adds resistance to the circuit and reduces the fault current that will flow in the circuit. Moreover, since conduit impedance increases with decrease in fault current, the conduit impedance would be 0.129 ohms/1000 feet at a current of 1,400 amperes.

Allowing for the arc impedance in the circuit described and the increase in conduit impedance at 1400 amps (about 50 percent) and other variable factors including a further increase of conduit impedance at the still lower current, the ground-fault current flow is more likely to be in the order of about 300 amperes. With a 300-am-

> For all sizes of busways, the matter of good electrical connection at the joints must be checked carefully.

pere fault current, the 400-ampere high-interrupting capacity current-limiting fuse obviously would not clear the fault.

For a long feeder such as this, it may be necessary to increase conductor sizes to account for voltage drop. Section 250.122 requires that the equipment grounding conductor also be increased in size in proportion to size of the ungrounded conductors. For example, if the 3-inch conduit run was only 100 feet long, the anticipated fault current would be about 3000 amperes, which would satisfactorily operate the overcurrent device. Both fault-current values were based on the impedance of only that part of the circuit that would be within the conduit, not the impedance of the entire circuit. The fault-current values met in practice would likely be less than those given here.

What can be done to get maximum safety? The system must be designed to see that performance of the grounding and bonding in the system fulfills all the requirements as set forth in 250.4(A)(5) and 250.4(B)(4) to obtain proper or effective grounding, bonding, and an effective path for fault current. This may mean adding an equipment-grounding conductor within the conduit in parallel with the conduit if the calculations indicate its need.

Where an equipment-grounding conductor is added inside a raceway, it is done to increase the capacity, that is, decrease the impedance of the raceway that is serving as an equipment grounding conductor. Where a metallic system is installed, the conductor enclosure also is a part of the equipment-grounding conductor.

The assistance of that parallel conduit path is valuable, but proper use can be made of it only if the equipment grounding conductor and the conduit are bonded together as frequently as is practical.

If the impedance of the ground-fault circuit is higher than will pass enough current to properly operate the overcurrent devices in a reasonable time, then a lower impedance can be obtained by adding an equipment grounding conductor within the conduit in parallel with the conduit. That conductor must be run within the conduit, that is, run with the circuit conductors. The equipment grounding conductor should never be run outside the conduit or raceway through which the conductors serving the equipment are run. Where run external to the conduit, it becomes quite ineffective in the grounding circuit, for virtually all the ground-fault current will return on the conduit. Further, the equipment grounding conductor must be run as close to the phase conductors as is practical, right to the point where it attaches to the neutral conductor at the service.

By determining the impedance of the circuit involved when a ground fault occurs, it can be determined whether the metallic enclosure will make an acceptable equipment grounding conductor or whether it will be necessary to supplement the enclosure with an equipment grounding conductor. Where an iron conduit is a part of the electric circuit, as it will be where a ground fault occurs, there will be a large increase in both the resistance and reactance of the circuit, which will vary considerably with the amount of fault current.

Laboratory tests have shown that where a single-phase current flows in a conductor within an iron conduit, the impedance of the circuit is approximately equal to the impedance of the conduit itself. The size of the conductor within the conduit has relatively little effect on the circuit impedance. Also, despite the fact that there are many parallel paths external to the conduit, the current flowing in all the parallel paths will be very small, and under normal conditions would be under 10 percent of the total fault current.

Two other factors are to be taken into account in estimating the ground-fault current flow. They are the effects of the conduit couplings in increasing the impedance of the circuit and the voltage drop across the point of fault. If conduit couplings are installed wrenchtight, as required by *Code*, the increase in impedance of the conduit with couplings is about 50 percent more than the impedance for a straight run without couplings. This is where the value and importance of installing a properly-sized equipment grounding conductor inside the conduit and bonding it to the conduit at frequent intervals is proven for ensuring low impedance and safety.

Impedance values for conduit can be obtained from manufacturers. By using these impedance values, adjusting for the couplings and allowing 50 volts drop across the fault, a reasonable value for the amount of fault current that will flow in the circuit can be determined.

Assume a 200-foot run of 3-inch conduit with 500-kcmil conductors on a 208Y/120-volt circuit protected by 400-ampere overcurrent devices. On a ground fault, the amount of current that will flow will thus be:

$$\frac{E \text{ (voltage)}}{Z \text{ (impedance)}} = I \text{ (Current)}$$

With a 50-volt drop at the fault and Z (impedance) = 0.02970, the current that will flow will be about 2350 amperes. The use of a 3-inch conduit as the equipment grounding conductor where

Where necessary, Increase size of conduit or install equipment grounding conductor and bond to raceway as often as practicable. Recommended not more than 100 feet apart.

Figure 11-10. Long Conduit Runs

400-ampere overcurrent devices are used is, thus, satisfactory for this run.

A simpler method of determining if the conduit or metallic enclosure will perform satisfactorily is to first calculate the minimum desired fault-current flow (5 times the overcurrent device rating or better yet to reach the instantaneous portion of the time/current curve) which, in this case, is 5 x 400 or 2000 amperes. Then, on the basis of 70 volts available for a 120-volt-to-ground circuit, calculate Z (impedance), which will be found to be 0.035. A straight run of 3-inch conduit has about 0.099 ohms impedance per thousand feet at 2000 amperes. To that figure add 50 percent to include a factor of safety. That will give an impedance of 0.01485 ohms per 100 feet. The impedance value that will allow 2000 amperes to flow in that circuit was found to be 0.035. Since 235 feet of 3-inch conduit carrying 2000 amperes will have an impedance of 0.035, it has been determined that to have a minimum current flow of 2000 amperes in a ground fault, up to 235 feet of 3-inch conduit can be installed where a 400-ampere overcurrent device is involved.

Table 11 in chapter twenty provides supporting data by the Georgia Institute of Technology relative to the maximum lengths of steel conduit or tubing that may safely be used as an equipment grounding conductor.

If a 4-inch conduit was used instead of 3-inch, and the overcurrent device rating did not change but remained at 400 amperes, the maximum length of conduit can be determined by reference to Table 11. With some interpolation and using the same calculations, it is found that 260 feet of 4-inch conduit could be installed and provide a satisfactory equipment grounding conductor for this circuit.

It can be determined for any circuit and any size conduit, for any size overcurrent device, the maximum safe length of conduit which will allow a fault current to pass which will be sufficient to facilitate the operation of the overcurrent device. Should the circuit length exceed the maximum safe length as calculated, and then it will be necessary to add a metallic (copper or aluminum) equipment grounding conductor in parallel with the conduit or increase the size of the conduit. Minimum equipment grounding conductor sizes are given in Table 250.122.

It would be neither practical nor desirable to substitute a copper or aluminum conductor for the conduit. Rather, add the copper or aluminum conductor, and connect it in parallel with the conduit to form an equipment grounding conductor of two conductors in parallel. The copper or aluminum conductor and the conduit should be connected together at convenient practical intervals, about every 100 feet or less. That will reduce the length of circuit through which the ground-fault current may flow on the conduit alone.

The *Code* permits an equipment grounding conductor to be bare or insulated.

The *Code* permits an equipment grounding conductor to be bare or insulated. However, if the conductor is bare, there may be some arcing between the bare conductor and the interior of the conduit at points other than the point at which the ground fault occurs. This is due to slight differences in impedance between the raceway and wire, resulting in potential differences which cause the arcing. Such arcing may damage the phase conductors without adding to the proper functioning of the ground-fault circuit. That makes a strong case for the use of insulated equipment grounding conductors where installed in a metallic enclosure.

If aluminum conduit was used instead of steel conduit for the same conditions cited above, (500-kcmil copper conductors, 3-inch conduit and a 400-ampere overcurrent device) the circuit run could be about 900 feet long, and the aluminum conduit would provide a satisfactory equipment grounding conductor. A 3-inch aluminum conduit has a dc resistance of about 0.0088 ohms/1000 feet and 500-kcmil copper cable has a dc resistance of 0.0222 ohms/1000 feet.

Flexible metal conduit is suitable as an equipment grounding conductor for not more than 6-foot lengths in the ground-fault return path and with not more than 20-ampere overcurrent protection of the contained conductors. As such, flexible metal conduit should include a metallic equipment grounding conductor. The *Code* requires that the various metal raceways shall be so constructed that adequate electrical and mechanical continuity of the complete system be secured. However, owing to the various joints involved, it is well for the engineer to investigate such conductor enclosures to be assured that their impedance is sufficiently low to function along the lines cited above in the event of a ground fault.

Adequate Size of Conductor

In general, the minimum size of equipment grounding conductor is given in Table 250.122 and is based on the rating of the overcurrent device ahead of the conductors. (An analysis of Table 250.122 is provided in Table 7 of chapter twenty.) The rule of thumb may be applied, but it should be checked by calculation that the equipment grounding conductor should not be less than

> Grounding conductors, circuit conductors, busbars, bonding jumpers, and so forth, are not intended to break current.

25 percent of the capacity of the phase conductors or the overcurrent device that supply the circuit. A note has been added to the table indicating that the size of equipment grounding conductor given in the table must be increased if necessary to comply with 250.4(A)(5). This adds emphasis to the heading of Table 250.122 because it indicates that equipment grounding conductors given in the table are the minimum size.

An analysis of Table 250.122 shows that the relation of the equipment grounding conductor to the size of the overcurrent device (based on the continuous rating of 75°C-rated wire). It is from 50 to 125 percent of the phase conductor for overcurrent devices up to 100-ampere rating. The rating varies from 33 to 25 percent for overcurrent devices rated up to 400 amperes and is from 22 percent for 600-ampere overcurrent devices to a low of only 8 percent for an overcurrent device of 6000 amperes.

Obviously, the equipment grounding conductor must be large enough to carry that amount of current for a given time, which is required to clear the overcurrent device with which it is associated, and not result in extensive damage. In addition, it is vital that the equipment grounding conductor be run with the circuit conductors or

encloses the circuit conductors. This was covered extensively in chapter nine.

Conductor Withstand Rating

Section 110.9 states, "Equipment intended to interrupt current at fault levels shall have an interrupting rating sufficient for the nominal circuit voltage and the current which is available at the line terminals of the equipment."[11] This includes fuses, circuit breakers, disconnect switches, contactors, and similar equipment.

Section 110.10 reads in part, "The overcurrent protective devices, the total impedance, the component short-circuit current ratings, and other characteristics of the circuit to be protected shall be selected and coordinated to permit the circuit protective devices used to clear a fault to do so without extensive damage to the electrical components of the circuit. This fault shall be assumed to be either between two or more of the circuit conductors, or between any circuit conductor and the grounding conductor or enclosing raceway."[12]

Grounding conductors, circuit conductors, busbars, bonding jumpers, and so forth, are not intended to break current. These conductors must be large enough to safely carry any short-circuit and/or ground-fault current for the time it takes the overcurrent protective device to clear the fault. This is clearly stated in 110.10, 240.1 FPN, 250.4, 250.90 and 250.96. Section 310.10 also discusses in detail the temperature limitations for conductors.

The integrity of equipment grounding conductors, grounding electrode conductors, main

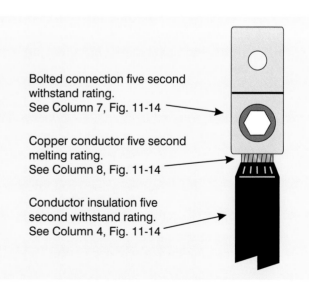

Bolted connection five second withstand rating. See Column 7, Fig. 11-14

Copper conductor five second melting rating. See Column 8, Fig. 11-14

Conductor insulation five second withstand rating. See Column 4, Fig. 11-14

Figure 11-11. Wire, Insulation and Connection Withstand Ratings

bonding jumpers and other circuit conductors must be assured by sizing them properly. Equipment-grounding conductors that are too small are of little value in clearing a fault and may, in fact, give one a false sense of security. Grounding conductors must not burn off under ground-fault conditions, leaving the equipment live thus, in many cases, creating a shock hazard that can cause serious injury or could even be fatal. Safety cannot be compromised. In fact, the first sentence of the *National Electrical Code* states that, "the purpose of this *Code* is the practical safeguarding of persons and property from hazards arising from the use of electricity."[13]

Bolted Connections

It can be shown by calculation, using values from the Insulated Cable Engineers Association Publication P 32-382 (1994), that an insulated copper conductor with a bolted connection may safely carry one ampere for every 42.25 circular mils for five seconds without destroying its va-

Short-Circuit Current Withstand Chart for Copper Cables with Thermoplastic Insulation
Allowable Short-Circuit Currents for Insulated Copper Conductors*

Figure 11-12. Courtesy of Insulated Cable Engineers Association

lidity. That will be the short-time rating or I²t (amperes x amperes x time) value of the conductor. Then, from the time-current characteristic curves of various approved overcurrent devices, the amount of current will be necessary to clear the overcurrent device in five seconds can be found.

It is desirable to limit the heat of the faulted circuit to reduce damage to adjacent insulated conductors.

Using that formula, the size of the equipment grounding conductors which will be proportional to those given in Table 250.122, as analyzed in Table 7 of chapter twenty, can be determined. Grounding conductors are permitted to be bare, and, in most cases, are pulled into the same raceway as the insulated phase conductors. This presents a potential problem. Where a grounding conductor is carrying ground-fault current, an extreme rise in its temperature can cause the insulation on the adjacent phase conductors to melt, causing further damage. Again, the potential for equipment damage and electrical shock hazard to personnel is increased. It is desirable to limit the heat of the faulted circuit to reduce damage to adjacent insulated conductors.

Thus, as discussed in this chapter, for copper conductors, the clearing time and short-circuit current must be limited to:
- One ampere...
- for five-seconds...
- for every 42.25 circular mils.

This can be expressed by the formula ampere squared seconds (I²t).

For example, from Table 8 of *NEC* Chapter 9, an 8 AWG conductor has a cross-sectional area of 16,510 circular mils. By dividing the circular mil area of the conductor by 42.25, the conductor's five-second withstand rating can be calculated.

Stated another way, this conductor has an I²t five-second withstand rating of: 391 x 391 x 5 = 764,405 ampere squared seconds.

From this five-second withstand rating value, it is easy to calculate the conductor's withstand rating for other values of time and/or for other values of current.

Example 1: How many amperes will the 8 AWG copper conductor be able to safely carry if the impedance of the circuit along with the operating characteristics of the overcurrent device protecting the circuit results in a 2 cycle (0.0333 seconds) opening time? Example 1 is solved as follows:

$$I^2t = 764,405 \text{ ampere squared seconds}$$

$$I^2 = \frac{764,405}{t}$$

$$I = \sqrt{\frac{764,405}{0.0333}}$$

$$I = 4,791 \text{ amperes}$$

Example 2: How many amperes will the 8 AWG copper conductor be able to safely carry if the impedance of the circuit along with the operating characteristics of the overcurrent device protecting the circuit results in a 1/4 cycle (0.0042) opening time? Example 2 is solved as follows:

$$I^2t = 764,405 \text{ ampere squared seconds}$$

$$I^2 = \frac{764,405}{t}$$

$$I = \sqrt{\frac{764,405}{0.0042}}$$

$$I = 13,491 \text{ amperes}$$

Note that in the example above, because a much faster total clearing time is achieved, the allowable fault current that the conductor will be subjected to can be increased. This is a result of substituting different time values in the I^2t formula.

Generally, where current-limiting overcurrent devices are protecting the circuit, the equipment grounding conductor sizes are determined directly from Table 250.122. Where available fault currents are high, and the overcurrent protective device takes longer than $1/4$ cycle to clear the fault, it is suggested that the grounding conductor be sized per figure 11-12, to be on the safe side.

Figure 11-14 provides information to assist the installer in the proper selection of equipment grounding conductors. Among other information, it includes:

1. safe values for 75°C thermoplastic insulated conductors,
2. safe values for bolted connections,
3. unsafe (melting) values for the copper conductor itself.

Because the weakest link in any system is the insulation short-circuit withstand rating as found in columns 4, 5, and 6 of figure 11-14, it is recommended that column 4 be the deciding factor where selecting equipment grounding conductors. The Insulated Cable Engineers Association data, figure 11-14, column 4 calculates out to the previously discussed: Do not exceed one ampere... for five-seconds...for every 42.25 circular mils of copper conductor.

This chart also shows the 8 AWG conductor used in the above text examples. Where you are absolutely certain that the equipment grounding conductor will not come into contact with any of the current-carrying insulated circuit conductors, the withstand rating of the grounding conductor, for every 29.1 circular mils of copper conductor, cannot exceed one ampere for five seconds. See column seven, figure 11-14.

This is the standard previously referred to in

Figure 11-13. Courtesy Bussman Division, Cooper Industries.

Copper 75° C. Thermoplastic Insulated Conductor, Bare Conductor, Bolted Connection Five Second Withstand Rating In Amperes and Melting of Copper Wire

1	2	3	4	5	6	7	8
Wire size	Area in circular mils	Area in square mm	ICEA Amperes	IEC Amperes	IEE Amperes	Bolted connection 250°C. Amperes	Melting of conductor 1,083°C. Amperes
14	4,110	2.080	97	107	107	141	254
12	6,530	3.310	155	170	170	224	403
10	10,380	5.261	246	271	271	357	641
8	16,510	8.367	391	430	430	567	1,020
6	26,240	13.300	621	684	684	902	1,621
4	41,740	21.150	988	1,088	1,088	1,435	2,578
3	52,620	26.670	1,245	1,372	1,372	1,808	3,251
2	66,360	33.620	1,571	1,729	1,729	2,281	4,099
1	83 690	42.410	1,981	2,181	2,181	2,876	5,170
1/0	105 600	53.490	2,499	2,751	2,751	3 629	6,523
2/0	133,100	67.430	3,150	3,468	3,468	4,574	8,222
3/0	167,800	85.010	3,972	4,372	4,372	5,767	10,366
4/0	211,600	107.200	5,009	5,513	5,513	7,272	13,071
250	250,000	126.700	5,918	6,516	6,516	8,592	15,443
300	300,000	152.000	7,101	7,818	7,818	10,310	18,532
350	350,000	177.300	8,285	9,119	9,119	12,029	21,621
400	400,000	202.700	9,467	10,425	10,425	13,747	24,709
500	500,000	253.300	11,834	13,027	13,027	17,184	30,887
600	600,000	304.000	14,201	15,636	15,636	20,621	37,064
700	700,000	354.700	16,568	18,243	18,243	24,057	43,241
750	750,000	380.000	17,752	19,544	19,544	25,776	46,330
800	800,000	405.400	18,935	20,850	20,850	27,494	49,419
900	900,000	456.000	21,302	23,453	23,453	30,931	55,596
1,000	1,000,000	506.700	23,669	26,060	26,060	34,368	61,773

Column 4 - Insulated Cable Engineers Association publication P32-382. One ampere for five seconds for every 42.25 circular mils of conductor area.

Column 5 - International Electrotechnical Commission publication 364-4-43.

Column 6 - Institute of Electrical Engineers publication 434-6.

Column 7 - Calculated from data in Electrical Engineers Handbook (75°C ambient). One ampere for five seconds for every 29.1 circular mils of conductor area.

Column 8 - Calculated from data in Electrical Engineers Handbook (75°C ambient). One ampere for five seconds for every 16.19 circular mils of conductor area.

Figure 11-14.

this text as the "Do not exceed one ampere for five seconds for every thirty circular mils of conductor area."

This value is only to be used where bare equipment grounding conductors are used in such a manner that they will not come in contact with insulated conductors. In this application, the limiting element of the circuit is the bolted connection of the lug.

The short-time rating of the equipment grounding conductor bears an approximately constant relation to the size of the overcurrent device.

Column 8 of figure 11-14 gives the current in amperes at which the melting temperature of copper conductors is reached. Of course, you never want to reach the current shown as the equipment grounding conductor will burn off leaving the equipment ungrounded and a possible shock hazard.

Conclusion on Equipment Grounding Conductor
For a grounded system, it is vital to safety that a low-impedance equipment grounding conductor path be provided in addition to a good grounding electrode with as low a resistance as practical. This allows sufficient current to flow to clear a ground fault automatically in a limited time, which would be as quickly as is practical, without undue interruption of service.

The I^2t values found in column 7 of figure 11-14 are based on the adequacy of a copper conductor and its bolted joints to carry the current values without destroying its validity. The values are obtained from an IEEE committee report in "A Guide to Safety in AC Substation Grounding." The figure expressed in amperes per circular mil is one ampere for every 29.1 circular mils cross section. The time of five seconds was used to provide a safety factor and was considered a reasonable approach for distribution systems of 600 volts or less protected by high-interrupting-

capacity current-limiting fuses and having equipment ground-fault protection.

As previously stated, where grounding conductors might come into contact with insulated phase conductors, use the values found in column four of figure 11-14. This column is based on one ampere for every 42.25 circular mils of conductor for five seconds. Figure 11-14 has all the calculations done and is much easier to use rather than performing complicated calculations.

The short-time rating of the equipment grounding conductor bears an approximately constant relation to the size of the overcurrent device. The I^2t values of the conductors given in Table 250.122 are between about 13 and 28 times their nominal continuous rating based on one ampere for every 42.25 circular mils cross section.

[1] *Electric Power Distribution for Industrial Plants.* The Institute of Electrical and Electronics Engineers, Inc., © 1954 AIEE, 445 Hoes Lane, P.O. Box 1331, Piscataway, NJ 08855-1331.
[2] "Some Fundamentals of Equipment-Grounding Circuit Design" by R. H. Kaufmann, Paper 54-244 presented at the AIEE Summer and Pacific General Meeting, Los Angeles, California, June 24-25, 1954, © 1954 AIEE, (now IEEE), 445 Hoes Lane, PO Box 1331, Piscataway, NJ 08855-1331.
[3] "Some Fundamentals of Equipment-Grounding Circuit Design" by R. H. Kaufmann, Paper 54-244 presented at the AIEE Summer and Pacific General Meeting, Los Angeles, California, June 24-25, 1954, © 1954 AIEE, (now IEEE), 445 Hoes Lane, PO Box 1331, Piscataway, NJ 08855-1331.
[4] Some Fundamentals of Equipment-Grounding Circuit Design" by R. H. Kaufmann, Paper 54-244 presented at the AIEE Summer and Pacific General Meeting, Los Angeles, California, June 24-25, 1954, © 1954 AIEE, (now IEEE), 445 Hoes Lane, PO Box 1331, Piscataway, NJ 08855-1331.
[5] "Some Fundamentals of Equipment-Grounding Circuit Design" by R. H. Kaufmann, Paper 54-244 presented at the AIEE Summer and Pacific General Meeting, Los Angeles, California, June 24-25, 1954, © 1954 AIEE, (now IEEE), 445 Hoes Lane, PO Box 1331, Piscataway, NJ 08855-1331.
[6] Fuses (JCQR), *General Information for Electrical Equipment 2000*, (Underwriters Laboratories, Northbrook, IL, 2001), p. 44.
[7] Circuit Breakers, Moulded-Case and Circuit Breaker Enclosures (DIVQ), *General Information for Electrical Equipment 2000*, (Underwriters Laboratories, Northbrook, IL, 2001), p. 11.
[8] Circuit Breaker Current Limiters (DIRW), *General Information for Electrical Equipment 2000*, (Underwriters Laboratories, Northbrook, IL, 2001), p. 9.
[9] 1991 IEEE Industry Applications Society Annual Meeting, Volume II. The Institute of Electrical and Electronics Engineers, Inc., © 1954 AIEE (now IEEE), 445 Hoes Lane, PO Box 1331, Piscataway, NJ 08855-1331.
[10] Modeling and Testing of Steel EMT, IMC and Rigid (GRC) Conduit School of Electrical and Computer Engineering Georgia Institute of Technology, Atlanta, Georgia 30332.
[11] NFPA 70, *National Electrical Code* 2002, 110.9, (National Fire Protection Association, Quincy, MA, 2002), p. 70-41.
[12] NFPA 70, 110.10, p. 70-41.
[13] NFPA 70, Front page, p. 70-1.

Chapter Eleven: The questions included here were developed using material included in this chapter. The answers can be found by reviewing the text. It is also important that students make use of the 2002 *NEC*, where many of the answers can be found.

1. A conducting connection, whether intentional or accidental, between any of the conductors of an electrical system whether it be from line to line or line to the grounded conductor, is defined as a ____.
   a. ground fault
   b. phase fault
   c. short circuit
   d. unidentified fault

2. A conducting connection, whether intentional or accidental, between any of the conductors of an electrical system and the conducting material which encloses the conductors, or any conducting material that is grounded, or that may become grounded is defined as a ____.
   a. identified fault
   b. ground fault
   c. short circuit
   d. failure of the system

3. A short circuit may be from one phase conductor to another phase conductor, or from one phase conductor to the grounded conductor or ____.
   a. unidentified conductor
   b. enclosure
   c. neutral
   d. equipment grounding conductor

4. The grounding system must provide a permanent and electrically continuous path; must have ample carrying capacity to conduct safely any fault currents likely to be imposed on it; and must have an impedance sufficiently low to limit the voltage to ground and to facilitate the operation of the circuit protective devices in the circuit. This is best described as being ____.
   a. improved
   b. effective
   c. sufficient
   d. required

5. In general, the rule of thumb may be applied, but should be checked by calculation, that the equipment grounding conductor should not be less than ____ percent of the capacity of the phase conductors that supply the circuit.
   a. 25
   b. 20
   c. 15
   d. 18

6. There is no overcurrent protection on the line side of service conductors, there is only ____ protection on the primary of the transformer.
   a. ground-fault
   b. short-circuit
   c. overload
   d. equipment ground fault

7. The grounded service conductor must be run to each service disconnecting means and bonded to the enclosure even if the service supplies only ____.
   a. 240-volt loads
   b. 208-volt loads
   c. 480-volt loads
   d. line-to-line loads

8. A 8 AWG THW copper conductor has an allowable ampacity of 50 amperes according to Table 310.16. This conductor also has a safe (ICEA) five-second withstand rating of ____ amperes, but will melt if subjected to ____ amperes for five seconds.
   a. 8367   10,000
   b. 391   1020
   c. 30   500
   d. 4110   254

9. When ICEA conductor short-circuit withstand rating tables are not available, it is possible to calculate the short-circuit current withstand ratings for a conductor. For example, to determine the safe withstand rating for copper conductors with 75°C insulation, use ____ ampere for every ____ seconds for every ____ circular mils of cross sectional area of the conductor.

    a.   one, five, 42.25
    b.   one, five, 16.9
    c.   one, five, 29.1
    d.   one, five, 10.2

10. Table 250.122 shows the minimum size equipment grounding conductors. Where high values of fault current are available, the grounding conductors may have to be ____ in size to be capable of safely carrying the available fault current.

    a.   decreased
    b.   increased

11. Referring to Figure 11-12, find the minimum conductor sizes required for the fault current values indicated.

    a.   10,000 amperes for one cycle.   Size _____

    b.   20,000 amperes for two cycles.   Size _____

    c.   4200 amperes for one cycle.   Size _____

    d.   10,000 amperes for eight cycles. Size _____

Key Notes

Grounding Separately Derived Systems

In many distribution systems of 600 volts or less for commercial or industrial occupancies, it is common to have separately derived systems at another voltage level lower or higher than the electrical system supplied by the service. Several *Code* requirements must be met when installing a separately derived system. First, if there is a separately derived system, this means there will be a new source established producing a voltage and providing a capacity from which current would be drawn. This capacity is usually referred to as the kVA or kW rating of the system. The system itself must have a capacity to supply the intended load to be served. Overcurrent protection sizing and the protection of conductor taps to the secondary of the separately derived system are also required. An another important element is what grounding and bonding rules must be applied to satisfy the requirements. This chapter focuses on those requirements.

The first item that must be established is that the system is separately derived. An example of a system or transformer that is not separately derived is an autotransformer. If a generator is installed on a premises wiring system, it must be grounded as a separately derived system if there is a switching action in the grounded conductor through the transfer equipment. If there is no switching action in the grounded conductor through the transfer switch or equipment, then the grounded conductor remains grounded through the service grounding connection point located at the service and the generator-produced system is not separately derived. Section 700.8(B) requires a sign at the service equipment when the grounding electrode system at the service is also used for the grounding electrode for a generator. In this case, the generator grounding is established through the use of equipment grounding conductor and an insulated neutral or grounded conductor that is isolated from grounded metal parts at that location to comply with 250.24(A)(5).

Definition

"**Separately Derived System.** A premises wiring system whose power is derived from a battery, a solar photovoltaic system, or from a generator, transformer, or converter windings, and that has no direct electrical connection, including a solidly connected grounded circuit conductor, to supply conductors originating in another system."[1]

Transformer-Type Separately Derived System

Figure 12-1 is a diagram of a transformer-type separately derived system (for simplicity,

"... no direct electrical connection, including a solidly-connected grounded circuit conductor, to supply conductors originating in another system."

Figure 12-1. Transformer-Type Separately Derived System

grounding and bonding conductors are not shown.) The supply or primary of the transformer is at one voltage level, and the secondary is often at another voltage level, either lower or higher. As shown, there is no direct electrical

> A short circuit or ground fault on the secondary of the transformer is often seen as a load by the transformer primary winding.

connection between the transformer primary and secondary so the installation meets the definition of a separately derived system. Where 3-phase systems are employed, the primary is usually connected delta so a neutral conductor is not needed. The secondary may be connected delta or wye as desired.

Grounding Primary Side Equipment

An equipment grounding conductor must be supplied with the primary circuit to provide a low-impedance fault-current path from the transformer case to the main service or source of supply. The equipment grounding conductor is permitted to consist of any of the means included in 250.118 including wires and the wiring method, where appropriate. The overcurrent device on the primary of the transformer will then clear a short circuit or ground fault up to, and including, the transformer primary windings.

The equipment grounding conductor connection to the transformer enclosure, plus the bonding jumper from the secondary does not constitute a direct electrical connection from the primary system to the secondary system. The equipment grounding conductor is not a system conductor as the ungrounded and grounded conductors are.

A short circuit or ground fault on the secondary of the transformer is often seen as a load by the transformer primary winding. The amount of current flowing in the fault on the secondary side will determine whether or not the primary overcurrent device will open.

Systems and Grounding Rules

Whether a separately derived system requires grounding in accordance with 250.30, it is established in 250.20. Sometimes there is a choice as to whether the system is required or permitted to be grounded. If the system can be grounded so that the voltage to ground from any of the system conductors does not exceed 150 volts, then

The means of attachment of the bonding jumper must be by the exothermic welding process, listed pressure connectors, listed clamps, or other listed means.

generally it must be grounded. Section 250.20(B)(1), (2), and (3) determines when the system produced must be grounded and a choice is not permissible. If that system is required to be grounded, or if the system is grounded by choice, in other words, if it is not required by 250.20 but is grounded anyway as permitted, then the separately derived system must be grounded according to the applicable requirements of 250.30. The grounding requirements are as follows in the order they appear in the *NEC*:

(1) Bonding Jumper.
(2) Grounding Electrode Conductor.
   (a) Single Separately Derived System.
   (b) Multiple Separately Derived Systems.
(3) Grounding Electrode Conductor Taps.
(4) Grounding Electrode.
(5) Equipment Bonding Jumper Size.
(6) Grounded Conductor.[2]

Bonding Jumper

The bonding jumper is the first component addressed in 250.30. The bonding jumper is the vital link for the fault-current path at the source and is the connection between the grounded conductor and the equipment grounding conductors at the source or equipment containing the first overcurrent protective device. It functions the same as the main bonding jumper at the service location.

Those in the electrical field often refer to the *bonding jumper* of a separately derived system as the *main bonding jumper* used at the service equipment. Its function is essentially the same, but a common language of communication must be used to apply the requirements of the *Code* prop-

erly. A bonding jumper is used at separately derived systems.

The bonding jumper must be installed in accordance with the requirements of 250.28(A) through (D). The sizing is to be based on the derived phase conductors supplied by the separately derived system. A brief look at the requirements in 250.28 identifies the material, construction, attachment, and sizing requirements for the bonding jumper. The bonding jumper is required to be copper or other corrosion-resistant material, and is permitted to be in the form of a wire, bus, screw, or similar conductor. Bonding jumpers that are of the screw types must be identified by the color green that identifies the screw after installation. This type of bonding jumper might be employed when the bonding jumper and grounding conductor connection to the system are installed at the first system overcurrent protective device, such as a

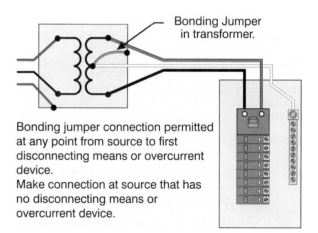

Bonding jumper connection permitted at any point from source to first disconnecting means or overcurrent device.
Make connection at source that has no disconnecting means or overcurrent device.

Figure 12-2. Bonding Jumper Location and Function

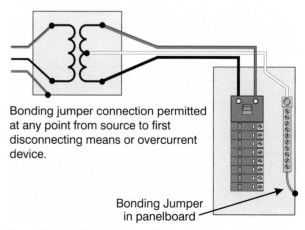

Bonding jumper connection permitted at any point from source to first disconnecting means or overcurrent device.

Figure 12-3. Bonding Jumper Location and Function

panelboard. The means of attachment of the bonding jumper must be by the exothermic welding process, listed pressure connectors, listed clamps, or other listed means. The more common location for the bonding jumper and grounding electrode conductor connection to the system is at the source enclosure.[3] However, the *Code* permits these connections at any point from the source to the first system disconnecting means or overcurrent protective device, or it must be made at the source where there is no disconnecting means or overcurrent device located in the equipment supplied by the separately derived system. The bonding jumper is required to be connected to the separately derived system at the same location as the grounding electrode conductor connection.

Bonding Jumper Location Exceptions

Section 250.30(A), Exception recognizes the provision for connecting the bonding jumper for a separately derived system according to the requirements for a high-impedance grounded system. These rules are given in 250.36 and 250.186. These systems are grounded only through an impedance device (often a resistor) usually at the

> The more common location for the bonding jumper and grounding electrode conductor connection to the system is at the source enclosure.

panelboard or motor control center and not at the transformer. These systems are designed to limit the fault current of the first ground fault to a predetermined value and often incorporate an indication or alarm system. Additional information on high-impedance grounded systems is provided in chapter four.

Section 250.30(A)(2)(a) permits the grounding electrode connection to be made to an equipment grounding terminal bar, rather than to the neutral bar where the main bonding jumper is a wire or busbar, as permitted by 250.24(A)(4). This provides for residual-type equipment ground-fault protection systems where, often, a current transformer is located on the bonding jumper to mea-

This application typically is used for residual-type equipment ground-fault protection systems. (Current transformer typically installed on main bonding jumper.)

Figure 12-4. Bonding Jumper Exception

sure ground-fault currents flowing back to the source. Another common application of this exception is in switchboards where the main bonding jumper is a wire or busbar.

Equipment Bonding Jumper for Separately Derived Systems

The installation and sizing requirements for equipment bonding jumpers installed with the derived phase conductors from the source to the first disconnect or overcurrent protective device have been clarified in the 2002 *NEC*. When the grounding connections for separately derived systems in accordance with 250.30(A) are made at the source location, most installations include the routing of the derived phase conductors, the grounded conductor of the system, and an equipment bonding jumper for bonding the metal enclosure of the derived system source to the metal enclosure at the first disconnect or overcurrent protective device are installed. This conductor is now recognized by the code as an *equipment bonding jumper* and must be sized in accordance with the requirements of 250.28 (see figure 12-5). These conductors derived from the secondary of a transformer, generator or other separately derived system are considered unprotected or unfused conductors or line-side conductors.

Unless the secondary conductors are protected against overcurrent in a manner specified in 240.4(F) for transformer secondary conductors, they are not protected by the line-side overcurrent device, but obtain their overcurrent protection through proper application of the appropriate tap rules in 240.21.

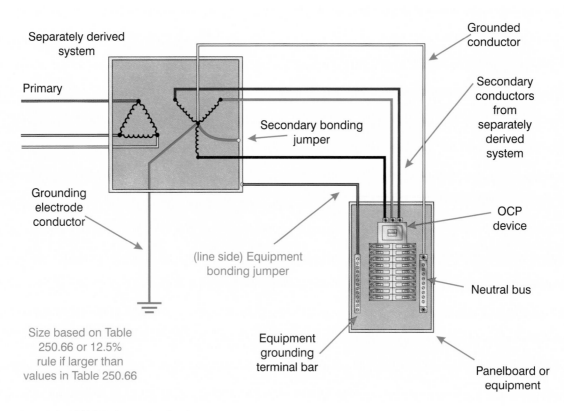

Figure 12-5. 250.30(A)(5) Equipment Bonding Jumper

The sizing requirements for this equipment bonding jumper installed from the source enclosure to the first system disconnect or overcurrent device are similar to the equipment bonding jumper sizing requirements for equipment bonding jumpers on the supply side of service equipment disconnecting means. They must be sized based on the circular mil area of the derived phase conductors and the values in Table 250.66. Where the size of the total circular mil area of the derived phase conductors exceeds the values given in Table 250.66, the 12 ½ percentage rule must be applied. An example would be a 750-kVA transformer supplying a 2500-ampere switchboard. If the derived secondary phase conductors were 10 sets of 4 – 500-kcmil copper XHHW conductors per phase, the total circular mil area of one set would equal 5,000,000 circular mils. 5,000,000 x 12.5 percent = 625,000 circular mils. Take that value back to Table 8 in *NEC* Chapter 9 and the next higher size is required. The minimum size for the equipment bonding jumper installed from the secondary of the derived system to the first disconnecting means or overcurrent device would be required to be not smaller than 700-kcmil copper. If the equipment

bonding jumpers and phase conductors are installed in separate raceways, the equipment bonding jumper should be sized based on the size of the derived phase conductors installed in each raceway, but no smaller than 1/0 AWG.

These conductors must work in parallel to carry any fault current that might be imposed on them, and the rules for parallel conductors must be applied in accordance with 310.4.

Use of Derived Neutral for Bonding

There is an alternative method of bonding that utilizes the grounded conductor and allows it to be bonded to the source enclosure and the first system disconnect or overcurrent device enclosure where doing so does not create a parallel path for current that would be returning to the source over the grounded conductor (see 250.30(A)(1) Exception 1, or refer to figure 12-6).

Section 250.142(A)(3) permits the grounded circuit conductor to be used for grounding equipment "On the supply side or within the enclosure of the main disconnecting means or overcurrent devices of a separately derived system where permitted by 250.30(A)(1)." As used in 250.142(A)(3), the term, *on the supply side* no doubt means "up to or within the enclosure for the disconnecting means or overcurrent devices" as it is widely accepted to make bonding connections of the neutral or grounded circuit conductor within the disconnecting means enclosure for the separately derived system and upstream to the transformer or generator.

As shown in figure 12-6 and 12-7, the grounded circuit conductor serves two purposes. First, it allows line-to-neutral or grounded conductor loads to be supplied, and the conductor carries the unbalanced loads from ungrounded conductors. Secondly, it serves as the equipment grounding conductor and will carry line-to-ground fault currents back to the source. This grounding scheme parallels that for services and for supply to additional buildings on the premises served from a common service.

Where this scheme is used, it is not necessary to install an equipment grounding conductor between the source and the disconnecting means or overcurrent protection enclosure. This will allow nonmetallic raceways to be used. Where nonmetallic raceways are used, a parallel path for neutral current is not created by the wiring method.

Keep in mind that this method of bonding the grounded conductor at both ends is permitted only where doing so does not create a parallel path for grounded conductor current. A parallel path for neutral current exists where a metal raceway is used between the source of the separately derived system and the metal enclosure for the disconnecting means or overcurrent device. A parallel path can also be established by other means such as through metal pipes, cable trays and structural steel. The neutral current will divide between the available paths depending upon the impedance of the paths. The lowest impedance path will obviously carry the most current.

Where the bonding jumper is installed at both

Permitted if no parallel path for grounded circuit current. As shown, wiring method does not serve as equipment

Figure 12-6. Use of Derived Neutral for Bonding

Separately derived system

Primary

Bonding jumper

OCP device

Grounded conductor

Panelboard

Neutral bus

- Routed with derived phase conductors
- Not smaller than grounding electrode conductor
- Size based on Table 250.66 or 12.5% rule if larger than values in Table 250.66
- Same rules for parallel conductors that are specified in Section 250.24(B)(2)

Grounding terminal bar

Grounding electrode conductor

Figure 12-7. 250.30(A)(6) Grounded Conductor Minimum Size For Separately Derived Systems the source location and the first disconnecting means as permitted in 250.30(A)(1) Exception 1, the grounded conductor must be adequately sized to carry any fault current likely to be imposed. Section 250.30(A)(6) covers the sizing and routing requirements for grounded conductors on the secondary of derived systems connected in this manner. Basically the sizing shall be no smaller than the required grounding electrode conductor in 250.66 and where the derived phase conductors exceed 1100-kcmil copper or 1750-kcmil aluminum or copper-clad aluminum, the grounded conductor shall not be smaller than $12^1/_2$ percent of the total kcmil of the derived phase conductors. The grounded conductor of a 3-phase, 3-wire delta-connected separately derived system shall have an ampacity not less than the ungrounded derived phase conductors.

The *NEC* in recent editions has continued its migration away from the use of the grounded conductor for grounding non-current-carrying metal parts of equipment on the downstream side of the service grounding connection point or the grounding connection point for a separately derived system.

Careful consideration should be exercised when considering the use of the grounded conductor to make grounding connections at points other than the service equipment or source of separately derived system. Current will always try to seek its source, be it normal current flow on the grounded or neutral conductor or be it ground-fault current. Parallel paths for this current flow can create additional shock hazards for persons and also introduce additional impedance into the path for fault current (see chapter one for in-depth discussions of electrical circuits and paths for current flow).

The Grounding Electrode Conductor(s)

"**Grounding Electrode Conductor.** The conductor used to connect the grounding electrode(s) to the equipment grounding conductor, to the grounded conductor, or to both, at the service, at each building or structure where supplied from a common service, or at the source of a separately derived system."[4]

A separately derived system required to be grounded must have a grounding electrode conductor(s) to establish the connection of the grounded system conductor and metal equipment supplied by the derived system to earth.

Separately derived systems are permitted to be grounded individually with an individual grounding electrode conductor, or, under certain conditions, multiple separately derived systems are permitted to be grounded to a common, single grounding electrode conductor to which grounding electrode conductor taps must be connected.

The grounding electrode conductor for a separately derived system must be sized at a minimum in accordance with the sizes specified in Table 250.66 and based on the circular mil area of the derived phase conductors. This conductor must connect the grounded conductor of the derived system to the grounding electrode and must be installed in accordance with the requirements of 250.64. This section includes all applicable installation requirements for grounding electrode conductors, including physical protection from damage and protection from magnetic fields. The connection of the grounding electrode conductor to the grounded conductor of the separately derived system is required to be made at the

Grounding and Bonding at transformer.

Grounding electrode connection permitted at any point from source to first disconnecting means or overcurrent device.
Generally at same point bonding jumper is connected.

Figure 12-8. Grounding Electrode Conductor Connections

Grounding electrode connection and bonding jumper in panelboard.

Figure 12-9. Grounding Electrode Conductor Connections

same location as the bonding jumper connection unless the installation conforms to the requirements of 250.24(A)(4). In this case, the grounding electrode conductor is permitted to be connected to an equipment grounding terminal bar installed at the source or first system disconnect or overcurrent protective device and a bonding jumper sized in accordance with 250.28(D) is installed from the grounded conductor termination point to the equipment grounding terminal bus. This is typically done for a residual or ground strap type of equipment ground-fault protection systems (see chapter fifteen for more information on ground-fault protection systems).

Grounding Electrode Conductors for Multiple Separately Derived Systems

An alternative method for establishing and installing grounding electrode conductors for multiple separately derived systems is now included in the rules. It is now permitted and formally recognized by the *NEC* to install a common grounding electrode conductor to which multiple separately derived systems shall be connected by individual grounding electrode (tap) conductors that are sized based on the derived phase conductors of each individual separately derived system. An example of where this alternative

method of grounding electrode conductor installation may be effective would be a typical high-rise building that has no effective grounded building steel but has an effectively grounded metal water piping system that enters the building in the basement of the structure. Section 250.30(A)(4) requires the effectively grounded metal water pipe to be used as the grounding electrode for the separately derived systems installed in that building, and the grounding electrode conductors to be installed and connected to the water pipe at a location within 1.5 m (5 ft) of where the water pipe enters the building. This requirement presented some challenges in the field, which led to the new alternative methods, permitted in this section.

The rules that must be followed when installing grounding electrode conductors for multiple separately derived systems in accordance with these new provisions are as follows:

Where more than one separately derived system is grounded to a common grounding electrode conductor, the common grounding electrode conductor shall be sized based on the total circular mil area of the derived phase conductors from separately derived systems. The largest size grounding electrode conductor required by 250.66 and Table 250.66 is 3/0 AWG copper or 250-kcmil aluminum or copper-clad aluminum. With mul-

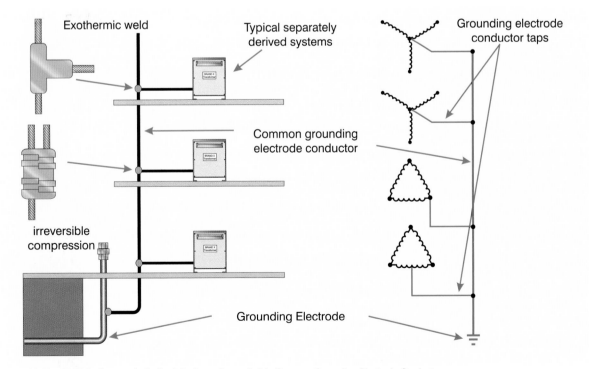

Figure 12-10. Multiple Separately Derived Systems Connected to Common Grounding Electrode Conductor

Common grounding
electrode conductor

Tap conductors sized based on the size
of the derived phase conductors of that
separately derived system

Connections by exothermic weld,
irreversible compression, or to copper
bus bars not less than ¼ inch x 2 inches

Grounding electrode
conductor taps

Common grounding electrode conductor size based on
total kcmil of derived phase conductors

Common grounding electrode conductor run to grounding
electrode system to remain continuous without a splice

Figure 12-11. 250.30(A)(3)(b) Connections

tiple systems connected to this common grounding electrode conductor, the size of this conductor would end up reaching the minimum values in Table 250.66 rather quickly. If longer lengths for this common grounding electrode conductor are needed, which is often the case, the common grounding electrode conductor usually is increased in size based on the conditions and the design of the system.

It is common to see a high-rise building that has no effectively grounded building steel but utilizes the building grounding electrode system established in the basement of the structure and route a 500-kcmil conductor up through the core of the structure.

Individual grounding electrode conductor taps would then be installed from each derived system to this common large grounding electrode conductor. This has been an industry practice for some time to accomplish the grounding for separately derived systems installed in high-rise buildings. However, these practices, although installed to exceed the minimum requirements, were never really covered by any written *Code*s. The 2002 *NEC* now includes the rules when alternative methods are employed for grounding multiple separately derived systems.

Once the size of the common (main) grounding electrode conductor is established, then sizing the individual grounding electrode conductor taps is based on the derived phase conductors supplied by each separately derived system. The grounding electrode conductor tap connects the grounded conductor of the derived system to the common (main) grounding electrode conductor, which is connected to the grounding electrode. Installation for both the common grounding electrode conductor and the grounding electrode conductor taps must be in accordance with the installation requirements for grounding electrode conductors specified in 250.64. This new alternative for grounding electrode conductor taps to a common grounding electrode conductor parallels the alternative methods for grounding electrode conductor taps for services in accordance with 250.64(D).

The common grounding electrode conductor must be installed in a continuous length without a splice or joint, unless splicing is accomplished by irreversible compression connectors or by the exothermic welding process. The grounding electrode conductor taps must be connected to the common grounding electrode conductor at an accessible location by one of three methods:

Grounding electrode is required to be:

as near as practicable, and preferably in same area as grounding electrode conductor connection, and the nearest one of the following:

an effectively grounded structural metal member of the structure, or

an effectively grounded metal water pipe within 5 ft from the point of entrance into the building.
In qualifying industrial and commercial buildings where the water pipe is exposed for entire length, connection to water pipe permitted at any point.

Other grounding electrodes permitted where above electrodes are not available

Figure 12-12. 250.30(A)(4) Grounding Electrode

1. The tap conductor is permitted to be connected to the common grounding electrode conductor by an irreversible compression connector that is listed for the purpose; or

2. the tap conductor is permitted to be connected to the common grounding electrode conductor by the exothermic welding process; or

3. the grounding electrode tap conductor is permitted to be connected to a copper busbar sized not less than 6mm x 50 mm ($\frac{1}{4}$ x 2 in.) to which the common grounding electrode conductor is connected to the busbar in a manner that keeps it continuous without a splice or joint.[5]

The grounding electrode tap conductors must comply with the installation requirements of 250.64(A), (B), (C), and (E).[6]

Grounding Electrode for the Separately Derived System

Although there is no formal definition for a grounding electrode, it can be described as simply the conducting element that establishes a connection to the earth for a service or separately derived system. The rules regarding grounding electrodes permitted to be used for grounding a separately derived system are located and described in Part III of Article 250.

Specific requirements apply to the allowable choices in regard to which electrode should be used for the separately derived system. The first requirement is that whichever grounding electrode is chosen, it must be located as near as practicable and preferably in the same area as the grounding

An effectively grounded structural metal member of the building or structure is the first electrode specified.

electrode conductor connection to the system. The second requirement is that the grounding electrode shall be the nearest one of the following grounding electrodes that are available.

An effectively grounded structural metal member of the building or structure is the first electrode specified. An effectively grounded metal water pipe within 1.5 m (5 ft) from the point of entry to the building is the second grounding electrode specified, and any of the other electrodes as specified in 250.52 are permitted where neither of the above are available.

There is an exception to the requirement of a connection to an effective grounded metal wa-

EGC permitted as GEC for separately
derived system where sized large enough

Figure 12-13. Separately Derived Systems in Listed Equipment

ter pipe within 1.5 m (5 ft) from the point of entry of the water pipe to the building. This connection shall be permitted at any point on the water pipe system in industrial and commercial buildings where conditions of maintenance and supervision ensure that only qualified persons will service the installation and the entire length of the interior metal water pipe that is being used for the grounding electrode is exposed. The term *exposed* is defined in Article 100 as "on or attached to the surface or behind panels designed to allow access."

Where separately derived systems are an integral part of listed equipment, such as in unit substations, the grounding electrode used for the service equipment, or equipment that is suitable for use as service equipment installed in accordance with the requirements of Article 225 for feeders, shall be permitted to serve as the grounding electrode for the separately derived system. The internal equipment grounding bus in such equipment is permitted to serve as the grounding electrode conductor if it is large enough, provided the grounding electrode conductor from the service or feeder to the grounding electrode is also large enough for the separately derived system (see figure 12-13).

Ungrounded Separately Derived Systems

Rules are provided in 250.30(B) for grounding ungrounded separately derived systems. Basically, a grounding electrode conductor connects the metal equipment of the separately derived system to a grounding electrode similar to the rules for grounded systems. However, the sys-

tem itself is not grounded. Where the separately derived system itself is not grounded, the non-current-carrying metal parts of equipment and enclosures for conductors, metal raceways, and so forth, are required to be grounded. There must be a grounding electrode conductor installed from the metal enclosures or equipment supplied by the derived ungrounded system to a grounding electrode as specified in 250.30(B)(2). This grounding electrode conductor is permitted to be connected to the metal enclosures or equipment at any point on the separately derived system from the source to the first system disconnecting means. The grounding electrode for the ungrounded system must also be as near as practicable and preferably in the same area as the grounding electrode conductor connection to the

> It is necessary that an effective path for ground-fault current be permanent and electrically continuous, of adequate capacity, and low impedance from the equipment supplied back to the separately derived system.

enclosures or equipment. It shall be the nearest grounding electrode as specified in 250.30(A)(4).

Ground-fault Return Path

As with all electrical systems, it is vital to be certain that, for separately derived systems, an effective ground-fault return path exists from the furthermost point on the electrical system back to the source.

Fault current downstream from the separately derived system follows the same laws of physics as does the fault current on the line side. It attempts to return to its source, which is the secondary windings of the transformer, and seeks the lowest impedance path.

As can be seen in figure 12-14, fault current downstream from the separately derived system will return to the secondary windings of the transformer, not to the primary or to the service. The transformer primary sees the fault current on the secondary as a load, and the overcurrent protec-

Bonding jumper installed
at transformer enclosure

Permanent and electrically continuous

Ample capacity for ground-fault current

Lowest practicable impedance

Ensure ground path
complies with 250.4(A)(5)

Bonding jumper installed
at panelboard enclosure

Ensure ground path
complies with 250.4(A)(5)

Figure 12-14. Effective Ground-Fault Return Path

tion devices on the line side of the transformer will respond according to their rating.

It is necessary that an effective path for ground-fault current be permanent and electrically continuous, of adequate capacity, and low impedance from the equipment supplied back to the separately derived system. Often, it is best to make a one-line diagram of the installed system to be certain a complete and low-impedance path is provided.

If the bonding jumper is installed in the panelboard or fusible switch rather than in the transformer, and flexible metal conduit is used as the wiring method, an equipment bonding jumper must be installed between the transformer enclosure and the overcurrent device enclosure. In this case a parallel path for fault current will exist through the equipment bonding jumper and the wiring method.

Equipment Grounding Conductors

Where permitted, the equipment grounding conductors of both the primary and the separately derived system may be obtained by using the metal conductor enclosures, such as conduit or

electrical metallic tubing. An equipment grounding conductor may be used, if necessary, to be certain the circuit is of low enough impedance to be effective. In both the primary and the separately derived system, a low-impedance path to the grounded conductor or neutral of each system must be provided. This is vital to ensure that overcurrent devices will function to remove ground faults that may occur.

The important connections here require an equipment grounding conductor for both the primary and secondary that are separately derived systems independent of each other and terminating at the respective points of supply at their grounded conductor or neutral. Both equipment grounding conductors will meet at some common point, as at the point where the bonding jumper is connected, and they both may connect to the common grounding electrode.

Interconnecting the equipment grounding conductor and the equipment bonding jumper of the two systems does not violate the definition of separately derived system. This is because the equipment grounding conductor is not a system or power conductor nor a grounded conductor.

In addition, the equipment grounding conductor usually does not make a direct electrical connection of the neutral to ground. By following, on a one-line diagram, the bonding jumper and equipment grounding conductors of the two systems, it can be seen that the systems are, in fact, connected together by these conductors.

Bonding of Structural Steel and Water Piping

Section 250.30(A)(3)(d) requires that "where exposed structural steel that is interconnected to form the building frame or interior metal piping exists in the area served by the separately de-

rived system, it shall be bonded to the grounding electrode conductor."[7] This effectively eliminates any possible differences of potential that may exist between the grounded conductor of the new source and either item and also provides an effective path directly to the source for any fault current likely to be imposed on either the steel or the water pipe. Bonding requirements for the metallic water piping system that serves the same area as the separately derived system are found in 250.104(C). See Figure 12-15.

Generator-Type Separately Derived System

Figure 12-16 illustrates a separately derived system from a generator. Note that there is no system connection, including that of a solidly grounded neutral, between the two systems. An easy way to determine whether or not a generator must be grounded as a separately derived system in accordance with 250.30 is to examine the transfer equipment. If the grounded conductor (usually a neutral), in addition to the phase conductors, is switched at the transfer equipment, the generator is required to be grounded as a separately derived system, (see figures 12-16 and 12-17).

If the neutral is not switched by the switching action, but is solidly connected, then it is not a separately derived system. Where the generator is a separately derived system, the neu-

Separately Derived System

Bond interior metal water pipe at the nearest available point in the area served by the separately derived system, to the grounded conductor -

- Connect at same point GEC is connected.
- Size bonding jumper per Table 250.66.

Figure 12-15. Bonding of Metal Water Piping

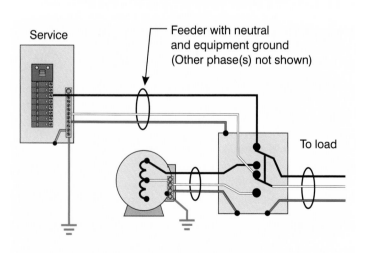

Service

Feeder with neutral and equipment ground (Other phase(s) not shown)

To load

Figure 12-16. Generator-Type Separately Derived Systems

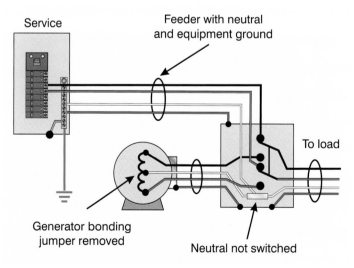

Service

Feeder with neutral and equipment ground

To load

Generator bonding jumper removed

Neutral not switched

Figure 12-17. Generator - Not Separately Derived

tral bonding jumper must be installed, either at the generator or at the first disconnecting means or overcurrent device or any point between. In addition, a grounding electrode conductor must be installed between the neutral and a grounding electrode.

Section 700.8(B), regarding grounding of emergency systems, requires that where the grounded circuit conductor connected to the emergency source is connected to a grounding electrode conductor at a location remote from the emergency source, a sign must be placed at the grounding location that shall identify all emergency and normal sources connected at that location.

Identical requirements are in 701.9(B) for legally required standby systems and in 702.8(B) for optional standby systems (see figure 12-20).

Alternate power supply systems that include grounded (neutral) conductors that are not switched through a transfer switch present a safety concern. Electricians that are unfamiliar

An equipment grounding conductor is installed throughout the system with the circuit conductors between the service, transfer switch, and other non-current-carrying parts of the installation

with the system grounding scheme may inadvertently disconnect the grounded (neutral) conductor when working on the normal source. If the emergency system is operating, the neutral conductor from the transfer switch to the location where it is grounded will function as a grounding electrode conductor, and disconnecting it may present a serious shock hazard.

Generator Not Separately Derived

Where the generator is not a separately derived system, the neutral bonding jumper must be removed from the generator, and the neutral must not be grounded at the generator or at any point between the generator up to the service.

In this case, the system is grounded by its solid connection to the neutral of the premises wiring system. An equipment grounding conductor is installed throughout the system with the circuit conductors between the service, transfer switch, and other non-current-carrying parts of the installation (see figure 12-17).

Ground-Fault Systems

There are situations where the location of the bonding jumper is determined by the conditions of the installation and the type of equipment installed. For example, a large step transformer separately derived system in the

Photo 12-1. Generator-Type Separately Derived Systems

Photo 12-2. Generator-Type Separately Derived Systems

voltage configuration of 240-volt 3-phase primary stepping up to 277/480-volt 3-phase, 4-wire secondary that feeds equipment rated at 1000 amperes or greater may require ground-fault protection for equipment (see 240.13 and 215.10).

Larger switchboards that include GFP equipment often require the bonding jumper to be installed at the equipment and not at the source to ensure the performance of the GFP equipment and conform to the listing and installation instructions of the equipment [see 110.3(B)].

A word of caution. It is imperative that care be exercised where generators supply systems that have equipment ground-fault protection. In most cases, the engineer will specify that the neutral be switched by the transfer switch to avoid grounding the neutral downstream from the ground-fault protection equipment. This is vital to prevent desensitizing the protection system. In this case, the generator is a separately derived system, and it must be grounded and bonded separately.

Section 700.26 indicates that equipment ground-

fault protection with automatic disconnecting means is not required for emergency systems. However, indication of a ground fault on the emergency system must be given in accordance with the rules of 700.7(D). It is required that, where practicable, audible and visual signal devices be provided that will indicate a ground fault in a solidly-grounded wye-connected emergency system of more than 150 volts to ground and circuit protective devices rated 1000 amperes or more. A similar exception from the requirement for ground-fault protection of equipment is provided in 701.17 for legally required standby systems.

[1] NFPA 70, *National Electrical Code* 2002, Article 100, (National Fire Protection Association, Quincy, MA, 2002), p. 70-38.
[2] NFPA 70, 250.30, p. 70-101 & 70-102.
[3] NFPA 70, 250.28, p. 70-100 & 70-101.
[4] NFPA 70, Article 100, p. 70-36.
[5] NFPA 70, 250.30(A)(3)(b), p. 70-101.
[6] NFPA 70, 250.30(A)(3)(c), p. 70-101.
[7] NFPA 70, 250.30(A)(3)(d), p. 70-101 & 70-102.

Figure 12-18. Signs Required for Standby Power Source

Chapter Twelve: The questions included here were developed using material included in this chapter. The answers can be found by reviewing the text. It is also important that students make use of the 2002 *NEC*, where many of the answers can be found.

1. A premises wiring system whose power is derived from generator, transformer, or converter windings without any direct electrical connection, including a solidly-connected grounded circuit conductor to supply conductors originating in another system, describes some ____.
    a.   medium voltage systems
    b.   utility supplied systems
    c.   separately derived systems
    d.   high voltage systems

2. Phase conductors used for a separately derived system sized at 500 kcmil aluminum, require a grounding electrode conductor not smaller than No. ____.
    a.   4 AWG copper or 2 AWG aluminum
    b.   2 AWG copper or 1/0 AWG aluminum
    c.   6 AWG copper or 4 AWG aluminum
    d.   8 AWG copper or 6 AWG aluminum

3. The connection of the grounding electrode conductor is permitted to be made at the transformer, or the ____ disconnecting means or overcurrent device or any point between.
    a.   first
    b.   second
    c.   any
    d.   line-side

4. The grounding electrode for a separately derived system is required to be located as near as practicable to, and preferably in, the ____ as the separately derived system.
    a.   same enclosure
    b.   same cabinet
    c.   same area
    d.   same pull box

5. A grounding electrode conductor for a separately derived system is required to be connected to the nearest available, effectively grounded ____ in the structure.
    a.   underground metal gas pipe
    b.   structural metal member
    c.   service equipment enclosure
    d.   metal raceway

6. Where the grounding electrode for a separately derived system, as in question No. 5, is not available, the grounding electrode is required to be ____.
    a.   the nearest available effectively grounded metal enclosure
    b.   the nearest available effectively grounded metallic gas pipe
    c.   an effectively grounded metal water pipe within 5 ft. of entry if available.
    d.   the nearest available effectively grounded plate electrode

7. Where installed in or at the building, the same ____ is required to be used for the grounding of all ac systems.
    a.   grounding electrode
    b.   grounded raceway
    c.   electrode enclosure
    d.   bonding jumper

8. The ____ grounding conductors of both the primary and the separately derived system are permitted to be obtained by using the conductor enclosures, where permitted by *Code*.
    a.   isolated
    b.   identified
    c.   system
    d.   equipment

9. If the neutral and all phase conductors are switched through transfer switching associated with an on-site generator, then the generator ____ required to be grounded as a separately derived system.
    a.   is
    b.   is not
    c.   may be
    d.   cannot be

10. Where a generator is a separately derived system, the ____ bonding jumper must be installed, either at the generator or at the first disconnecting means or overcurrent device.
    a.   equipment
    b.   neutral
    c.   service
    d.   10 AWG copper

11. Where the generator is not a separately derived system, the ____ bonding jumper must be removed from the generator and the neutral must not be grounded. In this case, the system is grounded by its solid connection to the neutral of the premises wiring system.
    a.   10 AWG
    b.   equipment
    c.   service
    d.   neutral

12. The equipment bonding jumper installed between the separately derived system enclosure and the first overcurrent device enclosure is required to be sized based on _____.
    a.   Section 250.30(A)(5), 250.28
    b.   Table 250.122
    c.   Section 250.122
    d.   Section 220.22

13. The minimum size copper grounded conductor required for a separately derived system where the derived phase conductors are 300 kcmil copper conductors (1 per phase) and there is very little neutral load on the system is_____
    a.   6 AWG
    b.   1/0 AWG
    c.   2 AWG
    d.   2/0 AWG

Key Notes

CHAPTER THIRTEEN

Grounding at More Than One Building

Section 250.32 provides requirements for grounding of electrical systems and equipment at buildings or structures that are supplied from a common service. Included are rules for grounding electrodes, for grounded systems, ungrounded systems, and grounding at buildings where the disconnecting means are located in a separate building or structure on the same premises.

OBJECTIVES

After studying this chapter, the reader will be able to understand

- General requirements for grounding at more than one building.

- Sizing of grounding conductors.

- Bonding grounding electrodes together.

- Disconnecting means for separate buildings.

- Objectionable currents.

- Requirements for mobile homes, recreational vehicles, and agricultural buildings.

Definition

"**Grounding Electrode Conductor:** The conductor used to connect the grounding electrode(s) to the equipment grounding conductor, to the grounded conductor, or to both, at the service, at each building or structure where supplied from a common service, or at the source of a separately derived system."[1]

As indicated in the definition, this conductor includes both the grounding electrode conductor used at services and separately derived systems as well as the conductor used to connect the equipment or the grounded circuit conductor to a grounding electrode(s) at a second or additional building or structure.

Grounding Electrode Required

The general rule in 250.32(A) is that at each building or structure served from a common service by one or more feeders or branch circuits, a grounding electrode meeting the requirements of Article 250 Part III must be connected in a manner specified in 250.32(B) or (C). Where there are no existing grounding electrodes at the building or structure, the grounding electrode(s) required in Part III of Article 250 must be installed.

Grounding electrodes specified in 250.52 must be installed or used as required in 250.32(A) include:

(A) Electrodes Permitted for Grounding.
(1) Metal Underground Water Pipe.

(2) Metal Frame of the Building or Structure.
(3) Concrete-Encased Electrode.

"Where none of these electrodes are available at the building or structure, one or more of the following electrodes specified in 250.52(A)(4) through (A)(7) shall be installed and used."[2]

(4) Ground ring.
(5) Rod and Pipe Electrodes.
(6) Plate Electrodes.
(7) Other Local Metal Underground Systems or Structures.

(B) Electrodes Not Permitted for Grounding. The following shall not be used as grounding electrodes:

(1) Metal underground gas piping systems
(2) Aluminum electrodes"[3]

The exception to 250.32(A) indicates that a *"grounding electrode at separate buildings or structures shall not be required where only one branch circuit supplies the building or structure and the branch circuit includes an equipment grounding conductor for grounding the conductive non-current-carrying parts of all equipment."*[4] See 250.53 and chapter six for additional information on installing grounding electrodes.

Grounding Requirements for Grounded Systems

Two general methods are provided in 250.32(B) for grounding electrical systems and/or equipment at additional buildings or structures on the

Figure 13-1. Insulated Neutral Second Building 250.32(B)(1)

premises. Buildings or structures supplied by a feeder(s) or more than one branch circuit that contains an equipment grounding conductor must comply with the rules in 250.32(B)(1). Buildings or structures supplied by a feeder(s) or more than one branch circuit that does not contain an equipment grounding conductor and meet the additional requirements must comply with the rules in 250.32(B)(2).

Grounding with an Equipment Grounding Conductor

In this method, an equipment grounding conductor is run with the feeder(s) or branch circuit(s),

Figure 13-2. Insulated Neutral at Second Building

along with the ungrounded and grounded conductors to the additional building or structure. "Any installed grounded conductor shall not be connected to the equipment grounding conductor or to the grounding electrode(s)."[5] This method is similar, or identical, to installing a feeder to a feeder panel in the same building or service where the service is located (see figures 13-1 and 13-2).

The equipment grounding conductor, that is run from the building or structure where the service is located, is connected to a terminal bar on the supply side of or inside the building or structure disconnecting means where it is grounded by connecting the grounding electrode to the equipment grounding terminal bar. Equipment grounding conductors of the type included in 250.118 are acceptable and include wires as well as some conduits or other raceways.

The number of conductors for various systems that must be installed is summarized in Table 13-1.

Equipment Grounding Conductor Installed			
System	Grounded	Ungrounded	Equip.Ground
120V	1	1	1
120/240V	1	2	1
208/120V	1	3	1
480/277V	1	3	1

Table 13-1.

Figure 13-3. Grounding at Separate Buildings 250.32(B)(2)

The grounded (often a neutral) conductor, installed as a part of the feeder or branch circuit from the building or structure where the service is located to the second or additional building or structure, must be an insulated conductor [see 310.2(A)].

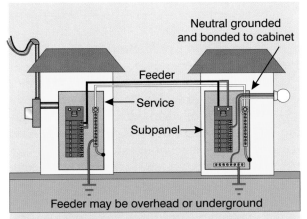

This method permitted only where
1. An equipment grounding conductor is not run with the supply to the building or structure
2. There are no continuous metallic paths bonded to the grounding system in both buildings or structures involved
3. Ground fault protection of equipment has not been installed on the common ac service

Figure 13-4. Grounding Neutral at Second Building

By using this method, the grounded (often a neutral) conductor(s) must be connected to the terminal bar for the grounded conductors, and the equipment grounding conductor(s) must be connected to the equipment grounding terminal bar where the grounding (electrode) is connected inside the building or structure disconnecting means enclosure.

The equipment grounding conductor that is run to the additional building or structure must be sized from Table 250.122 based on the rating of the overcurrent device at the beginning of the feeder or branch circuit.[6]

Grounding without an Equipment Grounding Conductor

Section 250.32(B)(2) permits the grounded circuit conductor (often a neutral) to be grounded again at the disconnecting means for the building or structure only where all the following conditions are complied with, where:

"(1) an equipment grounding conductor is not run with the supply to the building or structure,

(2) there are no continuous metallic paths bonded to the grounding system in both buildings or structures involved, and

(3) ground-fault protection of equipment has not been installed on the common ac service...."[7]

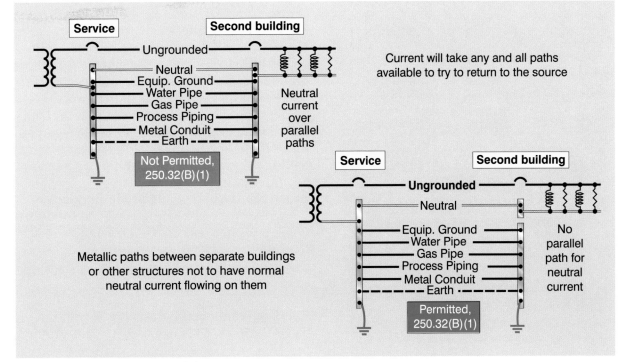

Figure 13-5. Objectionable Currents – Paths

In this case, the electrical system at the additional building or structure is treated like a service for grounding purposes, although the building or structure is supplied from a service in another building or structure by a feeder or branch circuit. The grounded circuit conductor is bonded to the disconnecting means enclosure for the building or structure. The grounding electrode connection to the grounded (often a neutral) conductor must be made on the supply side of or within the building or structure disconnecting means.

In this method, both the grounded circuit conductor (usually a neutral) and the non-current-carrying equipment are connected to a grounding electrode (system) at the additional building or structure. The number and type of conductors that must be taken from the building or structure where the service are located to the second building or structure is summarized in the following Table 13-2 (see figures 13-3 and 13-4).

Both the grounded (usually a neutral) conductor(s) and the equipment grounding conductor(s) are permitted to be connected to the terminal bar where the grounding electrode conductor connects to the grounded conductor.

Equipment Grounding Not Conductor Installed			
System	Grounded	Ungrounded	Equip.Ground
120V	1	1	0
120/240V	1	2	0
208/120V	1	3	0
480/277V	1	3	0

Table 13-2.

Equipment grounding conductor sized based on Table 250.122

Feeder

Service

Subpanel→

Grounding electrode conductor at both buildings sized based on Table 250.66

Figure 13-6. Insulated Neutral at Second Building

Note that the grounded conductor from the building or structure where the service is located to the second or additional building or structure must generally be an insulated conductor [see 310.2(A)]. Section 338.10(B)(2) permits the bare conductor of Type SE cable to be used only as an equipment grounding conductor for feeders or branch circuits and not as a neutral between buildings.

Size of Grounded Conductor
Where an equipment grounding conductor is not run from the service to the additional building or structure, the grounded circuit conductor must be sized no smaller than an equipment grounding conductor from Table 250.122. In this case, the grounded circuit conductor serves three purposes: first, to permit line-to-neutral loads to be utilized; second, to carry unbalanced loads back to the source; and, third, as an equipment grounding conductor. As such, it must be sized for the calculated load according to 220.22 and not smaller than an equipment grounding conductor from Table 250.122.

Grounding Electrode Conductor
A grounding electrode conductor, sized in accordance with 250.66, is used to connect the grounded circuit conductor or equipment grounding conductor and equipment to the grounding electrode system that exists or is installed at the additional building or structure. As covered above, where no grounding electrodes exist at the additional building or structure, one or more grounding electrodes must be installed in compliance with Part III of Article 250.[8]

This connection must be made inside the building or structure disconnecting means to the equipment grounding terminal bar or to the terminal bar for the grounded conductor as appropriate.

Sizing the Grounding Electrode Conductor
"The size of the grounding electrode conductor to the grounding electrode(s) shall not be smaller than given in 250.66, based on the largest ungrounded supply conductor. The installation shall comply with Part III of this article."[9]

Bonding Grounding Electrodes Together
Section 250.58 does not require that grounding electrodes at separate buildings or structures be bonded back to the service. A study

of the system will show that there is not really two separate grounding electrode systems. The grounding electrode system for the electrical system at the individual buildings or structures is connected together either by the grounded conductor or by the equipment grounding conductor. This is illustrated in figures 13-1 and 13-3.

Figure 13-7. Ungrounded System

Ungrounded Systems

"Where an ac system is connected to a grounding electrode in or at a building as specified in 250.24 and 250.32, the same electrode shall be used to ground conductor enclosures and equipment in or on that building."[10] A grounding electrode system in compliance with Part III of Article 250 must be installed at the building or structure where none exists (see figure 13-7).

The exception to 250.32(A) excludes buildings or structures supplied by only one branch circuit that includes an equipment grounding conductor from the requirement to install a grounding electrode system at the building or structure.

Disconnecting Means Located in Separate Building or Structure on the Same Premises

Special rules have been provided for large capacity, multibuilding industrial installations under single management in 250.32(D). These occupancies have trained and qualified personnel and have established procedures for safe switching of electrical feeders or circuits. As a result, the disconnecting means are permitted to be at other

Figure 13-8. Disconnecting Means Remote from Outbuildings

locations on the premises rather than at the building served. Often, the switching is managed by automatic or manual means from a control room or station.

The special rules for grounding the electrical system at these separate buildings or structures are as follows:

(1) The connection of the grounded circuit conductor to the grounding electrode at a separate building or structure shall not be made.

> **The basic requirement in 550.32(A) is that mobile home service equipment be located remote from the structure but within sight from and not more than 9.0 m (30 ft) from the mobile home.**

(2) An equipment grounding conductor for grounding any non-current-carrying equipment, interior metal piping systems, and building or structural metal frames is run with the circuit conductors to a separate building or structure and bonding to existing grounding electrode(s) required in Part III of this article, or, where there are no existing electrodes, the grounding electrode(s) required in Part III of this article shall be installed where a separate building or structure is supplied by more than one branch circuit.

(3) Bonding the equipment grounding conductor to the grounding electrode at a separate building or structure shall be made in a junction box, panelboard, or similar enclosure located immediately inside or outside the separate building or structure.[11]

Mobile Homes and Recreational Vehicles

The basic requirement in 550.32(A) is that mobile home service equipment be located remote from the structure but within sight from and not more than 9.0 m (30 ft) from the mobile home. There,

the grounded (usually a neutral) conductor is grounded to a grounding electrode. Feeders consisting of four insulated and color-coded conductors, one of which is an equipment grounding conductor, must be run to the mobile home distribution panelboard [see 550.33(A)].

"The service equipment shall be permitted to be located elsewhere on the premises, provided that a disconnecting means suitable for service equipment is located in sight from and not more than 9.0 m (30 ft) from the exterior wall of the mobile home it serves."[12] This exception permits the two options for grounding to be used as described earlier in this chapter.

A three-conductor feeder is permitted to be run from the service located remote from the mobile home disconnecting means where the neutral is grounded, and a feeder consisting of four insulated conductors in compliance with 550.33(A) are run from the mobile home disconnecting means to the mobile home. This permits the mobile home service equipment to be located at a common location, such as a separate laundry building as is required by some serving electric utilities. The grounding of the grounded circuit conductor at the disconnecting means must comply with 250.32(B) as covered above.

Section 550.32(B) permits service equipment to be installed directly on the manufactured home (not mobile home — see the definitions in 550.2) under four conditions.

(1) The manufacturer shall include in its writ-

Remote service

Disconnecting means <=30 ft.
Grounding per 250.32

3- or 4-Wire Feeder

4-Wire Feeder

Figure 13-9. Service Equipment Located Remote From Mobile/Manufactured Home

ten installation instructions information indicating that the home shall be secure in place by an anchoring system or installed on and secured to a permanent foundation.

(2) The installation of the service equipment shall comply with Article 230.

(3) Means shall be provided for the connection of a grounding electrode conductor to the service equipment and routing it outside the structure.

(4) Bonding and grounding of the service shall be in accordance with Article 250.[13]

Where this concept is chosen, the service is grounded to the grounding electrode at the manufactured home, and no service disconnecting means remote from the manufactured home is required.

For additional information, see Part 3280, Manufactured Home Construction and Safety Standards, of the federal Department of Housing and Urban Development for requirements related to manufactured homes.

Agricultural Buildings

The grounding of agricultural buildings is covered in Article 547. Sections 547.5(F), 547.9(A)(6), 547.9(B)(1), 547.9(C), and 547.10 all include grounding and bonding requirements that amend or add to the general requirements of Article 250. Where a building or structure houses livestock, an insulated or covered copper conductor must be installed where the equipment grounding conductor is run underground from one building or structure to another. See chapter sixteen for additional information on this subject.

Two Buildings in One Structure

Figure 13-10 shows two buildings in one structure that are separated by a fire wall. The two buildings

> Where separate services supply a building and are required to be connected to a grounding electrode, the same grounding electrode shall be used.

may be served by a single service or by multiple services in accordance with 230.2(A), (B), (C) or (D).

As previously discussed, 250.58 requires that all grounding electrodes at a building be bonded together. "Where separate services supply a building and are required to be connected to a grounding electrode, the same grounding electrode shall be used. Two or more grounding electrodes that are effectively bonded together shall be considered as a single grounding electrode system in this sense."[14] The connecting conductor must be properly sized according to 250.66.

Since the fire wall creates two buildings, bonding of the electrodes together is desirable but not required by *Code*. The electrodes can be bonded together by connecting them together by a properly sized bonding jumper.

---

[1] NFPA 70, *National Electrical Code* 2002, Article 100, (National Fire Protection Association, Quincy, MA, 2002), p. 70-36.

[2] NFPA 70, 250.50, p. 104.

[3] NFPA 70, 250.52, p. 105.

[4] NFPA 70, 250.32(A), p. 102.

[5] NFPA 70, 250.32(B)(1), p. 103.

[6] NFPA 70, 250.32(B)(1), p. 103.

[7] NFPA 70, 250.32(B)(2), p. 103.

[8] NFPA 70, 250.32(E), p. 103.

[9] NFPA 70, 250.32(E), p. 103.

[10] NFPA 70, 250.58, p. 70-106.

[11] NFPA 70, 250.32(D), p. 70-103.

[12] NFPA 70, 550.32(A), p. 70-448.

[13] NFPA 70, 550.32(B), p. 70-448.

[14] NFPA 70-250.58, p. 106.

Fire wall for building separation

Bonding grounding electrodes together permitted but not required. 250.58

Figure 13-10. Two Buildings in One Structure

Chapter Thirteen: The questions included here were developed using material included in this chapter. The answers can be found by reviewing the text. It is also important that students make use of the 2002 *NEC*, where many of the answers can be found.

1. A conductor used to connect the grounding electrode(s) to the equipment grounding conductor, to the grounded conductor, or to both, at the service, at each building or structure where supplied from a common service, or at the source of a separately derived system of a wiring system is defined as a ____.
   a.   main bonding jumper
   b.   grounding electrode conductor
   c.   grounded conductor
   d.   neutral conductor

2. A grounding electrode is required at a separate building or structure under which of the following conditions?
   a.   one grounded system feeder is installed
   b.   one ungrounded system feeder is installed
   c.   more than one circuit is installed
   d.   all of the above

3. Which of the following methods is NOT acceptable for grounding the electrical system at separate buildings or structures?
   a.   grounding the grounded system conductor
   b.   grounding the equipment grounding conductor
   c.   grounding the ungrounded conductor

4. Where a grounding electrode(s) is not available at the second building or structure, ___.
   a.   one or more must be installed.
   b.   installing one or more is optional.
   c.   an equipment ground must be installed
   d.   a grounded conductor must be installed

5. An accepted method for grounding at separate buildings recognizes the installation of an ____ where it is run with the feeder or circuit ungrounded (hot) conductors.
   a.   approved conductor
   b.   identified bonding jumper
   c.   equipment grounding conductor
   d.   acceptable conductor

6. All of the following are acceptable as a grounding electrode at a separate building or structure EXCEPT
   a.   metric designator 16 (trade size $^1/_2$) conduit
   b.   16 mm ($^5/_8$ in.) iron or steel not less than 8 feet long
   c.   metric designator 21 ($^3/_4$-in.) galvanized conduit
   d.   plate electrode

7. Where an equipment grounding conductor is run with the feeder or circuit from a first building to a separate building, the size is based upon the ampere rating of the overcurrent protective device ____ the feeder or circuit.
   a.   on the load side of
   b.   in the transformer of
   c.   on the line side of
   d.   in the service disconnect of

8. A grounding electrode conductor is used to connect the ____ circuit conductor, or equipment grounding conductor and equipment, to the grounding electrode system at the separate building(s) or structure(s).
   a.   feeder
   b.   main bonding
   c.   grounded
   d.   branch

9. Under certain circumstances, the grounded circuit conductor ____ to be used for grounding equipment on the line (supply) side of the disconnecting means for separate buildings.
   a.   is permitted
   b.   is not permitted
   c.   is permitted by special permission
   d.   is required

10. Currents that introduce noise or data errors in electronic equipment, such as data processing equipment, are ____ considered objectionable currents.

    a.   not to be
    b.   considered to be
    c.   always to be
    d.   sometimes

11. Where the additional building or structure houses livestock, an insulated or covered ____ conductor must be installed where run underground.

    a.   bonding
    b.   copper clad
    c.   aluminum
    d.   copper

Ground-Fault Circuit Interrupters

OBJECTIVES

After studying this chapter, the reader will be able to understand

- Requirements for replacement of ungrounded receptacles.
- Requirements for locations of GFCI-protected receptacles.
- Ground-fault circuit-interrupter principles of operation.
- Ratings of GFCI devices.
- Markings for GFCI devices.
- GFCI application and consideration.

The Underwriters Laboratories (UL) requirement for Class A ground-fault circuit interrupters (GFCIs) is that tripping shall occur when the continuous 60-hertz differential current exceeds 6 mA but shall not trip at less than 4 milliamperes (5 mA ± 1 mA). Some people contend that 5 mA is too low and should be increased to 10 mA or higher.

Several eminent investigators, including C.F. Dalziel, F.P. Kouwenhoven, O.R. Langworthy and others, have prepared papers on the dangers of electric shock hazards. They define *let-go current* as the maximum current at which a person is able to release a conductor by commanding those muscles directly stimulated by the shock. Currents over the let-go levels are said to freeze the victim to the circuit.[1]

C.F. Dalziel's paper, titled "Electric Shock Hazard," published in *the Institute of Electrical and Electronics Engineers (IEEE) Spectrum*, Vol. 9, February 1972, summarizes the studies that estimate shock currents based on the effective impedance of the body under various conditions. According to Dalziel, the reasonably safe electric current for normal healthy adults is the let-go current from which 99.5 percent of the population can extricate themselves from the circuit by releasing the conductor. On page 44, Dalziel states, "[s]o far, it has been impossible to obtain reliable [let-go] values for children; they just cry at the higher values." However, the IIT Research Institute report cites a value for children evidently based on engineering judgment by Dalziel and others. The following summarizes the let-go currents for 99.5 percent of the population:

1. for children   4.5 mA;
2. for women   6.0 mA;
3. for men   9.0 mA.

Based on information provided in the report by IIT Research Institute, 4.5 milliamperes for children may be the appropriate GFCI trip level relative to the 6 mA and 9 mA thresholds for men and women, but the rationale for this selection is not very evident in Dalziel's or other publications.

Based on Dalziel's data, UL's estimates of the percentage of the population that would be protected against inability to let-go at various current levels is shown in the following table.

For general purposes, a let-go limit for GFCIs higher than 6 mA appears inappropriate because too large a fraction of the population would be left unprotected. UL has decided to continue to designate 5 mA (± 1 mA) as the limit based on tolerated reaction and physiological effects. Moreover, UL retains the 5 mA limit because the 4.5 mA level is only an estimate that is not based on data and because the 5 mA limit has withstood the test of time with no evidence of being

inadequate and has appeared in a number of standards for many years.[2]

The first sensation of electricity can be felt by most people at currents considerably less than 0.5 mA, 60 Hz (frequency in Hertz). Current near 0.5 mA may produce an involuntary startled reaction such as to cause a person to drop a skillet of hot grease or cause a workman to fall from a ladder. As the current increases, involuntary muscular contractions increase, accompanied by current-generated heat.[3]

Higher currents of longer duration than one second can cause the heart to go into ventricular fibrillation, considered to be the most dangerous effect of electric shock. Once fibrillation begins, it practically never stops spontaneously. Death is almost certain within minutes. The rhyth-

Percentage of the Population Estimated to Be Protected Against Inability to Let Go for Several Levels of Shock Current

Level of Shock Current	6mA(rms)	10mA(rms)	20mA (rms)	30mA(rms)
Men	100%	98.5%	7.5%	0%
Women	99.5%	60%	0%	0%
Children*	92.5%	7.5%	0%	0%

*half of let-go threshold for men

Table 14-1.

Implied Safe Voltage[4] Based on Several Published Values of Body Resistance and Selected Body Current Safety Criteria as Published by Dalziel

Criterion	Body Resistance 300 ohms	Body Resistance 500 ohms	Body Resistance 1500 ohms	Body Resistance 3000 ohms
"let-go" 4.5 mA for children	1.35 volts	2.25 volts	6.75 volts	13.5 volts
"let-go" 9 mA for adult males	2.7 volts	4.5 volts	13.5 volts	27 volts
fibrillation at 23 mA 5 sec. pulse of 60-Hz current for 18 kg children	6.9 volts	11.5 volts	34.5 volts	69 volts
fibrillation at 52 mA 5 sec. pulse of 60-Hz current for 50 kg adult	15.6 volts	26 volts	78 volts	156 volts

Table 14-2.

mic contractions of the heart become disordered, its pumping action stops, and the pulse soon ceases altogether. Fibrillation in adults can occur at 52 mA and in children at 23 mA. According to V.G. Biegelmeier, the onset of fibrillation in a 50-kilogram (110-pound) adult occurs within the range of 50 mA to 200 mA when the duration of the shock exceeds two seconds.[4] Table 14-2 shows

The trend toward grounding equipment and appliances has been gradual, characterized as deliberate but cautious.

implied safe voltages based on these values.

The table's 3000-Ω (ohm) body resistance column, for example, indicates that 156 volts would result in a shock current of 52 milliamperes.

Extensive work on this subject has been reported in "Effects of Current on Human Beings and Livestock" IEC 479-1 by the International Electrotechnical Commission.[5] Several charts and graphs with background material are provided. Measurements were made on 50 living persons at a touch voltage of 15 volts and on 100 persons at a touch voltage of 25 volts. The total body impedance of one living person was measured with touch voltages of up to 200 volts. Measurements were also made on a large number of corpses.

Three-Wire Grounded System vs. GFCI

For many years, grounding was emphasized as the primary means for the protection of the electrical system, equipment and personnel from fires and injury, as well as for operating and maintenance advantages. The grounded neutral as a protective element was recognized more and more with each succeeding edition of the *NEC*. The *Code* emphasizes effective grounding in order to ensure that wiring faults to ground became *overcurrents* as required to activate the overcurrent device. The wiring and equipment were designed to be protected, and the thought that the grounded system also provided adequate protection for people became ingrained and accepted.

The trend toward grounding equipment and appliances has been gradual, characterized as deliberate but cautious. The belief that ground-

ing provided adequate protection against electric shock and fire hazards became so ingrained that consumers generally have not recognized its limitations, and they find it difficult to accept other more effective means of protection from electric shock and electric fire hazards.

Ground faults occur when electrical current flowing in a circuit returns:
1. through the equipment grounding wire,
2. through conductive material other than the electrical system ground (metal, water, plumbing pipes, and so forth),
3. through a person, or
4. through a combination of these ground return paths.

When a person becomes a pathway for electricity flowing to ground, the person will incur an electrical shock, may be seriously injured or may be electrocuted depending on:
1. amount of current or amperes flowing,
2. duration or time the current flows,
3. size of the person, and
4. pathway the current follows through the body.

On the other hand, arcing ground faults can occur just about anywhere on an electrical system, resulting in a fire. In either case, the ground faults may be of too low a magnitude to activate the overcurrent device and interrupt the flow of electricity.

A person may become a pathway to ground in one of two ways: in series contact or in parallel contact. In series contact, the person is the only current path to ground and the ground wire is not involved in the circuit. There are many ways in which contact can occur. One example of series contact was that of an infant that stuck a hairpin into a receptacle slot while sitting on a floor heating vent. The infant was electrocuted. Another case involved a man operating a metal-encased electric drill that incorporated a 3-wire grounded cord. He used a *two-to-three* wire adapter but did not connect the adapter pigtail to the wall plate grounding screw. Inadvertently, the pigtail touched the plug blade, thus energizing the drill case and electrocuting the man. In this incident, the equipment ground contributed to the electrocution by providing a current path to energize the drill case. In both cases, the 3-wire grounded system was totally ineffective since the ground wire was not involved in the current path, and the current through the body was not large enough to trip the overcurrent protective device.

In parallel contact, the victim becomes a path to ground in parallel with the equipment grounding conductor. One scenario of parallel contact occurs when the metal case of an appliance becomes charged electrically by some internal fault condition resulting in current leakage to ground via the grounding conductor. The leakage to ground, however, may not be sufficient to activate the branch-circuit overcurrent protective device. A person who touches the charged case and, at the same time, contacts a grounded surface such as a water faucet or pipe will be subjected to an electric shock. In such parallel contact situations, the effectiveness of the grounding conductor in preventing electrocution of the victim is dependent on several variables, including the following:

1. whether or not the ground-fault current reaches the instantaneous trip level of the overcurrent protective device (which is relatively high—over 15 amperes),
2. how fast the overcurrent device reacts,
3. the voltage level from faulted enclosure to ground, and
4. the impedance of the grounding path (composed of connections, contacts and the ground wire).

An effective grounding system depends upon the integrity of many series connections which must be properly made and maintained. The higher the resistance or impedance of the ground path, the less effectual will be the protection provided by the 3-wire grounded system. Higher ground impedances may be due to long wire lengths, small wire sizes, loose and/or corroded ground wire clamps and connections and other causes. A detailed analysis of the relationship of the impedance ratio of the line circuit to ground circuit to shock current levels is described in the previously cited Consumer Product Safety Commission paper titled "Three-Wire Grounding Systems vs. GFCI" and the cited IITRI report. [6]

A similar study by Mr. R.H. Lee of Dupont Company corroborates IITRI's analysis of circuit impedances.[7] His paper deals with the hazard vs. safeguard of a 3-phase grounded power distribution system.

An assessment of the effectiveness of the 3-wire grounded system for providing protection against electric shock hazards was conducted by Mr. A.W. Smoot of Underwriters Laboratories.[8] He analyzed 164 fatal electric shock accidents occurring over a 3 ¾ year period in and around homes. His study indicated the limitations of the 3-wire system and was submitted to the *National Electrical Code* Committee to support proposed amendments to the 1971 *NEC*.

Principles of Operation

"The GFCI sensing system continuously monitors the current balance in the ungrounded 'hot' conductor and the neutral conductor. If the current in the neutral wire becomes less than the current in the 'hot' wire, a ground fault could exist. A portion of the current returns to the supply by some path other than the neutral wire. With a current imbalance as low as 4-6 mA, the GFCI will interrupt the circuit and this will be

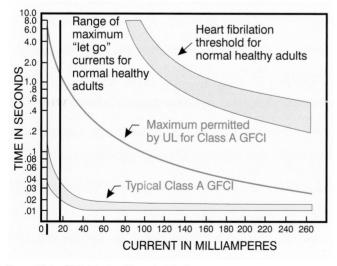

Figure 14-1. GFCI Tripping Characteristics[3]

*Receptacle-type GFCI devices switch both the phase and neutral.

Figure 14-2. Ground-Fault Circuit-Interrupter Principles of Operation shown by a trip or 'off' indicator on the device (see figure 14-2).

"The GFCI does not limit the magnitude of the ground-fault current. It limits the time that a current of given magnitude can flow. The trip level-time combinations are based on physiological data established for avoiding injury to normal healthy persons. These trip level-time combinations may be too high for persons with heart problems, such as those wearing a pacemaker or under treatment in health care facilities." [9]

The principle of operation of ground-fault circuit interrupters provides a significant advancement in safety for both equipment that is grounded by an equipment grounding conductor as well as for equipment that is ungrounded. Since the GFCI detects an imbalance of current in both the supply and return paths, it protects equipment supplied by both a 2-wire circuit and a 2-wire with ground circuit. This is the reason some *NEC* sections will allow a grounding-type receptacle to be supplied on a 2-wire circuit that has GFCI protection.

Several kitchen appliances, as well as portable heaters, are manufactured with a 2-wire supply cord and, thus, are ungrounded. A significant advancement in safety is realized where these appliances are supplied from receptacle outlets that have GFCI protection.

GFCI Required for Replacement Receptacles

The general requirement in 406.3(A) is that "receptacles installed on 15- and 20-ampere branch circuits shall be of the grounding type. Grounding-type receptacles shall be installed only on circuits of the voltage class and current for which they are rated, except as provided in Tables 210.21(B)(2) and (B)(3)." [10]

"Ground-fault circuit-interrupter protected receptacles shall be provided where replacements are made at receptacle outlets that are required to be so protected elsewhere in this *Code*." [11]

This requires that installers be aware of the rules that call for GFCI-protected receptacles in areas covered by 210.8 for dwellings, nondwellings, as well as many other locations in the *Code* for other facilities. This includes on 15- and 20-ampere, 125-volt receptacles installed in dwelling unit kitchens, bathrooms, garages, outdoor receptacles, and so forth. Also, receptacles that are replaced in locations where GFCI protection is required applies to dwellings and to other than dwelling units such as commercial repair garages, elevators and elevator pits, health care facilities and bathrooms in commercial and industrial facilities, as provided in 210.8.

Ground-fault circuit interrupters of various types, the most common of which are the circuit-breaker and receptacle-type, are permitted to be used to provide the protection required unless the particular type of device is specified. For example, 620.85 requires that the ground-fault circuit-interrupter protection "in pits, on elevator car tops, and in escalator and moving walk wellways shall be of the ground-fault circuit-interrupter type." [12]

Specific marking requirements exist where nongrounding-type receptacles are replaced with grounding-type receptacles where an equipment ground does not exist at the outlet. These receptacles must be marked "No Equipment Ground" so the user will be informed that even though the receptacle has an equipment-ground slot, an equipment grounding conductor is not connected to the device. An equipment grounding conductor is not permitted to be connected from the GFCI-type receptacle to any outlet supplied from the GFCI receptacle.

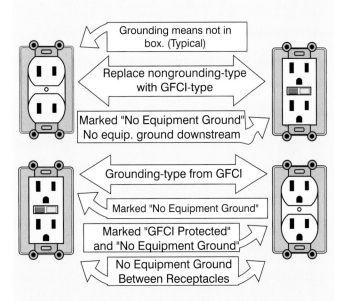

Grounding means not in box. (Typical)

Replace nongrounding-type with GFCI-type

Marked "No Equipment Ground" No equip. ground downstream

Grounding-type from GFCI

Marked "No Equipment Ground"

Marked "GFCI Protected" and "No Equipment Ground"

No Equipment Ground Between Receptacles

GFCI protection may be circuit breaker or receptacle type.

Figure 14-3. Replacement of Receptacles Where Now Required To Be GFCI Protected

Location	NEC Location
Accessory building dwellings	210.8(A)(2)
Agricultural buildings	547.5(G)
Basements, dwelling units	210.8(A)(5)
Bathrooms, dwellings	210.8(A)(1)
Bathrooms, nondwellings	210.8(B)(1)
Dwelling Unit Boathouses	210.8(A)(8)
Carnivals, Circuses, etc.	525.23
Construction sites	527.6
Crawl spaces, dwellings	210.8(A)(4)
Electric Vehicle Charging	625.22
Elevators, etc.	620.85
Existing	406.3(D)(2)
Feeders	215.9
Fountains	680.51
Garages, dwellings	210.8(A)(2)
Garages, commercial	511.12
Health care facilities	517.20(A), 517.21
Heating, AC, Ref. Outlet 210.8(B)(2)	210.63,
Hydromassage bathtubs	680.71
Kitchens in dwellings	210.8(A)(6)
Kitchens in nondwellings	210.8(B)(3)
Marinas and Boatyards	555.19(B)
Mobile Homes	550.13(B)
	550.32(E)
Outdoors, dwellings	210.8(A)(3)
Park Trailers	552.41(C)
Rooftops, nondwellings	210.8(B)(2)
RV Parks	551.41(C)
Signs, Outdoor Portable	600.10(C)(2)
Signs, with fountains	680.57(B)
Spas and Hot Tubs	680.43(A)(2) & (3)
	680.44
	680.41(A)(2)
Swimming Pools	680.5
	680.6
	680.22(A)(1) & (5)
	680.22(B)(4)
	680.23(A)(3)
	680.32
Therapeutic Pools and Tubs	680.62(A)
Wet bar sinks, dwellings	210.8(A)(7)

Table 14-3. GFCI Requirement Locations.

Where grounding-type receptacles are installed on a circuit that does not include an equipment grounding conductor but are protected on the line side by a ground-fault circuit-interrupter device, the receptacle(s) must be marked "GFCI Protected" and "No Equipment Ground." This will inform the user that even though the receptacle has a grounding terminal, it is not in fact grounded but is provided with GFCI protection. The grounding-type receptacle is protected by the GFCI device, but an equipment grounding conductor is not present as may be required for some equipment such as computers. An equipment grounding conductor is not permitted to be connected between the grounding-type receptacles as this would present incorrect information to the users. As explained earlier, the GFCI device will protect the user from line-to-ground faults even though an equipment grounding conductor is not connected to the devices.

Section 250.130(C) provides a method for installing an equipment grounding conductor for receptacles that are being replaced at a location where an equipment grounding conductor is not present in the box.

"The equipment grounding conductor of a grounding-type receptacle or a branch-circuit extension shall be permitted to be connected to any of the following:

(1) Any accessible point on the grounding electrode system as described in 250.50
(2) Any accessible point on the grounding electrode conductor
(3) The equipment grounding terminal bar within the enclosure where the branch circuit for the receptacle or branch circuit originates
(4) For grounded systems, the grounded service conductor within the service equipment enclosure.
(5) For ungrounded systems, the grounding terminal bar within the service equipment enclosure"[13]

Receptacles in Bathrooms

Section 210.8(A) requires that "all 125-volt, single-phase, 15- and 20-ampere receptacles installed in dwelling unit bathrooms have ground-fault circuit-interrupter protection for personnel."

The word bathroom is defined in Article 100 as "An area including a basin with one or more of the following: a toilet, a tub, or a shower." It is important to recognize that the definition does not use the word *room* but rather *an area* to describe the location where these receptacles must be provided with GFCI protection. This makes it clear that GFCI protection of these receptacles is required even though the receptacle(s) may be in separate rooms but are in the same area. For example, a wall with a door or doorway separates a basin from the area where the toilet and tub or shower is installed. No distance measurement is given beyond which GFCI protection of these receptacles is not required. This

definition applies to all bathrooms regardless of type of facility, due to the requirement in 210.8(B) for other than dwelling units (see figure 14-4 and 14-5 for a graphic illustration of bathroom areas).

Receptacles in Dwelling Unit Garages and Accessory Buildings

Section 210.8(A)(2) generally requires GFCI protection for all 125-volt, single-phase, 15- and 20-ampere receptacles installed in garages and in accessory buildings that have a floor located at or

An area with a basin and toilet.

An area with a basin and toilet.

An area with a basin and tub.

An area with a basin, toilet and tub.

Figure 14-4. GFCI Protection Requirements in Bathrooms

An area with a basin and shower.

An area with a basin, toilet and a shower.

An area with a basin, toilet and tub and laundry area.

Figure 14-5. GFCI Protection Requirements in Bathrooms

1. Duplex receptacles require GFCI protection unless they supply two appliances located in dedicated space that are not easily moved from one place to another in normal use and are cord- and plug-connected per Sections 400.7(A)(6), (A)(7) or (a)(8).

2. Single receptacle requires GFCI protection unless it supplies a single dedicated appliance that is not easily moved in normal use and is cord- and plug-connected per Sections 400.7(A)(6), (A)(7) or (A)(8).

3. Receptacle for laundry equipment requires GFCI protection unless it is a single contact device.

4. Receptacle that is not readily accessible does not require GFCI protection. Typical for garage door opener. (Exception No. 1)

5. Additional receptacle required if other receptacles in garage are not readily accessible or are located for appliances that are within dedicated spaces.

Figure 14-6. GFCI – Garages and Accessory Buildings – 210.8(A)(2)

Garage and accessory buildings with floor at or below grade

below grade level, that are not intended as habitable rooms but are used for storage or work areas.

Exception No. 1 excludes those receptacles that are not *readily accessible*. Readily Accessible is defined in Article 100 as "Capable of being reached quickly for operation, renewal, or inspections without requiring those to whom ready access is requisite to climb over or remove obstacles or to resort to portable ladders, and so forth." No specific height above the garage floor is given as a qualifier. In most cases, receptacles that are located on the ceiling for garage door operators are not considered readily accessible.

Exception No. 2 to 210.8(A)(2) excludes from GFCI requirements *"A single receptacle or a duplex receptacle for two appliances located within dedicated space for each appliance that, in normal use, is not easily moved from one place to another and that is cord-and-plug connected in accordance with 400.7(A)(6), (A)(7), or (A)(8)."* This will clarify that duplex or multi-receptacles installed at outlets in a dwelling unit garage must have ground-fault circuit-interrupter protection for personnel, unless a duplex receptacle is used for the supply to two appliances that meets all the conditions.

Dwelling Unit Outdoor Receptacles

Section 210.8(A)(3) generally requires all 125-volt, single-phase, 15- and 20-ampere receptacles installed outdoors to have GFCI protection for personnel. This requirement applies to these receptacles that are installed outdoors, for any purpose and at any height above grade or platform. The receptacles do not have to be accessible from grade level before the requirement applies.

The Exception for 210.8(A)(3) excludes from

the GFCI requirement those *"Receptacles that are not readily accessible and are supplied by a dedicated branch circuit for electric snow-melting or deicing equipment shall be permitted to be installed in accordance with the applicable provisions of Article 426."* Section 426.28 requires ground-fault protection of equipment (not GFCI protection for personnel) for resistance heating elements used for electric snow-melting or deicing unless the cable is "mineral-insulated, metal-sheathed cable that is embedded in a noncombustible medium."

Dwelling Unit Crawl Space Receptacles

Section 210.8(A)(4) requires that receptacles that are of the 125-volt, single-phase, 15- and 20-ampere type and located in crawl spaces that are at or below grade level have GFCI protection for personnel.

These receptacles are often installed for servicing equipment such as heating, air-conditioning or refrigerating equipment that is located in crawl spaces.

Dwelling Unit Unfinished Basements

Section 210.8(A)(5) requires that receptacles that are of the 125-volt, single-phase, 15- and 20-ampere type and located in unfinished basements have GFCI protection for personnel.

"Unfinished Basements — for purposes of this section, unfinished basements are defined as portions or areas of the basement not intended as habitable rooms and limited to storage areas, work areas, and the like."[14] It is likely that build-

Figure 14-7. GFCIs – Dwelling Unit Outdoor Receptacles

Figure 14-8. GFCIs Dwelling Unit Crawl Space Receptacles

1. Duplex receptacles require GFCI protection unless they supply two appliances located in dedicated space that are not easily moved from one place to another in normal use and are cord- and plug-connected per Sections 400.7(A)(6), (A)(7) or (A)(8).

2. Single receptacle requires GFCI protection unless it supplies a single dedicated appliance that is not easily moved in normal use and is cord- and plug-connected per Sections 400.7(A)(6), (A)(7) or (A)(8).

3. Receptacle for laundry equipment requires GFCI protection unless it is a single contact device.

4. Receptacle that is not readily accessible does not require GFCI protection. Ex. No. 1. (Defined in Article 100)

5. Additional receptacle required if other receptacles in unfinished basement are not readily accessible or are located for appliances that are within dedicated spaces.

Unfinished basement

Figure 14-9. GFCIs Dwelling Unit Basements – 210.8(A)(5)

ing codes will be consulted for determining whether rooms in basements are permitted to be classified as habitable rooms.

Exception No. 1 excludes from the GFCI requirement those receptacles that are not *readily accessible*.

Exception No. 2 to 210.8(A)(5) excludes from GFCI requirements "*A single receptacle or a duplex receptacle for two appliances located within dedicated space for each appliance that, in normal use, is not easily moved from one place to another and that is cord-and-plug connected in accordance with 400.7(A)(6), (A)(7), or (A)(8).*" This clarifies that duplex or multi-receptacles installed at outlets in an unfinished basement must have ground-fault circuit-interrupter protection for personnel, unless a duplex receptacle is used for the supply to two appliances that meets all the conditions.

GFCI protection not required as receptacle does not serve counter-top surface

Receptacle outlets that serve countertop surfaces require GFCI protection for personnel.

Figure 14-10. GFCI Requirements for Dwelling Unit Kitchens

Kitchen Counter Receptacles

All 125-volt, single-phase, 15- and 20-ampere receptacles that serve countertop surfaces in kitchens are required by 210.8(A)(6) to have ground-fault circuit-interrupter protection for personnel.

This requirement does not apply to those receptacles that do not serve countertop surfaces but are installed for other purposes such as for a waste disposal, a trash compactor or a dishwasher. Also excluded from this requirement are wall-mounted receptacles that are installed for equip-

ment such as electric clocks or receptacles that are installed for appliances such as refrigerators where the receptacle is located behind the appliance and does not serve the counter surface.

Other Than Dwelling Units

Section 210.8(B)(3) requires all 125-volt, single-phase, 15- and 20-ampere receptacles installed in kitchens to have ground-fault circuit-interrupter protection for personnel. The one difference between the kitchen receptacle GFCI requirement

Single phase 125-volt 15- and 20-ampere receptacles located within kitchen locations in occupancies other than dwelling units require GFCI protection

Restaurant kitchen (Typical)

GFCI protection required

Figure 14-11. 210.8(B)(3) Other Than Dwelling Units

● GFCI protection required for receptacles located at boathouses

GFCI

Residential dwelling unit typical

Residential boathouse

Boat

Earth

GFCI required

Figure 14-12. 210.8(A)(8) Boathouses

in other than dwelling units is that the requirement applies to all the 125-volt 15- and 20-ampere receptacles in the kitchen, and not just the ones intended to serve the counter top surfaces as covered in 210.8(A)(6)

> The rule is the same regardless of whether the power is from a temporary construction service, from an existing permanent power source or from one or more receptacles installed as a part of the permanent wiring of the building or structure.

Dwelling Unit Wet Bar Sinks

Section 210.8(A)(7) requires that all 125-volt, single-phase, 15- and 20-ampere receptacles installed to serve the countertop surfaces and are located within 1.8 m (6 ft) of the outside edge of a wet bar sink .

"Wet bar sinks — where the receptacles are installed to serve the countertop surfaces and are located within 1.8 m (6 ft) of the outside edge of the wet bar sink."[15]

Rooftop Heating, Air-Conditioning and Refrigeration Equipment Receptacle Outlet

Receptacles installed on rooftops must have ground-fault circuit-interrupter protection for personnel in accordance with 210.8(A)(3) and (B)(2). Note that a receptacle outlet for servicing this equipment is required for both dwellings and other than dwelling units. The receptacle must have GFCI protection if installed at these locations. This receptacle is for the purpose of servicing heating, air conditioning and refrigeration equipment and is required to be located on the same level as the equipment, within 7.5 m (25 ft) of the equipment, and must not be connected to the load side of the equipment disconnecting means.

For other than dwelling units, 210.8(B)(2) Exception excludes from the GFCI requirement those *"Receptacles that are not readily accessible and are supplied from a dedicated branch circuit for electric snow-melting or deicing equipment shall be permitted to be installed in accordance with the applicable provisions of Article 426."*

Section 426.28 requires "Ground-fault protection of equipment shall be provided for fixed outdoor electric deicing and snow-melting equipment, except for equipment that employs mineral-insulated, metal-sheathed cable embedded in a noncombustible medium."

GFCI Requirements on Construction Sites

Section 527.6 contains requirements for ground-fault protection for personnel using temporary power at buildings, structures or on equipment or similar activities while performing any of the following tasks: "construction, remodeling, maintenance, repair, or demolition of buildings, structures, equipment, or similar activities."

Ground-fault circuit-interrupter protection is generally required for all 125-volt, single-phase, 15-, 20-, and 30-ampere receptacles used for these purposes. The rule is the same regardless of whether the power is from a temporary construction service, from an existing permanent power source or from one or more receptacles installed

15- or 20-ampere 125 V receptacle - GFCI protection required
Maximum 25 ft. from HACR equipment and on same level

Not connected on load side of disconnect

Figure 14-13. HACR Rooftop Receptacles – All

125-volt, 15-, 20- or 30-ampere receptacle outlets

GFCI protection required if used for temporary power during construction, remodeling, maintenance, repair, or demolition of buildings, structures, equipment or similar activities.

For receptacles not covered above, the "Assured Equipment Grounding Conductor Program" as provided in Section 527.6(B) is permitted to be used.

Figure 14-14. GFCI Requirements for Construction Sites

as a part of the permanent wiring of the building or structure.

A cord set incorporating listed ground-fault circuit-interrupter protection is permitted to be used by personnel in lieu of permanently or temporarily installed receptacle outlets. This may be the most practical means of complying with this requirement, especially for maintenance and repair activities.

The Exception to 527.6 states that *"In industrial establishments only, where conditions of maintenance and supervision ensure that only qualified personnel are involved, an assured equipment grounding conductor program as specified in 527.6(B)(2) shall be permitted for only those receptacle outlets used to supply equipment that would create a greater hazard if power was interrupted or having a design that is not compatible with GFCI protection."*

For receptacles like 250-volt, 15- or 20-ampere or 125/250-volt, 30- or 50-ampere, the assured equipment grounding conductor program, as provided for in Section 527.6(B)(2), is permitted only if the requirements and conditions of 527.6(A)(1) Exception are complied with. This program requires a written procedure enforced by one or more individuals at the construction site. Detailed requirements for testing of the equipment grounding conductor continuity on a specific time

schedule are provided. A record of this testing program must be maintained and provided to the authority having jurisdiction.

[1] A. Albert Biss, Ground-Fault Circuit-Interrupter (GFCI) Technical Report, (Washington, D.C.: U.S. Consumer Product Safety Commission, February 28, 1992).

[2] Walter Skuggevig, 5-Milliampere Trip Level for GFCIs, (Northbrook, IL: Underwriters Laboratories, March 1989).

[3] IITRI report, the voltage and current values are assumed to be "root-mean-square" RMS values.

[4] Electric Shock Prevention, p. 25.

[5] Technical Report IEC 479-1, Effects of current on human beings and livestock. International Electrotechnical Commission. Available from Global Engineering Documents, 15 Inverness Way East, Englewood, Colorado 80112.

[6] A. Albert Biss, Three-Wire Grounding Systems vs. GFCI, (Washington, D.C.: U.S. Consumer Product Safety Commission, December 1985).

[7] Ralph H. Lee, "Electrical Grounding: Safe or Hazardous?," *Chemical Engineering*, July 28, 1969.

[8] A.W. Smoot, Analysis of Accidents, (Northbrook, IL: Underwriters Laboratories, 1971).

[9] Application Guide for Ground-Fault Circuit-Interrupters, Standards Publication No. 280-1990, National Electrical Manufacturers Association, 1300 North 17th Street, Suite 1847, Rosslyn, VA 22209. Reprinted with permission.

[10] NFPA 70, *National Electrical Code* 2002, 406.3(A), (National Fire Protection Association, Quincy, MA, 2002), p. 70-248.

[11] NFPA 70, 406.3(D)(2), p. 70-248.

[12] NFPA 70, 620.85, p. 70-497.

[13] NFPA 70, 250.130(C), p. 70-115.

[14] NFPA 70, 210.8(A)(5), p. 70-53.

[15] NFPA 70, 210.8(A)(7), p. 70-53.

Chapter Fourteen: The questions included here were developed using material included in this chapter. The answers can be found by reviewing the text. It is also important that students make use of the 2002 *NEC*, where many of the answers can be found.

1. Where existing 125-volt, single-phase, 15- and 20-ampere rated receptacles are replaced, or where new receptacles are installed in locations that are required by the *Code* to be GFCI protected, they must be ___.
   a. of the grounding type
   b. GFCI protected
   c. on dedicated circuit
   d. polarized

2. Circuit breaker or receptacle-type ground-fault circuit interrupters ____ be used to protect the receptacle, where nongrounding types of receptacles are replaced with grounding type receptacles where an equipment ground does not exist in the receptacle enclosure.
   a. are permitted to
   b. cannot
   c. must always
   d. by special permission can

3. Where used for a single cord- and plug-connected appliance that occupies a dedicated space in a residential garage, GFCI protection is not required where a ____ receptacle is installed and dedicated for use by that appliance.
   a. duplex
   b. single
   c. isolated
   d. 15-ampere rated

4. Portions or areas of a dwelling unit basement that are not intended to be used as habitable rooms and limited to storage areas, work areas and the like, defines:
   a. a basement
   b. a finished basement
   c. an unfinished basement
   d. a storage space

5. Where installed in dwelling unit kitchens, all 125-volt, single-phase, 15- and 20-ampere receptacles that are installed ____ must be ground-fault circuit interrupter protected.
   a. within 6 feet of the sink
   b. on the dishwasher circuit
   c. to serve counter surfaces
   d. to serve refrigeration equipment

6. The distance of the required receptacle on a rooftop, other than at one- and two-family dwellings, must be located to be not further than ____ feet from the heating, air-conditioning and refrigeration equipment.
   a. 10
   b. 15
   c. 25
   d. 50

7. Where installed in ____ of commercial, industrial and all other nondwelling-type occupancies, all 125-volt, single-phase, 15- and 20-ampere receptacles must be ground-fault circuit interrupter protected.
   a. kitchens
   b. basements
   c. closets
   d. bathrooms

8. Where installed on roofs of which of the followings locations, all 125-volt, single-phase, 15- and 20-ampere receptacles must be ground-fault circuit interrupter protected.
   a. one family dwelling units
   b. two family dwelling units
   c. dwelling unit garages
   d. all the above

9. With a current imbalance as low as _____ milli-amperes, a GFCI will interrupt the circuit and this will be shown by a trip or "off" indicator on the device.
    a.   3
    b.   4
    c.   5
    d.   6

10. A GFCI of the Class _____ type is designed so that it will automatically trip if the neutral conductor is grounded on the load side of the sensor.

    a.   A
    b.   B
    c.   C
    d.   D

11. The basic GFCI consists of a ground-fault detecting means that is coupled with a circuit-interrupting means. This detecting means measures the ground-fault current as the difference between outgoing and incoming load current on the _____ conductors.
    a.   protected
    b.   neutral
    c.   bonded
    d.   service

12. A condition which could result in injury or electrocution to a person, caused by the flow of electrical current through the body, is defined as:
    a.   a high resistance ground
    b.   an electrical shock hazard
    c.   a short circuit
    d.   ground fault

13. To determine compliance with specified mechanical and electrical life requirements, _____ is made.
    a.   an endurance test
    b.   a ground fault test
    c.   an operational test
    d.   an overcurrent test

14. GFCI protection is required for receptacles in which of the following locations.
    a.   family room receptacles in a dwelling
    b.   commercial office receptacles
    c.   restaurant kitchens
    d.   basements of commercial office buildings

Equipment Ground-Fault Protection Systems

OBJECTIVES

After studying this chapter, the reader will be able to understand

- Requirements for ground-fault protection of equipment and various systems.

- Requirements for feeder ground-fault protection.

- Ground faults in an ungrounded system.

- Testing of equipment ground-fault protection.

S ection 230.95 requires "ground-fault protection of all equipment shall be provided for solidly grounded wye electrical services of more than 150 volts to ground but not exceeding 600 volts phase-to-phase for each service disconnect rated 1000 amperes or more." As can be seen, this protection is required for nominal 480/277-volt, three-phase, 4-wire wye connected systems. These provisions do not apply to fire pumps or continuous industrial processes where a nonorderly shutdown would introduce additional or increased hazards.

"The maximum setting of the ground-fault protection shall be 1200 amperes, and the maximum time delay shall be one second for ground-fault currents equal to or greater than 3000 amperes."[1]

The magnitude of current which will flow in the event of a line-to-ground fault on a grounded system is usually determined by the reactance of the grounded apparatus, the reactance of the lines or cables leading to the fault and the resistance

Solidly grounded. > 150V to Gnd.
< = 600V Phase to Phase.
> = 1000A service disconnect rating

Figure 15-1. Ground-Fault Protection Required

Photo 15-1.

Photo 15-2.

and reactance of the ground return path including any intentional grounding resistance or reactance. For interconnected systems, calculation of the current may be rather complicated. For simpler cases, an approximation of the available fault current may be obtained.

This equipment protection is required due to a history of destructive burndowns of electrical equipment operating at this voltage level. An electric arc, which generates a tremendous amount of heat, is readily maintained at 277 volts to ground. Since it is not a solid ground (bolted) connection, and the current is limited by the resistance of the arc, often, not enough current will flow in the circuit to allow the overcurrent device ahead of the fault to open. A great deal of damage is done to the electrical equipment while the arc is burning (see photos 15-1 and 15-2).

There are basically two types of equipment ground-fault protection systems in use, although these systems may have different names in the industry. The most common types are zero-sequence which may have more than one form and residual type, which is sometimes referred to as ground-strap, or ground-return type. Both types are designed to protect downstream equipment from destructive arcing burndowns. Note that this equipment will not protect equipment or the system on its line side from line-to-ground faults as this fault current will not pass through the ground-fault equipment.

Definition

"Ground-Fault Protection of Equipment. A system intended to provide protection of equipment from damaging line-to-ground fault currents by

Neutral bar
Current trans
Main bonding jumper
Equipment ground bar

Relay

Fault in Equipment

Service Enclosure

Figure 15-2. Residual EGFP System

Photo 15-3. Bolted Pressure Switch (Fuse) EGFP

operating to cause a disconnecting means to open all ungrounded conductors of the faulted circuit. This protection is provided at current levels less than those required to protect conductors from damage through the operation of a supply circuit overcurrent device."[2]

Residual-Type System

The residual or ground-strap type of equipment ground-fault protection system consists of a current transformer, control relay, and, usually, a shunt trip circuit breaker. The main bonding jumper passes through the current transformer as shown in figure 15-2.

The advantage of this system is that it is probably the least expensive, but it is limited to application at the main service. It cannot be used downstream of the main bonding jumper.

With the connections as shown, the maximum ground-fault current will flow through the main bonding jumper and will, thus, be recognized by the current transformer. Although there are many parallel bypass circuits for ground-fault current to flow, the relative impedance of the main circuit is so low compared to all the parallel bypass circuits that a minimum of 90 percent of the total ground-fault current will flow in the main bonding jumper and, thus, be seen by the current transformer.

If a current transformer of properly-selected value is placed in the main bonding jumper and has its secondary feeding a magnetic current relay, the relay may be adjusted to close a contact at any desired value of fault. The current transformer value is selected on the basis of the smallest current transformer whose short-time rating will not exceed the maximum value of ground-fault current that may be anticipated.

Examination of the diagram in figure 15-2, describing the ground-fault protective device, will show that the main ground-fault current path is from the transformer to the service, through the feeder to the fault, back to the service over the equipment grounding conductor, through the main bonding jumper where it returns to the transformer by the neutral conductor. The parallel circuit from the grounding electrode at the service to the grounding electrode is both a high-resistance and a high-reactance circuit. As a result, little current will return through the earth.

Some small ground-fault current will be carried by the building steel if it is in the circuit. It is preferable to adequately ground the building steel using the same grounding electrode as used for the service. This prevents the building steel from rising to a dangerous potential above ground and often serves as a grounding electrode as required by 250.52(A)(2). Even though the building steel represents a parallel path for fault current to flow, most of the current will flow back to the transformer through the neutral because of the lower reactance of that path as compared to the reactance of the other parallel paths.

Zero-Sequence Transformer-Type System

Probably the most popular and common type of equipment ground-fault protection system is the zero-sequence type. This system is shown in figures 15-3 and 15-4. It consists of a current transformer that is placed around all the conductors of the circuit, including the grounded (neutral) conductor, a control relay, and, usually, a shunt trip circuit breaker. As shown, the current transformer must be placed downstream from the main bonding jumper. The equipment grounding conductor does not pass through the window.

The general difference between the types of zero sequence equipment shown in figures 15-3 and 15-4 is the current transformers are generally built into the circuit breaker shown in figure 15-4 with an external current transformer through which the neutral passes that may be field installed. Often, these current transformers are used by the circuit breaker as a part of its internal operational system. The equipment ground-fault protection system shown in figure 15-4 is sometimes referred to as a residual type as the ground-fault relay or system sums up the current through all four coils and considers any excess current as residual. Generally, the current transformer through which all conductors of the circuit pass as shown in figure 15-3 is installed by the manufacturer of the switchboard. Cables for feeders are field installed.

Figure 15-3. Zero Sequence EGFP System

Figure 15-4. Zero Sequence EGFP System

Under normal operation, the vector summation of all phase and neutral (if used) currents approaches zero. This is due to the canceling effect of the currents in the conductors. A sensor (differential current transformer) around the phase and neutral conductors detects the current imbalance when a ground fault occurs. This is due to the fault current passing outside the current transformer window(s), which sets up the imbalance which is detected by the system.

The output of the sensor is proportional to the magnitude of the ground-fault current. This output is fed to a ground-fault relay. The relays are usually field adjustable. Pickup ranges of from 4 to 1200 amperes are common. Time delay settings are available from instantaneous (1.5 cycles) to 36 cycles delay. When the ground-fault current exceeds a pre-selected level, the relay will activate the circuit-interrupting device, which usually is a shunt trip circuit breaker, to open the circuit.

Ground-Fault Protection Not Required

Exception No. 1 of 230.95 requires that *"The ground-fault protection provisions of this section shall not apply to a service disconnect for a continuous industrial process where a nonorderly shutdown will introduce additional or increased hazards."*

Likewise, ground-fault protection of equipment is not required for fire pump services (see 230.95 Exception No. 2). For an emergency system, 700.26 excludes the alternate source of power from the requirement to have "ground-fault protection of equipment with automatic disconnection means." This applies to fire pumps that are classified as an emergency system in accordance with 700.1. Indication of a ground fault of the emergency source is required to comply with 700.7(D).

"To indicate a ground fault in solidly grounded wye emergency systems of more than 150 volts to ground and circuit-protective devices rated 1000 amperes or more. The sensor for the ground-fault signal devices shall be located at, or ahead of, the main system disconnecting means for the emergency source, and the maximum setting of the signal devices shall be for a ground-fault current of 1200 amperes. Instructions on the course of action to be taken in event of indicated ground fault shall be located at or near the sensor location."[3]

A similar exclusion from the ground-fault protection of equipment requirement is provided for legally required standby systems in 701.17. No such exclusion is provided for optional standby systems installed in accordance with Article 702.

System Coordination

System coordination with circuit-breaker systems is easily accomplished by use of ground-fault protective devices. These ground-fault protective devices may be cascaded where the economics of the design warrant doing so. The time-delay settings may become lower and lower as the device gets further from the service, so that the last ground-fault protective device in the system may even be set for instantaneous trip. Similar coor-

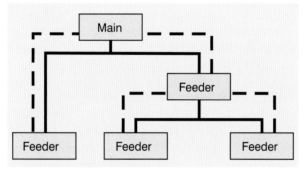

Figure 15-5. System Coordination

dination also can be obtained by locking out the relay or relays upstream from the sensor that sees the fault first.

Since it cannot be know where a fault-to-ground or enclosure may originate in the system, and since it is desirable to coordinate the protection and clear the fault nearest to its point of origin, it will be necessary to delay the action of tripping the main overcurrent device. The magnetic current relay is thus used to energize a time delay relay to provide proper coordination and maximum continuity of service consistent with safety.

When a fault-to-ground or enclosure occurs, the possibility should be considered that the fault may reach a high value. If so, the normal main overcurrent device should be set to a time value to allow it to function and clear the fault through that means. Even though the ground-fault protective relay should be set at as low a value as possible, enough time delay should be set to permit an overcurrent device near the point of origin of the fault to clear first. This ground-fault protective device is an adjunct to, but does not replace, the main overcurrent device. It gives protection below the rating of the main overcurrent device. Being affected only by ground-fault currents, it does not interfere with

Additional level of equipment ground-fault protection not permitted:

1. on the load side of an essential electrical system transfer switch, or
2. Between the on-site generating unit(s) and the essential electrical system transfer switch(es), or
3. On systems not solidly grounded wye systems more than 150 volts to ground and less than 600 volts phase-to-phase.

Figure 15-6. Second Level EGFP - Health Care Facilities

normal operation of the main overcurrent device. In effect then, the ground-fault protective device will keep all ground faults to a limited time, selected on the basis of experience, to give maximum safety without undue interruption of service.

Second Level Required in Health Care Facilities

Special rules for ground-fault protection of electrical systems apply to health care facilities. Sec-

- Where EGFP is installed on service, additional level of EGFP required downstream
- Limited application of essential electrical system
- Fully selective so feeder opens and not service
- Minimum six cycle separation required
- Tested when first installed

Figure 15-7. Health Care Facilities EGFP Systems

tion 517.17 requires that "Where ground-fault protection is provided for operation of the service disconnecting means or feeder disconnecting means as specified by 230.95 or 215.10, an additional step of ground-fault protection shall be provided in the next level of feeder disconnecting means downstream toward the load."

This rule clarifies that regardless of whether or not the health care facility is supplied by a service or feeder, where equipment ground-fault protection is provided at the service or feeder disconnecting means, a second level is required downstream.

The additional levels of ground-fault protection shall not be installed as follows:

(1) On the load side of an essential electrical system transfer switch

(2) Between the on-site generating unit(s) described in 517.35(B) and the essential system transfer switch(es)

(3) On electrical systems that are not solidly grounded wye systems with greater than 150 volts to ground, but not exceeding 600 volts phase-to-phase[4]

This feeder protection is required to be 100 percent selective so that where a ground fault

- 100% Selectivity required
- Minimum six cycle minimum separation between service and feeder GFP tripping bands
- Testing of GFP required in accordance with 517.17(C)

480/277 Volts
1000 Amps and greater

Figure 15-8. 517.17 Ground-Fault Protection

occurs downstream from the feeder overcurrent device, only the feeder overcurrent device will open and the service or feeder main will remain closed. This, of course, is to prevent the blackout of a facility caused when a main would open for a fault that can be isolated to a single feeder. To achieve this coordination, a six cycle, or

> The risk of destructive arcing burndowns is the same regardless of whether the system is a service or feeder.

greater, separation between the service and feeder tripping bands is required. Each level of the ground-fault protection system must be tested when first installed.

Feeder Ground-Fault Protection

Figure 15-9 illustrates a situation where equipment ground-fault protection is required for feeders. Section 215.10 requires this protection for feeders that are of the same voltage and current rating as similar services.

This covers the situation where the electrical system is delivered at more than 600 volts and has the service disconnect at that level. A transformer then reduces the voltage of the feeder to the level where equipment ground-fault protection is required. The risk of destructive arcing

Over 600 volt service
Secondary is:

Solidly grounded
More than 150 volts to ground
Less than 600 volts phase-to-phase

Figure 15-9. Equipment Ground Fault Protection for Feeders

burndowns is the same regardless of whether the system is a service or feeder.

Exception No. 1: The provisions of this section shall not apply to a disconnecting means for a continuous industrial process where a nonorderly shutdown will introduce additional or increased hazards. [Editor's note: see Article 685 – Integrated Electrical Systems for more information on these systems.]

Exception No. 2: The provisions of this section shall not apply to fire pumps. [Editor's note: see Article 695 and NFPA-20 for information on electrical requirements for fire pumps.]

Exception No. 3: The provisions of this section shall not apply if ground-fault protection of equipment is provided on the supply side of the feeder. [5] [Editor's note: However, care must be exercised to be certain that each new system created by a separately derived system has the protection required.]

Ground-Fault Protection of Building or Structure Main Disconnecting Means

For each building or structure disconnecting means meeting the criteria, 240.13 requires ground-fault protection of equipment to be installed in accordance with 230.95. This requirement applies for systems of identical voltage and amperage rating to that for services.

"The provisions of this section shall not apply to the disconnecting means for the following:

(1) Continuous industrial processes where a nonorderly shutdown will introduce additional or increased hazards [*Editor's note:* see Article 685 for additional information.]

(2) Installations where ground-fault protection is provided by other requirements for services or feeders

(3) Fire pumps installed in accordance with Article 695"[6]

It is most important that an overall engineered system approach is taken where more than one building or structure is on the premises that are fed from a common service. This is especially important where transfer switches and alternate power sources are installed in the premises wiring system. In most cases, four-

Solidly grounded wye, > 150V to ground
<= 600V Ø to Ø, Rated >= 1000 A

XYZ Main Plant

XYZ Plant 2

Equipment ground-fault protection required at building
or structure disconnecting means except for:
1. continuous industrial process ...
2. fire pumps.
3. where disconnecting means protected by
   service or feeder ground fault protection.

Figure 15-10. Equipment Ground Fault Protection for Feeders

pole transfer switches are installed so as to allow the system to function properly without grounding the neutral downstream from the service. It is critical to ground-fault protection systems functionality that the neutral be completely isolated from any grounding connections downstream from the service.

Combination Systems
An improvement in the protection of distribution systems of 600 volts or less may be obtained by using a ground-fault protective device as a supplement to high-interrupting capacity current-limiting overcurrent protection. Fault currents in low ranges may be recognized and virtually all fault currents can be held to a limited time duration with the resultant increase to the safety of such systems.

If circuit breakers are used as the disconnecting means and overcurrent protection on such secondary distribution systems and a ground-fault reaches a high enough value, the circuit breaker will open and clear all three phases, thus preventing a feedback into the ground fault from the two unaffected phases. When a circuit breaker is used, a ground-fault protection device also may be applied to automatically open the circuit breaker in a reasonable and limited time if a ground fault should develop and not be cleared by some overcurrent device near the point of fault.

If a manually-operated, fused load-break switch is used as the disconnecting means and overcurrent protection, and if a ground fault reaches a high enough value to clear one fuse, an alarm should be provided to indicate a blown fuse so all three phases may be cleared manually.

If an electrically-operated, fused load-break switch is used, it can be provided with an automatic blown fuse indicator which will immediately open the switch, and thus, all three phases, if any one or more fuses should blow. An electrically operated switch also permits the application of a ground-fault protective device which will automatically open the switch in a reasonable and limited time if a ground fault should develop and not be cleared by an overcurrent device near the point of the fault.

A vital part of the test is to remove the neutral disconnect link in the distribution equipment and test the neutral with a continuity tester or *megger* to be certain that it is clear from any grounding connections downstream from the main bonding jumper at the service.

For systems with large available short-circuit current, and which require the best degree of protection, the electrically-operated switch or circuit breaker can provide full protection where it is equipped with high-interrupting capacity current-limiting fuses, a ground-fault protective device and an automatic blown-fuse indicator.

Testing of EGFP System
Section 230.95(C) requires that ground-fault protection systems be performance-tested when first installed on site to ensure that they will operate properly. Experience has shown that the major-

ity of these systems do not operate properly, or at all, when first installed. This is often due to improper field wiring of the system.

The test must be performed in full compliance with the manufacturer's written instructions. These instructions must be furnished with the equipment. A vital part of the test is to remove the neutral disconnect link in the distribution equipment and test the neutral with a continuity tester or *megger* to be certain that it is clear from any grounding connections downstream from main bonding jumper at the service. Accidental or intentional grounding of the neutral past the service can desensitize ground-fault protection systems and render them ineffective.

"A written record of this test shall be made and shall be available to the authority having jurisdiction."[7] Safety requires that the equipment be properly installed, as does the threat of lawsuit where negligence of the installer can be proven.

Optional Protection

Figure 15-11 illustrates an installation where equipment ground-fault protection is not required even though the system voltage is the same as that where it is required. The difference is the rating of the disconnecting means is not 1000 amperes or greater, even though the combined rating exceeds 1000 amperes. For example, a service with a single 2000 ampere main requires equipment ground-fault protection at the stated voltage and ampere rating. If a service with four, 600-ampere overcurrent devices is installed rather than the single main, equipment ground-fault protection is not required. Obviously, the system can benefit from equipment ground-fault protection even though it is not required.

In the case where a 900-ampere fuse is installed in a 1200-ampere switch, the requirements are different. Equipment ground-fault protection is required since the fuse holder will accept a 1000-ampere or larger fuse. Where adjustable circuit breakers are used, the rating, for purposes of 230.95, is the actual overcurrent device installed in the circuit breaker or the maximum rating that it can be adjusted to.

[1] NFPA 70, *National Electrical Code* 2002, 230.95(A), (National Fire Protection Association, Quincy, MA, 2002), p. 70-83.
[2] NFPA 70, Article 100, p. 70-36.
[3] NFPA 70, 700.7(D), p. 70-556.
[4] NFPA 70, 517.17(A), p. 70-403.
[5] NFPA 70, 215.10, p. 70-61.
[6] NFPA 70, 240.13, p. 70-88.
[7] NFPA 70, 230.95(C), p. 70-84.

250 Amp

600 Amp

800 Amp

400 Amp

600 Amp

800 Amp

Solidly grounded wye, > 150V to ground <= 600V Ø to Ø

Figure 15-11. Optional Equipment Ground-Fault Protection (EGFP)

Chapter Fifteen: The questions included here were developed using material included in this chapter. The answers can be found by reviewing the text. It is also important that students make use of the 2002 *NEC*, where many of the answers can be found.

1. A system intended to provide protection of equipment from damaging line-to-ground fault currents, by operating to cause a disconnecting means to open all ungrounded conductors of the faulted circuit, is defined as ____.
   a.   ground-fault circuit-interrupter
   b.   ground-fault protection of equipment
   c.   leakage current detector
   d.   saturable core reactor

2. The *NEC* requires ground-fault protection of all solidly-grounded wye electrical services of more than 150 volts to ground, but not exceeding 600 volts phase-to-phase, for each service disconnect rated ____ amperes or more.
   a.   800
   b.   1000
   c.   600
   d.   400

3. Ground-fault protection of equipment is not applicable to ____ or continuous industrial processes where a nonorderly shutdown would introduce additional or increased hazards.
   a.   electronically-actuated fuses
   b.   phase converters
   c.   fire pumps
   d.   industrial buildings

4. Where a service is protected by a system of EGFP, the maximum setting of the EGFP devices is __ amps and the maximum time delay is 1 second for ground-fault currents equal to or greater than ____.
   a.   1300 - 2000
   b.   1200 - 3000
   c.   1400 - 1500
   d.   1600 - 1000

5. For feeders rated 1000 amps or more in a solidly-grounded wye system with greater than 150 volts to ground, but not exceeding 600 volts phase-to-phase, ground-fault protection ____ required.
   a.   is
   b.   is not
   c.   sensors are
   d.   indication is

6. The most common types of equipment ground-fault protection equipment is:
   a.   three-phase, three-wire
   b.   residual and bypass relaying
   c.   zero-sequence and residual
   d.   isolation transformer and relay

7. One of the most critical elements of a ground-fault protection system is:
   a.   the neutral must be isolated downstream
   b.   a choke coil is installed downstream
   c.   capacitors are installed downstream
   d.   all the above

8. The equipment ground-fault protection system must be tested:
   a.   by pushing the "push to test" button
   b.   on an annual basis
   c.   within 30 days of installation
   d.   when first installed on site

9. A second level of equipment ground-fault protection is required
   a.   for fire pump installations
   b.   for health care facilities
   c.   for emergency systems
   d.   for optional standby systems

10. Equipment ground-fault protection is required for electrical systems of more than 150 volts to ground and less than 600 volts phase-to-phase:
   a.   for feeders of more than 800 amperes
   b.   at a building disconnecting means rated at 1200 amperes or more
   c.   for feeders rated more than 1000 amperes
   d.   for a continuous industrial process

Special Location Grounding and Bonding

Protection against static electricity is a requirement of a number of industries and establishments. However, the grounding of equipment is not necessarily a solution to static problems. Each static problem requires its own study and solution. Humidity plays an important part. The higher the humidity, the less the chances of a static discharge. In some industries, increasing the humidity in the area of a static discharge has been found very effective. One example is in the printing industry.

OBJECTIVES

After studying this chapter, the reader will be able to understand

- Requirements for static and lightning protection.
- Bonding in hazardous (classified) locations.
- Requirements for agricultural buildings.
- Requirements for health care facilities.
- Requirements for swimming pools, hot tubs and spas.
- Requirements for electric signs.

Chapter Sixteen – Special Location Grounding and Bonding

Lightning Protection Systems

Lightning protection is an important factor at outdoor substations and at locations where thunderstorms are prevalent. Lightning discharges usually consist of very large currents of extremely

> The lightning protection system ground terminals shall be bonded to the building or structure grounding electrode system.

short duration. Protection is accomplished by deliberately providing a path of low resistance to earth, compared to other paths. There is no guarantee that lightning will necessarily follow the lower resistance path that has been provided, but at least the low-resistance path will reduce the likelihood of damage.

The steel fence around outdoor substations, when properly grounded, offers good lightning protection for the equipment, especially if the fence is higher than the equipment within the station. Lightning rods (air terminals) should project at least one foot above any part of the structure and equipment, and the path to ground should be as direct as possible.

Each air terminal should have at least two connections to ground with air terminals placed on each side of the structure. Surge arresters also should be used on the power lines and the equipment. The lightning down conductor system should be connected to earth through driven electrodes (ground terminals), as well as being bonded to the station ground bus. Lightning conductors should not be smaller than 4 AWG, increasing in size as the primary voltage of the system increases.

See NFPA-780, *Standard for the Installation of Lightning Protection Systems* for additional information on this subject.

Bonding of Lightning Protection Systems

"The lightning protection system ground terminals shall be bonded to the building or structure grounding electrode system."[1] This section no longer requires that metallic parts of electrical wiring system be bonded to the lightning protection system conductors where there is less than six feet of separation. However, specific requirements for bonding the systems together are found in NFPA 780, *Standard for the Lightning Protection Systems*.

Hazardous (Classified) Locations Bonding

The *Code* places some special requirements for grounding and bonding in hazardous (classified) locations. These requirements can be found in 501.16 for Class I locations, 502.16 for Class II locations, and 503.16 for Class III locations. For Class

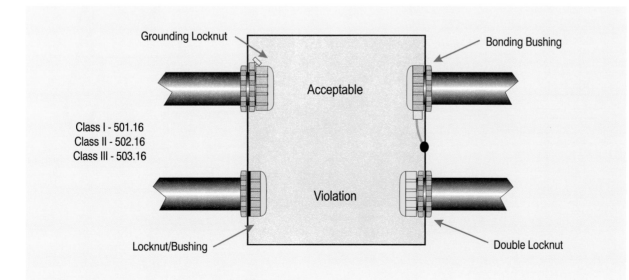

Figure 16-1. Bonding for Hazardous (Classified) Locations

249

Figure 16-2. Bonding at Hazardous (Classified) Locations

Figure 16-3. Bonding at Hazardous (Classified) Locations

I, Zones 0, 1 and 2 hazardous (classified) locations, see 505.25 which requires compliance with 501.16.

Grounding and bonding is essential for electrical safety in nonhazardous locations as well as in hazardous locations. In hazardous locations it is vital to have effective grounding and bonding to prevent an explosion. Under fault-current conditions when heavy currents are flowing through metal conduit, every connection point in the raceway system is a potential source of sparks and ignition. If there is an arcing fault to a metal enclosure in a hazardous (classified) location, the external surface temperature of the metal enclosure at the point of the arcing fault will start to rise to temperatures that could cause ignition of the flammable vapors or accumulations of combustible dust. Under these fault conditions it is essential that the overcurrent device operate as quickly as possible to prevent a hot spot on the enclosure or even arcs that may burn through the enclosure from igniting the atmosphere, dust, or fibers and flyings on the outside of the enclosure. It is extremely important that all threaded joints be made up wrenchtight to prevent sparking at those threaded joints. If joints are other than the threaded type, such as locknuts and bushings or double locknuts and bushings at boxes, enclosures, cabinets, and panelboards, it is essential that bonding be assured around those joints in the fault-current path to prevent sparking and assure a low-impedance path for the fault current.

Bonding requirements are found in Part V of Article 250. Section 250.90 requires that "bonding shall be provided where necessary to ensure electrical continuity and the capacity to conduct safely any fault current likely to be imposed." Section 250.100 also includes bonding requirements for hazardous locations and indicates that "Regardless of the voltage of the electrical system, the electrical continuity of non-current-carrying metal parts of equipment, raceways, and other enclosures in any hazardous (classified) location as defined in Article 500 shall be ensured by any of the methods specified for services in

It is extremely important that all threaded joints be made up wrenchtight to prevent sparking at those threaded joints.

250.92(B) that are approved for the wiring method used."[2]

By these special requirements, an effort is made to provide assured grounding and bonding to reduce the likelihood that a line-to-ground fault will cause arcing and sparking at connection points of metallic raceways and boxes or other enclosures. If such arcing or sparking were to occur in a hazardous (classified) location while a flammable gas is present in its explosive range, it is likely that the flammable atmosphere would be ignited.

Generally, these requirements are that locknuts on each side of the enclosure, or a locknut on the outside and a bushing on the inside, cannot be used for bonding. Bonding locknuts or bonding bushings with bonding jumpers be used to ensure the integrity of the bond and its capability of carrying the fault current that may be imposed, hopefully, without arcing or sparking at the connections.

The bonding means required in 501.16 are generally installed from the hazardous (classified) location to the service equipment, or point of grounding of a separately derived system that is the source of the circuit. "Such means of bonding shall apply to all intervening raceways, fittings, boxes, enclosures, and so forth between Class I locations and the point of grounding for service equipment or point of grounding of a separately derived

Service equipment or separately derived system

Bonding required in this building

Bonding requirements end here - if grounded conductor and grounding electrode conductor are connected together and branch-circuit protection is provided.

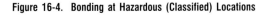

Figure 16-4. Bonding at Hazardous (Classified) Locations

system."[3] The goal is to ensure a substantial flame path for ground-fault currents and to facilitate fast operation of overcurrent protective devices supplying the circuit in the hazardous locations. Hazardous atmospheres can be ignited by hot temperatures on electrical enclosures. If a ground fault should occur inside an explosionproof enclosure, a hot spot at the point of fault on the enclosure could develop if the overcurrent device does not clear quickly. These more restrictive bonding rules in *NEC* Chapter 5 are in place for these reasons. The bonding in hazardous locations is applicable to all metal raceways and enclosures in the hazardous location, and all intervening metal raceways and enclosures of the circuit extending back to the applicable service or derived system. Since current and fault current seek the source, this method of bonding must be accomplished for the circuit to the source or service grounding point, where the main bonding jumper is installed.

The exception to 501.16 clarifies that "the specific bonding means shall only be required to the nearest point where the grounded circuit conductor and the grounding electrode are connected together on the line side of the building or structure disconnecting means as specified in 240.32(A), (B), and (C), provided the branch-circuit overcurrent protection is located on the load side of the disconnecting means."

Section 250.32 provides grounding requirements where more than one building or structure are on the same premises and are supplied from a common service. Where the grounded circuit conductor is not grounded at the building or structure, the rule in 501.16(A), 502.16(A) and 503.16(A) requires that the bonding extend from the hazardous location back to the service even if it is in another building. This requires that the feeder raceway system be bonded if it is metallic (see chapter thirteen for additional information on grounding electrical systems at additional buildings or structures on the premises).

Grounding in Hazardous Locations

"Where flexible metal conduit or liquidtight flexible metal conduit is used as permitted in 501.4(B) and is to be relied on to complete a sole equipment grounding path, it shall be installed with internal or external bonding jumpers in parallel with each conduit and complying with 250.102."[4] If installed outside the conduit, the bonding jumper is limited to 1.8 m (6 ft) in length per 250.102(E).

Exception: In Class I, Division 2 locations, the bonding jumper shall be permitted to be deleted where all the following conditions are met:

(a) Listed liquidtight flexible metal conduit 1.8 m (6 ft) or less in length, with fittings listed for grounding, is used.

(b) Overcurrent protection in the circuit is limited to 10 amperes or less.

(c) The load is not a power utilization load.[5]

See 502.16(B) for similar rules for Class II areas and 503.16(B) for Class III locations.

Agricultural Buildings

Sections 547.5(F) and 547.9 contain some specific requirements for grounding and bonding in agricultural buildings to which the article applies. The major concern is twofold. The first is for the integrity of the grounding path due to corrosive conditions that exist in these buildings. The second concern is for neutral-to-earth and stray voltage which, if excessive, cause behavior responses in livestock. These behavior responses in dairy cattle can lead to loss of production and health problems.

Section 547.9(C) requires that "Where livestock is housed, any portion of the equipment grounding conductor run underground to the building or structure shall be insulated or covered copper."[6]

Equipment Grounding

Where grounded equipment is installed in an agricultural building, a copper equipment grounding conductor must be installed to ground the equipment. This applies regardless of the type of wiring method employed. "If installed underground, the equipment grounding conductor shall be insulated or covered."[7]

"(L) Motor-Operated Water Pumps. Motor-operated water pumps, including the submersible type. [*Editor's note:* must be grounded.]

(M) Metal Well Casings. Where a submersible pump is used in a metal well casing, the well casing shall be bonded to the pump circuit equipment grounding conductor."[8]

One of the most important elements of a farm wiring system, especially where dairy cattle are involved, is to isolate the system neutral from the equipment grounding conductor and non-cur-

Feeder may be overhead or underground.
Equipment ground insulated or covered copper if underground. 547.9(C)

Figure 16-5. Grounding at Agricultural Buildings

- Generally, must comply with Article 250.
- Copper equipment grounding conductor required for all wiring methods.
- Bond metal well casing for submersible pump to pump circuit equipment grounding conductor.
- Ground motors per Article 430, Part XII.

Figure 16-6. Equipment Grounding in Agricultural Building

rent carrying metal parts of the electrical system at barns, milking parlors, and so forth. Separation of the neutral conductor will prevent voltage drop on the feeder neutral from becoming stray voltage in the building.

Isolation of the grounded conductor (usually the neutral) at an additional building or structure on the same premises as the electrical service is required by 250.32(B)(1). In the case of agricultural buildings, the conditions of 547.9(B) must be satisfied as well. Here two general methods are provided.

Grounding of electrical systems at agricultural

buildings is permitted to be performed in one of two ways in accordance with 547.9(B)(1) or (B)(2). The difference in the two methods permitted depends on whether the neutral is grounded at the site distribution point or at the agricultural building.

"Where the disconnecting means and overcurrent protection are located at the building or structures, the supply conductors shall be sized in accordance with Part IV of Article 220 and installed in accordance with the requirements of Part II of Article 225."[9] Article 220 gives the rules for calculating the load on conductors, feeders and services. Article 225 includes requirements for the disconnecting means at additional buildings and structures on the premises.

Where the disconnecting means and overcurrent protection is located at the agricultural building or structure, two methods for grounding are permitted. Where the second method is chosen, a number of conditions must be complied with.

The first method for grounding at the agricultural building where the neutral is grounded at the disconnecting means is covered in 547.9(B)(1).

(a) The grounded circuit conductor shall be permitted to be grounded to the building disconnecting means and to the grounding electrode system of that building or structure where all the requirements of 250.32(B)(2) are met.

Figure 16-7. Equipotential Bonding in Agricultural Buildings

(b) A separate equipment grounding conductor shall be run with the supply conductor's to the building(s) or structure(s)...[10] [Editor's note: see chapter thirteen for additional information on grounding at more than one building or structure supplied by a common electrical service.]

The second method for grounding at the agricultural building where the neutral is grounded at the disconnecting means is covered in 547.9(B)(1)(b). In this method, the neutral is not grounded at the agricultural building or structure. In addition to the requirement that an equipment grounding conductor be run with the supply conductors to the agricultural building or structure, the following conditions must be met:

(1) The equipment grounding conductor is the same size as the largest supply conductor, if of the same material, or is adjusted in size in accordance with the equivalent size columns of Table 250.122 if of different materials.

(2) The equipment grounding conductor is bonded to the grounded circuit conductor at the disconnecting means enclosure at the distribution point or at the source of a separately derived system.

(3) A grounding electrode system is provided in accordance with Part III of Article 250 and connected to the equipment grounding conductor at the building(s) or structure(s) disconnecting means.

(4) The grounded circuit conductor is not connected to a grounding electrode or to any equipment grounding conductor on the load side of the distribution point.[11]

As provided in 547.9(B)(2), "Where the disconnecting means and overcurrent protection for each set of feeder conductors are located at the distribution point, feeders to building(s) or structure(s) shall meet the requirements of 250.32 and Article 225, Parts I and II.

Stray (Tingle) Voltage

Voltage drop on a neutral conductor supplying a building or structure housing livestock can result in elevated levels of stray voltage if it is grounded at the building or structure. It is important to balance 120-volt loads to minimize neutral loads, operate motors at 240 volts wherever practical and size neutral conductors in service drops and feeders as large as practical.

Livestock behavioral responses, including production and health problems, can be caused by stray voltage (also referred to as *tingle-voltages*). These voltages appear between various portions of grounded or ungrounded metallic systems such as electrical equipment or piping systems and the earth or floor such that livestock can come between two different potentials. It is not uncommon to find voltage differences between adjacent concrete slabs or between a concrete slab and the adjacent earth.

Common causes of neutral-to-earth voltages (stray voltage if it is at livestock contact points) are currents on primary distribution systems, farm secondary neutral conductors and faulty wiring on the farm. Some electric utilities have installed primary-to-secondary neutral conductor isolators on their transformers in an attempt to solve the stray voltage problem. Neutral conductors that are too small for the load and length of run and loose or corroded splices and terminations will frequently elevate the neutral-to-earth voltage to a level that can adversely affect livestock.

Another common cause of stray voltage is ground faults where the fault current gets into the earth, concrete slabs or on metal equipment that may be contacted by livestock. Ground faults that can cause stray voltage may occur in water pumps, underground wires, sump pumps, manure pumps, electrically heated livestock watering fountains, electrically operated feeders and similar equipment. Proper grounding along with bonding this equipment together usually prevents the stray voltage from occurring even where there is a high-impedance fault. In addition, it is important to have a low-impedance ground-fault return path so overcurrent devices can clear ground faults.

Equipotential Plane

Section 547.10(A) requires that "Equipotential planes shall be installed in all concrete floor confinement areas of livestock buildings that contain metallic equipment that is accessible to animals and likely to become energized."

Also, "Outdoor confinement areas, such as feedlots, shall have equipotential planes installed around metallic equipment that is accessible to animals and likely to become energized. The equipotential plane shall encompass the area around the equipment where the animal stands while accessing the equipment."[12]

This is the best method of controlling the effects of stray voltages on livestock. This bonding does not correct or remove the faults that are causing the problem but keeps everything in the livestock area at the same potential. This, in essence, prevents the livestock from being aware of and being affected by the stray voltage.

Section 547.10(B) specifies areas that are not required to have an equipotential plane. "Equipotential planes shall not be required in dirt confinement areas containing metallic equipment that is accessible to animals and likely to become energized." However, all circuits are required to have GFCI protection if they provide electric power to equipment that is accessible to animals in dirt confinement areas.

The term *equipotential plane* is defined in 547.2 as, "An area where wire mesh or other conductive elements are embedded in or placed under concrete, bonded to all metal structures and fixed nonelectrical equipment that may become energized, and connected to the electrical grounding system to prevent a difference in voltage from developing within the plane."

Section 547.10(C) provides details on requirements for bonding of the equipotential plane. Wire mesh or other conductive elements, such as reinforcing steel rods, must be installed in the concrete floors in these areas and be bonded together, as well as to the building grounding electrode system. This is accomplished by bonding the reinforcing rods and wire mesh in the concrete together with a copper conductor that is insulated, covered or bare, not smaller than 8 AWG, and then connecting these bonding conductors to metal piping systems, stanchions and the building grounding electrode systems. This creates the required equipotential plane, meaning everything in the area is at the same potential. The equipotential plane prevents voltage differences between conductive bodies with which livestock can make contact. Connections must be made with pressure connectors or clamps of brass, copper, copper alloy, or an equally substantial approved means. Connections can also

All metal stanchions, bars, support beams ect. bonded to metal mesh or reinforcing bars are required to be bonded together to form an equipotential plane

Rebar or wire mesh in floor forms part of equipotential plane

Figure 16-8   Equipotential Plane

be made by exothermic welding. "Slatted floor sections that are supported by structures that are a part of an equipotential plane shall not require bonding."[13] These floor sections are typically precast concrete sections, which by their size and mass effectively become a part of the equipotential plane when supported by the rest of the structure. In addition, the installation of bonding conductors to these *floating* sections has proven difficult as these sections are removed for periodic wash downs and other cleaning.

Obviously, it is best to install this bonding system during the original construction of the building or portion of a building that houses livestock or serves as a milking parlor. Equipotential planes have been installed in existing buildings by sawing grooves in the concrete floors, installing bonding conductors and bonding all conductive elements together. However, this is always much more expensive than making the installation before the concrete is poured.

The FPN No. 1 following 547.10(C) provides information on methods for installing equipotential planes and voltage gradient ramps and reads, "Methods to establish equipotential planes are described in American Society of Agricultural Engineers (ASAE) EP473-2001 *Equipotential Planes in Animal Containment Areas*."[14]

Receptacles in Agricultural Buildings

Section 547.5(G) requires that "All 125-volt, single-phase, 15- and 20-ampere general-purpose receptacles installed in the following locations shall have ground-fault circuit-interrupter protection for personnel:

(1) In areas having an equipotential plane
(2) Outdoors
(3) Damp or wet locations"[15]

Note that Article 547 does not include a definition of *damp location* or *wet location* so the definitions of these terms in Article 100 applies. While the term *general purpose receptacles* is not defined, it no doubt refers to those receptacles that would be used for portable tools and portable power cords and not to receptacles installed for a specific purpose such as for fixed equipment like brooders, incubators, and so forth.

Health Care Facilities

Electrical systems in health care facilities are re-quired to comply with at least two safety standards, the *National Electrical Code*, NFPA 70 and the *Health Care Facilities Code®* NFPA-99. These codes provide the minimum requirements for installation as well as maintenance and testing of electrical systems in health care facilities.

Depending on the type of health care facility involved and the scope of the project, several other electrical codes and standards may be involved. These include *Centrifugal Fire Pumps*,® NFPA 20; *National Fire Alarm Code*,® NFPA 72; *Emergency and Standby Power Systems*,® NFPA 110; and *Installation of Lightning Protection Systems*,® NFPA 780.

Article 517 provides special requirements for grounding of equipment in certain health care facilities, particularly in patient care areas. Reasons for the extra or specialized requirements are given in the FPN following 517.11.

In a health care facility, it is difficult to prevent the occurrence of a conductive or capacitive path from the patient's body to some grounded object, because that path may be established accidentally or through instrumentation directly connected to the patient. Other electrically conductive surfaces that may make an additional contact with the patient, or instruments that may be connected to the patient, then become possible sources of currents that can traverse the patient's body. The hazard is increased as more apparatus is associated with the patient, and, therefore, more intensive precautions are needed. Control of electric shock hazard requires the limitation of electric current that might flow in an electric circuit involving the patient's body by raising the resistance of the conductive circuit that includes the patient, or by insulating exposed surfaces that might become energized, in addition to reducing the potential difference that can appear between exposed conductive surfaces in the patient vicinity, or by combinations of these methods. A special problem is presented by the patient with an externalized direct conductive path to the heart muscle. The patient may be electrocuted at current levels so low that additional protection in the design of appliances, insulation of the catheter, and control of medical practice is required.[16]

Grounding in Patient Care Areas

Section 517.13(A) requires "All branch circuits serving patient care areas shall be provided with a ground path for fault current by installation in a metal raceway system, or cable armor or sheath assembly. The metal raceway system, or cable armor, or sheath assembly, shall itself qualify as an equipment grounding conductor return path in accordance with 250.118. Type AC, Type MC, and Type MI cables shall have an outer metal armor or sheath that is identified as an acceptable grounding return path."

Section 517.13(B) requires that, "In an area used for patient care, the grounding terminals of all receptacles and all non-current-carrying conductive surfaces of fixed electric equipment likely to become energized that are subject to personal contact, operating at over 100 volts, shall be grounded by an insulated copper conductor. The grounding conductor shall be sized in accordance with Table 250.122 and installed in metal raceways or metal-clad cables with the branch-circuit conductors supplying these receptacles or fixed equipment."

Exception No. 1 provides that, *"Metal faceplates shall be permitted to be grounded by means of a metal mounting screw(s) securing the faceplate to a grounded outlet box or grounded wiring device."*[17]

Exception No. 2 provides that, *"Luminaires (light fixtures) more than 2.3 m (7½ ft) above the floor and switches located outside of the patient vicinity shall not be required to be grounded by an insulated equipment grounding conductor."*[18]

Switches located outside patient vicinity are not required to be grounded by insulated grounding conductor.

Figure 16-10. Grounding of Receptacles and Fixed Electric Equipment

The term *patient vicinity* is defined in 517.2 as follows, "In an area in which patients are normally cared for, the patient vicinity is the space with surfaces likely to be contacted by the patient or an attendant who can touch the patient. Typically in a patient room, this encloses a space within the room not less than 1.8 m (6 ft) beyond the perimeter of the bed in its nominal location, and extending vertically not less than 2.3 m (7 ½ ft) above the floor." As can be seen, luminaries that are higher than 7 ½ ft from the floor do not have to be grounded by the insulated equipment grounding conductor otherwise required in the patient care area. These luminaries do have to be grounded by the wiring method used to supply the luminaries. Switches that are located either inside or outside the patient room but outside the patient vicinity are also not required to be grounded by the insulated equipment grounding conductor (see figure 16-10).

Both a primary and secondary, or redundant, means of grounding is required in patient care areas of health care facilities (see 517.12). The metal raceway or outer metal jacket of a listed cable, that is suitable for grounding, provides the primary means of equipment grounding. An insulated copper equipment grounding conductor (the secondary, redundant equipment grounding means) are required to also be installed or be a part of listed cable for grounding receptacles and non-current-carrying metal portions of fixed electric equipment in the patient care areas

Raceway, cable armor, or sheath assembly must qualify as an equipment grounding return path as per 250.118

Insulated copper equipment grounding conductor

Metal outlet box

(A) Covers Wiring Methods
(B) Covers Insulated Equipment Grounding Conductor

Figure 16-9. 517.13(A) and (B) Grounding of Receptacles and Fixed Electric Equipment

Health Care Facilities
Patient Care Area Wiring Methods

1. Rigid metal conduit.
2. Intermediate metal conduit.
3. Electrical metallic tubing.
4. Listed flexible metal conduit, max. 6 feet long, max. 20A OC protection, fittings listed for grounding.
5. Armor of listed Type AC cable where the outer metal jacket is approved as a grounding means.
6. Listed Type MI cable and MC cable where the outer metal jacket is acceptable as a ground-return path.
7. Listed liquidtight flexible metal conduit maximum 6 feet long, and
   Max. 20 A OC protection ³/₈ to ½ inch or
   Max. 60 A OC protection ¾ - 1¼ inch

All wiring methods must contain an insulated, copper equipment grounding conductor.

Figure 16-11. Patient Care Area Wiring Methods

1. Galvanized steel armor.
2. Insulated bushing.
3. Green insulated ground conductor.
4. Nylon.
5. Thermoplastic insulation 90 C rated.
6. Copper conductors.
7. Bonding wire.

Figure 16-12. Health Care Facilities Type AC Cable. Courtesy AFC Cable Systems

that are subject to personal contact. The basic concept or idea here is the *Code* requires two grounding paths to be installed for electrical equipment in patient care areas so it is assured that at least one positive return path for fault current is always present if either should fail. This is often referred to as redundant grounding for the patient care area. The concept of this redundancy should be looked at from the standpoint that one is required to install two suitable grounding return paths, to always be assured to have at least one.

The *Code* is specific in requiring two separate types of grounding return paths in accordance with 250.118, one being the raceway or cable enclosing the circuit conductors, and the other, an insulated copper equipment grounding conductor. Installing two insulated copper grounding conductors in a raceway that does not qualify as

an equipment grounding conductor does it not meet the letter or intent of the *NEC* for branch-circuit grounding requirements in patient care locations. The wiring methods that are acceptable for providing the primary equipment grounding fault return path are the metal raceways or cable assembly included in 250.118 (see figure 16-11). Of course, where these wiring methods are installed, it is critical that good work practices be followed to ensure that the wiring method provides a reliable, continuous and low-impedance path for fault current.

Section 517.13(A) places additional restrictions on the use of cable wiring methods that are used for wiring in patient care areas. The outer jacket of the cable must qualify as an equipment grounding conductor path in accordance with 250.118.

Types AC, MC and MI cables are required in 517.13(A) to have an outer armor that itself qualifies "as an equipment grounding return path in accordance with 250.118." These requirements do not permit the combination of the cable outer jacket and the internal equipment grounding conductor to serve as providing the path for fault current required by this section.

Generally, the outer jacket of Type MC cable of the spiral-interlocking armor type is not suitable as an equipment grounding conductor and cannot be used in a patient care area of a health care facility that must comply with 517.13. Type MC cable with a smooth or corrugated continuous tube is suitable for these patient care areas if it contains a green-insulated equipment grounding conductor with a yellow stripe or surface marking or both to indicate that it is an additional equipment (or isolated) grounding conductor. This acceptable type of MC cable would not contain a supplemental bare or unstriped green-insulated grounding conductor. A supplemental bare or unstriped green-insulated equipment grounding conductor in Type MC cable of the smooth or corrugated continuous type is an indication that the outer armor by itself does not qualify as an equipment ground.

By its construction and listing by a qualified electrical testing laboratory, Type AC cable with a spiral interlocking metal jacket and a bonding strip or wire in intimate contact with the outer jacket is suitable as an equipment grounding conductor. Type AC cable with this outer cable armor assembly plus the internal insulated equipment grounding conductor constitutes two ground return paths and qualifies the cable for use in wiring in patient care areas.

Copper, Type MI cable, by nature of its construction, has an outer jacket that is suitable as an equipment grounding conductor. This cable is manufactured with bare conductors that are physically separated from each other by the mineral compound inside the metal sheath. The individual conductors are insulated by slip-on insulation at the time the cable is terminated.

While recognized in 250.118 as equipment grounding conductors, severe restrictions are placed on the use of flexible metal conduit and liquidtight flexible metal conduit due to their limited capabilities of carrying current and functioning as a ground-fault return path.

Additional restrictions are placed on the wiring of the emergency system in hospitals. While these requirements relate to mechanical protection of the circuits and not directly to grounding, the installer should be aware of these rules. See 517.30(B)(2) for designation of these branches, 517.2 for definitions, and 517.30(C)(3) for wiring requirements.

Testing of Grounding System in Patient Care Areas

NFPA 99, The *Standard for Health Care Facilities,* requires that the integrity of the grounding path provided by the wiring method in patient care areas of health care facilities be tested before acceptance of the initial installation and after any alterations or replacement of the electrical system is made (see 3-3.3.2.1). Both voltage and impedance measurements must be made of exposed conductive surfaces including the grounding contacts of receptacles in the patient-care vicinity.

Excluded from the testing requirements are small, wall-mounted conductive surfaces, not likely to become energized, such as surface-mounted towel and soap dispensers, mirrors and the like.

Also exempt are large, metal conductive surfaces not likely to become energized, such as window frames, door frames, and drains. A note following this section indicates that the grounding system, including both metallic raceway and equipment grounding conductor, is to be tested as an integral system. Removing equipment grounds from receptacles or equipment is not required or recommended.

The voltage and impedance measurements must be measured against a reference point. The reference point is permitted to be one of the following:

1. a reference grounding point;
2. a grounding point in the room under test that is electrically remote from the equipment under test, such as a metal water pipe; and
3. the grounding contact of a receptacle that is powered from a different branch circuit from the receptacle being tested.

The criteria for new construction that must be met are as follows:

1. the voltage limit is 20 mV;
2. the impedance limit is 0.1 ohm; and
3. for *quiet ground systems* the impedance limit is 0.2 ohm.[19]

Receptacle Testing in Patient Care Areas

In wet locations, fixed receptacles, equipment connected by cord and plug, and fixed electrical equipment must be tested:
(1) when first installed,
(2) where there is evidence of damage,
(3) after any repairs, or
(4) at intervals not exceeding 6 months.

Section 3-3.4.2.3 gives requirements for the testing interval for receptacles in patient care areas. It is required that:
(1) Testing be performed after initial installation, replacement, or servicing of the device.
(2) Additional testing be performed

Voltage and impedance measurements
From reference point:
1. reference grounding point
2. remote grounding point in room
3. grounding contact on receptacle
   From different branch circuit.

Criteria to be met for
new construction:
1. voltage max. 20 mV
2. impedance max. 0.1 ohm
3. Quiet ground max 0.2 ohm.

Figure 16-13. Testing Grounding System Patient Care Area

at intervals defined by documented performance data.

Receptacles that are not listed as hospital-grade must be tested at intervals not exceeding 12 months.

Record keeping requirements are found in 3-3.4.3. A record must be maintained of the tests required by the chapter and associated repairs or modification. At a minimum, this record must contain the date, the rooms tested, and an indication of which items have met or have failed to meet the performance requirements of this chapter.

Where an isolated power system is installed, 3-3.4.3.2 requires that a permanent record be kept of the results of each of the tests.

Tests that are required to be performed for each receptacle include:
(1) Visual inspection to confirm physical integrity.
(2) Continuity of the grounding circuit.
(3) Correct polarity.
(4) Retention force (except locking type) to be not less than 115 g (4 oz).[20]

Bonding of Panelboards in Patient Care Areas
Section 517.14 requires that the "equipment grounding terminal busses of the normal and essential branch-circuit panelboards serving the same individual patient vicinity be bonded together with an insulated continuous copper conductor not smaller than 10 AWG.

"Where more than two panelboards serve the same location, this conductor shall be continuous from panel to panel but shall be permitted to be broken in order to terminate on the ground bus in each panel."[21]

Typically, the circuits supplying receptacles serving the patient bed locations are supplied from different separately derived systems. This bonding is intended to ensure that little if any potential difference exists between exposed non-current-carrying metal portions of equipment in the patient-care area. It is not required that this bonding conductor be installed in a raceway between panelboards where it is protected by normal building construction such as studs and gypsum board.

This bonding is provided to equalize the potential of electrical equipment in the patient vicinity that might be supplied from two different panelboards. This is especially important should a line-to-ground fault occur in one piece of equipment while another piece of equipment in the same patient vicinity is in a stable condition. There might be a slight voltage rise on the faulted circuit until the overcurrent device on the line side of the fault

Figure 16-14. Panelboard Bonding – Patient Care Areas

clears. The 10 AWG bonding conductor helps keep the electrical equipment in the patient vicinity at the same potential thus reducing the shock hazard.

Patient Equipment Grounding Point

The *patient equipment grounding point* is an option for improving electrical safety in patient care areas. Though an optional feature [see 517.19(C)], a patient equipment grounding point is provided by many hospital equipment manufacturers and specified by consulting engineers. Where installed, it consists of a grounding terminal bus and may include one or more grounding jacks. "An equipment bonding jumper not smaller than 10 AWG copper shall be used to connect the grounding terminals of all grounding-type receptacles to the patient equipment grounding point."[22] Again, this is done to reduce any potential difference between conductive surfaces in the patient vicinity.

Panelboard Grounding, Critical Care Areas

Section 517.19(D) requires that where metal raceways or Type MC or MI cables are used for feeders, grounding of panelboards and switchboards serving critical care areas must be assured. Acceptable methods include as follows:

(1) A grounding bushing and a continuous copper bonding jumper, sized in accordance with 250.122, with the bonding jumper connected to the junction enclosure or the ground bus of the panel

(2) Connection of feeder raceways or Type MC or MI cable to threaded hubs or bosses on terminating enclosures

(3) Other approved devices such as bonding-type locknuts or bushings.[23]

Grounding bushing with bonding jumpers

Threaded bosses or hubs on enclosures

Bonding locknuts or bushings

Figure 16-15. Panelboard Grounding, Critical Care

Isolated Power Systems Permitted

Isolated power systems are ungrounded systems, and "shall be permitted to be used for critical care areas, and, if used, the isolated power system equipment shall be listed for the purpose and the system designed and installed so that it meets the provisions of and is in accordance with 517.160."[24] The exception to 517.19 states that "*The audible and visual indicators of the line isolation monitor shall be permitted to be located at the nursing station for the area being served.*"

An isolated power system usually consists of a one-to-one transformer that has an ungrounded secondary. A protective shield is placed between the primary and secondary windings. A line isolation monitor with visual and audible alarms is provided to warn of excessive line-to-ground leakage as well as to indicate that a line-to-ground fault has occurred. The primary purpose of an isolated power system is to allow equipment to function with the first line-to-ground fault without opening an overcurrent device. Faulty equipment can be repaired or replaced at the earliest opportunity but will remain in service until a second line-to-ground fault occurs which then becomes a line-to-line fault. This would cause the overcurrent device to operate and take the faulted equipment off the line. Reference grounding terminal bars are included in isolated power systems enclosures for termination of equipment grounding conductors of the branch circuits supplied by the isolated power system. These systems must be tested prior to being put into service and periodically thereafter.

There are specific requirements regarding the types of insulation on the branch-circuit conductors, length of the branch circuits, and color code for these conductors. Wire pulling compound is not permitted to be used for installation of these circuit conductors in raceways as it can have an adverse affect on the dielectric characteristics of the conductors and the system operation. It is important to follow the manufacturer's installation instructions to ensure compliance with the equipment listing requirements of 110.3(B) in addition to the requirements of 517.160.

Other specific requirements for isolated power systems are in 517.160.

Special Purpose Receptacle Grounding

Section 517.19(F) states that "Where an isolated

ungrounded power source is used and limits the first-fault current to a low magnitude, the grounding conductor associated with the secondary circuit shall be permitted to be run outside of the enclosure of the power conductors in the same circuit."

The FPN reads as follows: "Although it is permitted to run the grounding conductor outside of the conduit, it is safer to run it with the power conductors to provide better protection in case of a second ground fault."[25]

Wet Locations in Health Care Facilities

The term *wet locations* is defined under the Patient Care Area of 517.2 as "Those patient care areas that are normally subject to wet locations while patients are present. These include standing fluids on the floor or drenching of the work area, either of which condition is intimate to the patient or staff." This definition of wet locations modifies for the purposes of Article 517 the similar definition in Article 100.

The general requirement in 517.20 is that "All receptacles and fixed equipment within the area of the wet location shall have ground-fault circuit-interrupter protection for personnel if interruption of power can be tolerated, or be served by an isolated power system if such interruption cannot be tolerated."[26]

The Exception to 517.20 provides that "*Branch circuits supplying only listed, fixed, therapeutic and diagnostic equipment are permitted to be supplied from a normal grounded service, single- or 3-phase system, provided that*

"*(a) Wiring for grounded and isolated circuits does not occupy the same raceway, and*
(b) All conductive surfaces of the equipment are grounded."[27]

It is common that operating rooms and some special procedures rooms are designated as wet locations by the governing body of the facility, particularly in hospitals. Many complex and lengthy procedures are performed in these rooms, which cannot be interrupted by the operation of a ground-fault circuit interrupter. As a result, isolated power systems are often installed.

"Where an isolated power system is utilized, the equipment shall be listed for the purpose and be installed so that it meets the provisions of and is in accordance with 517.160."[28] This section provides the rules for isolated power systems.

Ground-Fault Circuit-Interrupter Protection for Personnel in Health Care Facilities

The general requirement for providing ground-fault circuit interrupters of all 125-volt, single-phase, 15- and 20-ampere receptacles in bathrooms is contained in 210.8(B)(1). GFCI protection of these receptacles is generally required. Note that the definition of the term *bathroom* is contained in Article 100, Part I and reads, "An area including a basin with one or more of the following: a toilet, a tub, or a shower." Generally, receptacles of the type mentioned above that are installed in areas in health care facilities that meet this definition are required to have ground-fault circuit-interrupter protection for personnel.

NEC 517.21 provides that. "Ground-fault circuit-interrupter protection for personnel shall not be required for receptacles installed in those critical care areas where the toilet and basin are installed within the patient room."[29] This section essentially provides an exception to the general requirement for GFCI protection of receptacles in 210.8(B)(1). Toilets of the traditional configuration are not generally installed in critical care patient rooms, as these patients are often too ill to get out of bed. Some special configurations of patient care equipment may include a patient toilet in the critical care room. GFCI protection of the receptacles is not required where, for example, only a basin is installed in the patient room. In this configuration, the room does not meet the definition of *bathroom* in Article 100.

GFCI protection of the receptacles identified above is required in other patient care rooms that meet the definition of bathroom, and typically

Figure 16-16. Ground-Fault Circuit-Interrupter Protection

have a basin and a toilet and/or a shower. Note that the definition of bathroom uses the term *area* rather than *room*. As a result, GFCI protection is typically required for receptacles at the basin and in the room where the toilet and/or shower are installed and not for the receptacles that are not located adjacent to the basin.

Swimming Pools, Fountains, and Similar Installations

It is important to understand the organization of Article 680 for proper application of its requirements. Article 680 is divided in several parts, namely Part I through Part VII. Part I includes general requirements and applies to all equipment and installations covered in Article 680. Part II covers permanently installed pools. Other parts cover the equipment or installation identified in the title of the part and do not necessarily apply to all equipment included in Article 680. For example, the rules in Part II for grounding panelboards do not apply to the wiring of a hydromassage bathtub covered in Part VII.

However, be aware of the requirements for wiring spas and hot tubs. Section 680.40 reads, "Electrical installations at spas or hot tubs shall comply with the provisions of Parts I and IV of this ar-

ticle." Generally, a spa or hot tub installed outdoors must be in accordance with 680.42 and is required to meet all the rules for swimming pools, while a spa or hot tub installed indoors is required to meet the rules in 680.43 which are different from those for swimming pools.

The scope of Article 680 has been revised for clarification and the article itself has been rewritten and reorganized for the 2002 *NEC*.

"**680.1 Scope.** The provisions of this article apply to the construction and installation of electrical wiring for and equipment in or adjacent to all swimming, wading, therapeutic, and decorative pools, fountains, hot tubs, spas, and hydromassage bath tubs, whether permanently installed or storable, and to any metallic auxiliary equipment such as pumps, filters, and similar equipment. The term *body of water* used throughout Part I applies to all bodies of water covered in this scope unless otherwise amended."[30]

GFCI Requirements

Section 680.5 contains specific requirements relative to ground-fault circuit interrupters for swimming pools. They are permitted to be self-contained units, circuit-breaker types, receptacle types, or other approved types.

"Conductors on the load side of a ground-fault circuit interrupter or of a transformer, used to comply with the provisions of 680.23(A)(8), shall not occupy raceways, boxes, or enclosures containing other conductors…[31] [*Editor's note*: …even if the other conductors are for pool-related equipment such as for a water pump.]" Grounding conductors are permitted in these raceways and enclosures.

Conductors protected by ground-fault circuit interrupters are permitted in a panelboard that contains circuits protected by other than ground-fault circuit interrupters.

Supply conductors to a feed-through-type ground-fault circuit interrupter are permitted in the same enclosure with the conductors supplied by the GFCI device.

GFCI Requirements for Receptacles

Section 680.22(A)(2) generally requires that receptacles on the property be located at least 3.0 m (10 ft) from the inside walls of

Photo 16-1. Typical Swimming Pool

Figure 16-17. Wiring For Underwater Luminaires (Fixtures)

Pool or outdoor spa or hot tub

GFCI required
GFCI not required
8'
GFCI not required
Sliding door
Window that opens
GFCI not required

Permitted between 5 and 10 feet if for permanently installed pool water pump motor. GFCI protection required.

Figure 16-18. GFCI Requirements for Receptacles

a pool (outside spa or hot tub) or fountain. A receptacle to supply power for a water-pump for a permanently installed pool or fountain, as permitted in 680.22(A)(1), must be installed between 1.5 m (5 ft) and 3.0 m (10 ft) from the inside walls of the pool or fountain. Where so located, the receptacle(s) must be of the single and of the locking and grounding types and be protected by a GFCI device.

Section 680.22(A)(3) requires "Where a permanently installed pool is installed at a dwelling unit(s), no fewer than one 125-volt 15- or 20-ampere receptacle on a general-purpose branch circuit shall be located not less than 3.0 m (10 ft) from and not more than 6.0 m (20 ft) from the inside wall of the pool. This receptacle shall be located not more than 2.0 m (6 ft 6 in.) above the floor, platform, or grade level

serving the pool. [32] [Editor's note: Where the word pool is used, it includes outdoor spas and hot tubs.]"

Section 680.22(A)(4) permits not more than one general purpose receptacle to be located between 1.5 m (5 ft) and 3.0 m (10 ft) from the inside walls of the pool for dwellings having restricted space for the pool. Restricted space is described as a "pool within 3.0 m (10 ft) of a dwelling and the dimensions of the lot preclude meeting the otherwise required clearances, not more than 1.5 m (5 ft) measured horizontally from the inside wall of the pool."

Section 680.22(A)(5) notes that "All 125-volt receptacles located within 6.0 m (20 ft) of the inside walls of a pool or fountain shall be protected by a ground-fault circuit interrupter." Section 680.22(A)(6) provides information on determining the distances from the pool. It reads, "In determining the dimensions in this section addressing receptacle spacings, the distance to be measured shall be the shortest path the supply cord of an appliance connected to the receptacle would follow without piercing a floor, wall, ceiling, doorway with hinged or sliding door, window opening, or other effective permanent barrier."[33]

The GFCI requirements for motors associated with permanently installed pools are found in Part II of Article 680, specifically 680.21.

Bonding Requirements
Bonding requirements for pool areas are found in 680.26. Section 680.26(A) is new and includes performance language relative to the purpose of bonding for these types of installations. The Code makes it clear that the bonding required for the pool must be installed to eliminate voltage gradients in the pool area as described in the article. Swimming pools, spas and hot tubs present special shock hazards to people due to mixing water enriched with chemicals, people and electricity together. Electric shock hazards will be reduced to an acceptable level if the measures prescribed in Article 680 are carefully followed. One of these measures is the bonding together of conductive portions of the pool and metal parts of electrical equipment associated with the pool. The goal is to provide a means of equalizing the potential of all equipment and parts so there will be no current flow between parts. This is accomplished by connecting all the parts together by an adequately-sized and properly-connected copper conductor.

Fence

- Receptacle outlet permitted within 10' restricted area in limited space

- Not less than 5'
- GFCI protected
- Existing or new receptacle outlet

POOL

3'

3'

8'

4'

Receptacle outlet not less than 5'

Dwelling Unit with zero lot-line (Typical)

Figure 16-19. 680.22(A)(4) Restricted Space

This concept is often called equipotential bonding or grounding [see 680.26].

These requirements emphasize the purpose of bonding in a swimming pool, spa or hot tub area — that of equalizing the potential (voltage) between various parts of the pool. By keeping the potential difference as low as practicable, the shock hazard is reduced significantly.

Since several pieces of electrical equipment commonly used with pools, hot tubs or spas that are bonded together are required to also be grounded with an insulated equipment grounding conductor, an interconnection between the grounded (neutral) conductor and bonding grid exists. This interconnection is often remote from the pool and is not intended to play a part in equipotential bonding.

Section 680.26(B) provides a list of items that are required to be bonded. Included are the following:

(1) Metallic Structural Components. All metallic parts of the pool structure, including the reinforcing metal of the pool shell, coping stones, and deck, shall be bonded. The usual steel tie wires shall be considered suitable for bonding

Listed pool junction box, supported properly

Equipment grounding conductors with branch circuits

Wet-niche luminaire (fixture)

Bonding with minimum 8 AWG CU

Figure 16-20. Grounding and Bonding

the reinforcing steel together, and welding or special clamping shall not be required. These tie wires shall be encapsulating nonconductive compound at the time of the manufacture and installation, it shall not be required to be bonded. Where reinforcing steel is encapsulated with a non-conductive compound, provisions shall be made for an alternate means to eliminate voltage gradients that would otherwise be provided by unencapsulated, bonded reinforcing steel.

265

- Reinforcing steel that is insulated by encapsulating nonconductive compound at the time of manufacture and installation shall not require bonding

Figure 16-21. 680.26(B)(1) Bonding-Metallic Structural Components

[*Editor's note:* The *Code* recognizes an 8 AWG solid copper conductor for bonding purposes which could be installed for this purpose.]

(2) Underwater Lighting. All forming shells and mounting brackets of no-niche luminaries (fixtures) shall be bonded unless a listed low-voltage lighting system with nonmetallic forming shells not requiring bonding is used.

(3) Metal Fittings. All metal fittings within or attached to the pool structure shall be bonded. Isolated parts that are not over 100 mm (4 in.) in any dimension and do not penetrate into the pool structure more than 25 mm (1 in.) shall not require bonding.

(4) Electrical Equipment. Metal parts of electric equipment associated with the pool water circulating system, including pump motors and metal parts of equipment associated with pool covers, including electric motors, shall be bonded. Metal parts of listed equipment incorporating an approved system of double insulation and providing a means for grounding internal nonaccessible, non-current-carrying metal parts shall not be bonded. Where a double-insulated water-pump motor is installed under the provisions of this rule, a solid 8 AWG copper conductor that is of sufficient length to make a bonding connection to a replacement motor shall be extended from the bonding grid to an accessible point in the motor vicinity. Where there is no connection between the swimming pool bonding grid and the equipment grounding system for the premises, this bonding conductor shall be connected to the equipment grounding conductor of the motor circuit.

(5) Metal Wiring Methods and Equipment. Metal-sheathed cables and raceways, metal piping, and all fixed metal parts except those separated from the pool by a permanent barrier shall be bonded that are within the following distances of the pool:

(1) Within 1.5 m (5 ft) horizontally of the inside walls of the pool

(2) Within 3.7 m (12 ft) measured vertically above the maximum water level of the pool, or any observation stands, towers, or platforms, or any diving structures.[34]

Common Bonding Grid

The parts identified in 680.26(B) "shall be connected to a common bonding grid with a solid copper conductor, insulated, covered or bare, not smaller than 8 AWG." The conductor can be insulated, covered or bare. Connections to the parts to be bonded and to the common bonding grid must be made by exothermic welding or by pressure connectors or clamps of that are labeled as being suitable for the purpose and are of the following material: stainless steel, brass, copper or copper alloy.

"The common bonding grid shall be permitted to be any of the following:

Figure 16-22. **Bonding for Pools and Outdoor Spas and Hot Tubs**

(1) The structural reinforcing steel of a concrete pool where the reinforcing rods are bonded together by the usual steel tie wires or the equivalent.
(2) The wall of a bolted or welded metal pool
(3) A solid copper conductor, insulated, covered, or bare, not smaller than 8 AWG
(4) Rigid metal conduit or intermediate metal conduit of brass or other identified corrosion-resistant metal conduit."[35]

Bonding of Wet-Niche Luminaires (Light Fixtures)

Where mounted in a pool or fountain structure, a wet-niche luminaire (lighting fixture) must be installed in a forming shell that is designed to support a wet-niche luminaire (lighting fixture) assembly. The fixture will be completely surrounded by water [see definition of Wet-Niche Luminaire (Lighting Fixture) in Article 680 and 680.23(B)(1)].

A forming shell must be installed for the mounting of all wet-niche underwater luminaries (lighting fixtures). The forming shell must also be equipped with provisions for threaded conduit entries.

Figure 16-23. **680.26(B)(4) Bonding Electrical Equipment**

Wiring methods permitted to be used to connect the forming shell to a suitable junction box or other permitted enclosure include rigid metal conduit, intermediate metal conduit, liquidtight flexible nonmetallic conduit or rigid nonmetallic conduit [see 680.23(B)(2)].

Rigid metal or intermediate metal conduit used to connect the wet-niche fixture housing must be made of brass or other approved corrosion resistant metal, such as stainless steel.

Where rigid nonmetallic or liquidtight flexible nonmetallic conduit is used, an 8 AWG insulated copper conductor must be installed along with the pool flexible cord assembly so that it can be terminated on a suitable lug in the forming shell, junction box, transformer enclosure or ground-fault circuit-interrupter enclosure. This conductor is referred to as an equipment grounding conductor in this section but the sizing must be 8 AWG copper minimum. This conductor performs as a bonding jumper to connect non-current-carrying electrical equipment together. Where a metallic method is used from the junction box to the forming shell, the bonding of the two enclosures is accomplished through the metal conduit. An equipment grounding conductor is included in the cord assembly of the wet-niche luminaire (lighting fixture) for returning any line-to-ground fault current back to the source.

At the point of termination within the forming shell, the 8 AWG copper conductor must be covered with, or encapsulated in, a listed potting compound [680.23(B)(2)(b)]. This compound provides protection from deteriorating effects often caused by the pool water. Where a listed potting compound is not used to encapsulate the 8 AWG bonding conductor inside the forming shell, brass or stainless steel conduit must be used between the pool junction box and the forming shell.

Where in contact with the pool water, metal parts of the luminaire (fixture) and forming shell must also be of brass or other approved corrosion-resistant metal. Forming shells used with nonmetallic conduit systems other than those that are part of a listed low-voltage lighting

system not requiring grounding, include provisions for the termination of an 8 AWG copper conductor [see 680.23(B)(1)]. The end of the flexible cord jacket and flexible cord conductor terminations located within a wet-niche fixture must be covered with, or encapsulated in, a suitable potting compound. This will help to prevent the siphoning of water into the luminaire (fixture) through the cord jacket or its contained conductors. This requirement is met by the manufacturer of the wet-niche luminaire (fixture). In addition, an equipment grounding conductor connection within a wet-niche fixture must be similarly treated to protect it from the deteriorating effect of pool water in the event of water entry into the luminaire (fixture).

The luminaire (fixture) must be bonded to and secured to the forming shell by a positive locking device that will ensure a low resistance contact. A special tool is required to remove the fixture from the forming shell. Bonding is not required for fixtures listed for the application, having no non-current-carrying metal parts.

Underwater wet-niche lighting fixtures include markings that will indicate the proper housing or housings with which they are to be used, and the fixture housings are marked to indicate the fixture or fixtures with which the housings are to be used. These luminaries (fixtures) are provided with a factory-installed permanently-attached flexible cord

Figure 16-24. 680.23(B)(2) Luminaries (Fixtures) Wiring Extended Directly to the Forming Shell

that extends at least 12 feet outside the luminaire (fixture) enclosure. This will permit removal of the fixture from the forming shell so that it can be lifted onto the pool or spa deck for servicing without lowering the water level or disconnecting the fixture from the branch-circuit conductors.

Luminaires (fixtures) with longer cords are available for installation where the junction box or splice enclosure is so located that a 12-foot-long cord will not permit the fixture to be removed from the forming shell and placement on the deck for servicing. To avoid possible cord damage, any cord length in excess of that necessary for servicing should be trimmed from the supply end rather than stored in the forming shell.

Junction Boxes and Enclosures

Special requirements are contained in 680.24(A) for "A junction box connected to a conduit that extends directly to a forming shell or mounting bracket for a no-niche luminaire (fixture) shall meet the requirements of this section.

(1) Construction. The junction box shall be listed and labeled for the purpose and shall comply with the following conditions:

(1) Be equipped with threaded entries or hubs or a nonmetallic hub listed for the purpose
(2) Be comprised of copper, brass, suitable plastic, or other approved corrosion-resistant material
(3) Be provided with electrical continuity between every connected metal conduit and the grounding terminals by means of copper, brass, or other approved corrosion-resistant metal that is integral with the box."[36]

680.24(D) requires that junction boxes have a number of grounding terminals that shall be at least one more than the number of conduit entries. For example, where there are two conduit entries, the grounding terminal bar is required to have not less than three terminals.

Installation of Luminaires (Fixtures) over 15 Volts

The junction box or other suitable enclosure must be located not less than 100 mm (4 in.), measured from the inside of the bottom of the box or other enclosure, above the ground level or pool deck, or not less than 200 mm (8 in.)

above the maximum pool water level, whichever provides the greatest elevation [see 680.24(A)(2)(a)]. It must also be located so that it is not less than 1.2 m (4 ft) from the inside wall of the pool unless it is separated from the pool by a solid fence, wall or other permanent barrier [see 680.24(A)(2)(b)]. If used on a lighting system operating at 15 volts or less, a flush deck box is permitted provided both of the following conditions are complied with:

(1) An approved [*Editor's note:* acceptable to the authority having jurisdiction, not listed or labeled] potting compound is used to fill the box to prevent the entrance of moisture.
(2) The flush deck box is located not less than 1.2 m (4 ft) from the inside wall of the pool.[37]

Grounding Requirements

The following swimming pool and outdoor spa and hot tub equipment is required to be grounded:

(1) Through-wall lighting assemblies and underwater luminaires (lighting fixtures), other than those low-voltage systems listed for the application without a grounding conductor
(2) All electric equipment located within 1.5 m (5 ft) of the inside wall of the specified body of water

Figure 16-25. 680.27(A)(2) Wiring Methods for Underwater Audio Equipment

(3) All electric equipment associated with the recirculating system of the specified body of water

(4) Junction boxes

(5) Transformer enclosures

(6) Ground-fault circuit interrupters

(7) Panelboards that are not part of the service equipment and that supply any electric equipment associated with the specified body of water [38]

Methods of Grounding

Section 680.23(F)(2) reiterates that "Through-wall lighting assemblies, wet-niche, dry-niche, and no-niche luminaries (lighting fixtures) shall be connected to an insulated copper equipment grounding conductor installed with the circuit conductors. The equipment grounding conductor shall be installed without joint or splice except as permitted in (a) and (b). The equipment grounding conductor shall be sized in accordance with Table 250.122 but shall not be smaller than 12 AWG."

The equipment grounding conductor installed between the wiring chamber of the secondary winding of a transformer and a junction box must be sized in accordance with the overcurrent device in this circuit (see Table 250.122). The equipment grounding conductor must be an insulated copper conductor and generally be installed with the circuit conductors in rigid metal conduit, intermediate metal conduit, liquidtight flexible nonmetallic conduit, or rigid nonmetallic conduit.

Electrical metallic tubing is permitted to be used for these conductors where installed on or within buildings. Electrical nonmetallic tubing is permitted to be used to protect circuit conductors where installed within buildings as ENT is generally limited to installation inside buildings as it is not suitable for direct-sunlight exposure [see 680.23(F)(1)].

Where connecting to transformers for pool luminaires (lights), liquidtight flexible metal conduit or liquidtight flexible nonmetallic conduit Type B (LFNC-B) is permitted to be used. Any one length is limited to not more than 1.8 m (6 ft) feet and a total of not more than 3.0 m (10 ft) is permitted. Liquidtight flexible nonmetallic conduit, Type B (LFNC-B), is permitted in lengths longer than 1.8 m (6 ft).

"The junction box, transformer enclosure, or other enclosure in the supply circuit to a wet-niche or no-niche luminaire (lighting fixture) and the field-wiring chamber of a dry-niche luminaire (lighting fixture) shall be grounded to the equipment grounding terminal of the panelboard. This terminal shall be directly connected to the panelboard enclosure."[39]

(a) If more than one underwater luminaire (lighting fixture) is supplied by the same branch circuit, the equipment grounding conductor, installed between the junction boxes, transformer enclosures, or other enclosures in the supply circuit to wet-niche luminaries (fixtures), or between the field-wiring compartments of dry-niche luminaries (fixtures), shall be permitted to be terminated on grounding terminals.

(b) If the underwater luminaire (lighting fixture) is supplied from a transformer, ground-fault circuit interrupter, clock-operated switch, or a manual snap switch that is located between the panelboard and a junction box connected to the conduit that extends directly to the underwater luminaire (lighting fixture), the equipment grounding conductor shall be permitted to terminate on grounding terminals on the transformer, ground-fault circuit interrupter, clock-operated switch enclosure, or an outlet box used to enclose a snap switch.[40]

Section 680.23(B)(3) requires "wet-niche luminaries (lighting fixtures) that are supplied by a flexible cord or cable shall have all their exposed non-current-carrying metal parts grounded by an insulated copper equipment grounding conductor that is an integral part of the cord or cable. This grounding conductor shall be connected to a grounding terminal in the supply junction box, transformer enclosure, or other enclosure. The grounding conductor shall not be smaller than the supply conductors and not smaller than 16 AWG."[41]

Pool-Associated Motors

All pool-associated motors must be connected to an equipment grounding conductor sized in accordance with Table 250.122 [see 680.21(A)]. This equipment grounding conductor must be a copper conductor and cannot be smaller than 12 AWG. "The branch circuits for pool-associated motors shall be installed in rigid metal conduit, intermediate metal conduit, rigid nonmetallic conduit, or Type MC cable listed for the location."[42] This Type MC cable will have an overall outer jacket of PVC material and an insulated equipment grounding conductor not smaller than 12 AWG. Electrical metallic tubing is permitted to be used to protect conductors where it is installed on or within buildings.

Where flexible connections are necessary at or adjacent to the motor, liquidtight flexible metal or liquidtight flexible nonmetallic conduit with approved fittings are permitted.

"In the interior of one-family dwellings, or in the interior of accessory buildings associated with a one-family dwelling, any of the wiring methods recognized in Chapter 3 of this *Code* shall be permitted that comply with the provisions of this paragraph. Where run in a raceway, the equipment grounding conductor shall be insulated. Where run in a cable assembly, the equipment grounding conductor shall be permitted to be uninsulated, but it shall be enclosed within the outer sheath of the cable assembly."[43]

Permanently installed pools are permitted to be provided with listed cord- and plug-connected pool pumps that are protected by a system of double insulation. They are to be provided with a means for grounding only the internal and nonaccessible non-current-carrying metal parts of the pump [680.21(A)(5)].

Panelboard Grounding

A panelboard, and where installed, a disconnecting means, that is not part of the service equipment or source of a separately derived system are required to have an equipment grounding conductor installed between its equipment grounding terminal, and the equipment grounding terminal located at the service equipment or source of a separately

Figure 16-26. Panelboard Grounding

derived system. This conductor must be sized in accordance with Table 250.122 and be an insulated conductor of copper, aluminum or copper-clad aluminum and can never be smaller than 12 AWG. This conductor must generally be installed with the feeder conductors in a rigid metal conduit, intermediate metal conduit, liquidtight flexible metal conduit, or rigid nonmetallic conduit.

Electrical metallic tubing is permitted to be used to protect conductors where installed on or within buildings in accordance with Article 358. Electrical nonmetallic tubing is permitted to be used to protect conductors where installed within buildings in accordance with Article 362. The equipment grounding conductor must be connected to an equipment grounding terminal of the panelboard and, where installed, to the enclosure for a disconnecting means.

Panelboards from Separately Derived Systems

There are several ways of grounding swimming pool panelboards that are supplied from a separately derived system. It is important to follow the rules for grounding the separately derived system as given in 250.30 and as covered extensively in chapter twelve.

Section 680.25(B) requires that the equipment grounding conductor from a separately derived system be sized in accordance with Table 250.66 based on the derived phase conductors but not smaller than an 8 AWG. The grounded conductor from the separately derived system is also required to be sized from Table 250.66 based on the derived phase conductors.

Existing Panelboards

The Exception to 680.25(A) states that *"An existing feeder between an existing remote panelboard and service equipment shall be permitted to run in flexible metal conduit or an approved cable assembly that includes an equipment grounding conductor within its outer sheath. The equipment grounding conductor shall comply with 250.24(A)(5)."* Section 250.24(A)(5) generally does not permit a connection of grounding conductor on the load side of the service disconnecting means.

The *NEC* does not define what is meant by the word *existing*. Often, inspection authorities consider an installation to be existing after it has had final approval or after a certificate of occupancy has been issued.

Swimming Pool Panelboards in Other Buildings

Grounding at the disconnecting means for remote buildings is covered in 250.32 and in chapter thirteen where the methods for complying with the *Code* rules are given. Normally, one can choose between grounding the neutral conductor again at the building and not taking the equipment grounding conductor to the building, and taking the equipment grounding conductor to the building, grounding it there and not grounding the neutral but floating it there. Specific limitations are given for re-grounding the grounded conductor (may be a neutral) at the remote building.

"**(2) Grounded Conductor.** Where (1) an equipment grounding conductor is not run with the supply to the building or structure, (2) there are no continuous metallic paths bonded to the grounding system in both buildings or structures involved, and (3) ground-fault protection of equipment has not been installed on the common ac service, the grounded circuit conductor run with the supply to the building or structure shall be connected to the building or structure disconnecting means and to the grounding electrode(s) and shall be used for grounding or bonding of equipment, structures, or frames required to be grounded or bonded."[44] Where an equipment grounding conductor is installed in accordance with 250.32(B)(1), and is included with a feeder to a separate building, it must be an insulated conductor in accordance with 680.25(B)(2).

Cord-Connected Equipment

Where fixed or stationary equipment is connected with a flexible cord to facilitate removal or disconnection for maintenance, repair, or storage as provided in 680.7, the equipment grounding conductor must be connected to a fixed metal part of the assembly. The removable part must be mounted on or bonded to the fixed metal part [see 680.7(C)].

All other electrical equipment must be grounded in accordance with Article 250 and be connected by an approved wiring method covered in *NEC* Chapter 3.

Spa or Hot Tub Installed Outdoors

"A spa or hot tub installed outdoors shall comply with the provisions in Part I and II of this article, except as permitted in 680.42(A) and (B), that would otherwise apply to pools installed

Figure 16-27. Grounding for Swimming Pool at Second Building

outdoors."[45] This requirement includes those for GFCI protection of receptacles and lighting fixtures as well as required clearances from receptacles and lighting fixtures.

Section 680.42(A) permits listed packaged units (spas or hot tubs) utilizing a factory-installed or assembled control panel or panelboard to be connected with not more than 1.8 m (6 ft) of liquidtight flexible conduit (metallic or nonmetallic) or be cord-and-plug connected with a cord not longer than 4.6 m (15 ft) if protected by a ground-fault circuit interrupter.

Section 680.42(B) permits bonding to be accomplished by metal-to-metal mounting on a common frame or base. This is a common construction of a *packaged spa or hot tub equipment assembly* which is commonly referred to as a *skid pack*.

Metal bands or hoops that are used to secure wooden staves must be bonded to a common bonding grid.

Spa or Hot Tub Installed Indoors

A spa or hot tub installed indoors must conform to the requirements of Article 680, Parts I and II except as modified by 680.43. They must be connected by wiring methods of *NEC* Chapter 3. The Exception to 680.43 reads, "*Listed spa and hot tub packaged units rated 20 amperes or less shall be permitted to be cord-and-plug connected to facilitate the removal or disconnection of the unit for maintenance and repair.*"

Section 680.43(A) states that "At least one 125-volt, 15- or 20-ampere receptacle on a general-purpose branch circuit shall be located not less than 1.5 m (5 ft) from and not exceeding 3.0 m (10 ft) from the inside wall of a spa or hot tub.

(1) Location. Receptacles shall be located at least 1.5 m (5 ft) measured horizontally from the inside walls of a spa or hot tub.

2) Protection, General. Receptacles rated 125 volts and 30 amperes or less and located within 3.0 m (10 ft) of the inside walls of a spa or hot tub shall be protected by a ground-fault circuit interrupter."[46]

Bonding of Spa or Hot Tub Installed Indoors

The following parts of a spa or hot tub are required by 680.43(D) to be bonded together:

(1) All metal fittings within or attached to the spa or hot tub structure

(2) Metal parts of electric equipment associated with the spa or hot tub water circulating system, including pump motors

(3) Metal conduit and metal piping that are within 1.5 m (5 ft) of the inside walls of the spa or hot tub and that are not separated from the spa or hot tub by a permanent barrier

(4) All metal surfaces that are within 1.5 m (5 ft) of the inside walls of a spa or hot tub and that are not separated from the spa or hot tub area by a permanent barrier. *Exception: Small conductive surfaces not likely to become energized, such as air and water jets and drain fittings, where not connected to metallic piping, towel bars, mirror frames, and similar nonelectric equipment, shall not be required to be bonded.*

(5) Electrical devices and controls that are not associated with the spas or hot tubs and that are located not less than 1.5 m (5 ft) from such units; otherwise they shall be bonded to the spa or hot tub system[47]

Methods of Bonding of Spa or Hot Tub Installed Indoors

"All metal parts associated with the spa or hot tub shall be bonded by any of the following methods:

(1) The interconnection of threaded metal piping and fittings

(2) Metal-to-metal mounting on a common frame or base

(3) The provisions of a copper bonding jumper, insulated, covered, or bare, not smaller than 8 AWG solid."[48]

GFCI Protection of Spa or Hot tub Installed Indoors or Outdoors

Section 680.44 contains GFCI requirements that apply to outdoor and indoor spa and hot tub equipment. GFCI protection of the outlet is generally required if it supplies:

1. self-contained spa or hot tub;
2. packaged spa or hot tub equipment assembly; and
3. a field-assembled spa or hot tub.

GFCI protection of the outlet is not required for a listed self-contained unit or listed packaged equipment assembly tht is marked to indicate that integral ground-fault circuit-interrupter protection is provided for all electrical parts within the

Chapter 3 wiring methods permitted to run to outside spa or hot tub

Spa or hot tub

Dwelling Unit

Equipment grounding conductor to be 12 AWG or larger, insulated or covered copper

Figure 16-28.   680.42(C) Interior Wiring to Outdoor Locations

a. Self contained spa or hot tub.
b. packaged spa or hot tub equipment assembly.
c. field- assembled spa or hot tub with heater load of 50 amperes or less.

Exception: Listed self-contained unit or listed packaged equipment assembly that are marked to indicate integral GFCI protection is provided for all electrical parts within the unit or assembly.

Electrical equipment for combination pool/hot tub or spa assembly commonly bonded does not require GFCI protection.

Figure 16-29. GFCI Protection of Spas and Hot Tubs

All metal piping systems, metal parts of electrical equipment, and pump motors to be bonded with 8 AWG solid copper conductor.

Double-insulated equipment exempted.

Figure 16-30. Bonding at Hydromassage Bathtubs

unit or assembly (pumps, air blowers, heaters, lights, controls, sanitizer generators, wiring, and so forth). The electrical supply for a field-assembled spa or hot tub is not required to be protected by a ground-fault circuit interrupter if the supply is rated greater than 250 volts or rated 3 phase or if the heater load is more than 50 ampheres. A combination pool/hot tub or spa assembly commonly bonded need not be protected by a GFCI.[49]

Bonding Hydromassage Bathtubs

"Hydromassage bathtubs as defined in Article 680.2 are required to comply with only Part VII" of Article 680.[50] This part has requirements for both GFCI protection and bonding of hydromassage bathtubs,

Hydromassage bathtubs and their associated electrical components must be protected by a GFCI. This GFCI is to be of the Class A type that trips at a range of 4 to 6 mA leakage. In addition, all 125-volt, single-phase receptacles within 5 ft (1.52 m) measured horizontally of the inside walls of a hydromassage tub must be protected by a GFCI(s).

Section 680.74 contains requirements on bonding of electrical equipment and metal piping sys-

tems associated with hydromassage bathtubs. It should be noted that these requirements apply to dwelling units as well as to other than dwelling units. This section requires that, "All metal piping systems, metal parts of electrical equipment, and pump motors associated with the hydromassage tub shall be bonded together using a copper bonding jumper, insulated, covered, or bare, not smaller than 8 AWG solid. Metal parts of listed equipment incorporating an approved system of double insulation and providing a means for grounding internal nonaccessible, non-current-carrying metal parts shall not be bonded."[51]

Electric Signs and Outline Lighting

Section 600.7 contains requirements for grounding of electric signs and metal equipment of outline lighting systems. The rules in Article 600 amend or supplement the general rules for grounding in Article 250 due to the rule of "Code Arrangement" in 90.3. Specific requirements in 600.7 are that, "Signs and metal equipment of outline lighting systems shall be grounded."

The equipment grounding conductor and proper bonding are essential elements for safety in electrical signs and neon installations. This safety component of the circuit acts as the silent servant waiting to perform its ever-important function. Electrical installations for sign circuits and neon installations are not exempt from these basic grounding safety requirements.

Metal parts of the building are not permitted to serve as a grounded or equipment grounding conductor. This means that the secondary circuit

Legend:
1. Insulated copper equipment grounding conductor sized in accordance with Table 250.122 but not smaller than 12 AWG.
2. Install in rigid metal conduit, intermediate metal conduit, liquidtight flexible nonmetallic conduit or rigid nonmetallic conduit.
3. Install in rigid metal conduit, intermediate metal conduit or rigid nonmetallic conduit or Type MC cable listed for application.
4. Electrical metallic tubing permitted inside buildings.
5. Electrical nonmetallic tubing permitted inside buildings per Article 362.
6. Liquidtight flexible metal or nonmetallic conduit permitted where flexibility required.
7. Wiring method of Chapter 3 with equipment grounding conductor insulated or covered by outer sheath not smaller than 12 AWG permitted in one-family dwelling and associated structures.
8. Flexible cord permitted.
9. Wet-niche or no-niche fixture supplied by flexible cord; insulated copper conductor not smaller than circuit conductors, not smaller than 16 AWG.
10. Rigid metal, intermediate metal, liquidtight flexible nonmetallic or rigid nonmetallic conduit. Metal conduit of brass or equal.

Figure 16-31. Summary of Grounding Requirements for Swimming Pools

conductors of a neon tubing system are required to both return to the power supply or transformer. Using the metal components of the building, such as flashing or steel framing members, as a connection from a neon electrode back to the source is unsafe, and can cause fires or injury to persons. Proper grounding and bonding of metal enclosures and associated metal parts ensure that these parts remain at earth potential. Other electrical resource material on electrical theory is also available that expands on this term.

Transformers installed and wired using the balanced mid-point reference wiring method require the secondary output conductors to be as short as possible, and the secondary return leads are required to terminate on a mid-point reference connection terminal provided for that purpose by the transformer manufacturer. This mid-point terminal is often connected to the case ground in the transformer internally by the manufacturer. Proper grounding and bonding connections of the entire branch-circuit wiring

methods and secondary circuit wiring methods are critical for proper operation and safety of these secondary circuits wired by this mid-point reference method (see figure 16-33).

Sections 300.3(B) and 250.134(B) indicate all conductors of the branch circuit, including any grounded conductor or equipment grounding conductor, supplying power to a sign or primary (line side, usually 120-volt input) of a neon transformer must be installed in the same raceway, cable, trench, wiring gutter, unless permitted otherwise by the *NEC*. This includes all conductors (wires) of the circuit, including the equipment grounding conductor, which can be in many forms. It can be in the form of a conductor, conduit, tubing, cable armor, or combination of cable armor and conductor (see 250.118). In Article 600, there are requirements for branch circuits supplying signs and outline lighting systems. The wiring method (cable, conduit or raceway) used to supply signs and outline lighting systems is required to terminate within the sign, outline

Figure 16-32. 600.32(A)(5) Metal Building Parts

lighting system enclosure, junction box, or a conduit body. This circuit that terminates at the sign or neon transformer or power supply enclosure contains the equipment grounding conductor for the circuit. This conductor should be terminated to the metal enclosure to accomplish the equipment grounding (see figure 16-34).

The termination of the equipment grounding conductor should be made by use of approved means. Sheet metal screws are not acceptable for attachment of equipment grounding conductors per 250.8. Proper grounding clips, screws, or lug type terminations are available for this purpose. This termination of the equipment grounding conductor establishes the connection to ground for those enclosures and puts them at the same zero voltage potential as the earth.

The *Code* also allows for listed flexible metal conduit or listed liquidtight flexible metal conduit (per Article 350) to be used as a bonding means in lengths not exceeding 30 m (100 ft) on the secondary side of the sign transformer or power supply. These flexible metal conduits are suitable as a bonding means in lengths up to 30 m (100 ft) because the current on the secondary side of a neon transformer is in the milliampere range and the

bonding purpose of the conduit here is not to clear an overcurrent device on the primary (supply side) of the neon transformer or power supply. One should keep in mind that there is a length limitation on secondary GTO conductors of 6.0 m (20 ft) when installed in metallic wiring methods and 15 m (50 ft) when installed in nonmetallic wiring methods. This limitation must minimize the capacitance effect on the GTO cable of the secondary circuit [see 600.32(J)(1)].

Metal parts and metal equipment associated with electric signs, neon installations, and outline lighting systems must be grounded. This is usually accomplished by the conduit connections, but often requires the use of bonding conductors.

Bonding Metal Parts and Equipment

Bonding of electrical equipment and enclosures simply means that the enclosures will be connected together in an appropriate manner to "ensure electrical continuity and the capacity to conduct safely any fault current likely to be imposed."[58] Normal current will always seek its source, which is the transformer or power supply secondary. The same is true for fault current. Any metal parts or components requiring bonding must be bonded back to the source (transformer/and or combination of transformer and enclosure). When a metal conduit is connected to a metal electrical junction box with a proper conduit connector or fittings, the two parts become one electrical metal object because they are bonded together.

"Small associated metal parts not exceeding 50 mm (2 in.) in any dimension, and not likely to become energized [*Editor's note:* such as the metal mounting means for tubing supports], and spaced at least 19 mm (¾ in.) from the neon tubing are not required to be bonded."[53] "Where listed

Figure 16-33. 600.32(H) Conductor Runs Neon Tubing and Midpoint Return

Figure 16-34. Bonding with Liquidtight Flexible Metal Conduit

liquidtight nonmetallic conduit is used for installing the secondary" high voltage GTO conductors "from the transformer or power supply" to the neon tubing and where there are associated

Figure 16-35. Bonding with Conduit and Flanged Receptacle

metal parts that require bonding, "a bonding conductor is required."[54] This "bonding conductor shall be installed separate and remote from the nonmetallic conduit."[55] It should be pointed out here that this wiring method is not electrical nonmetallic tubing, which was deleted from the 1999 *NEC* as an acceptable wiring method for GTO secondary conductors (see 600.32). The wiring method referred to here is liquidtight flexible metal conduit or rigid nonmetallic conduit.

Stresses on High Voltage GTO Cables

"Bonding conductors must be copper and not smaller than 14 AWG."[56] "Where listed nonmetallic conduit is used to enclose the secondary circuit conductor from transformer or power supply and a bonding conductor is required, the bonding conductor shall be installed separate and remote from the nonmetallic conduit."[57] This spacing requirement is established to reduce stresses that might be imposed on the GTO conductor because the magnetic flux lines would no longer be symmetrical around the conductor.

Figure 16-36. Bonding with Conductor

Installation of a bonding conductor or grounded and bonded electrically conductive parts remote from nonmetallic raceways and enclosures that contain high voltage secondary GTO cables is important to minimize the effects unbalanced stress on the high voltage secondary conductor installed in a nonmagnetic wiring method. Two requirements in 600.7 were introduced in the 1999 *NEC* to deal with this situation. When a conductor is carrying AC current, it generates a magnetic field, which surrounds the conductor as the current flows during normal operation. This electromagnetic field produces electromagnetic lines of force.

When the conductor is installed in a metal wiring method, these electro-magnetic lines of force will work to compress into the metal wiring method encircling the secondary conductor because the metal conduit introduces less resistance to the magnetic lines of force than air. Even though the magnetic lines of force (flux) are more concentrated, they are maintained in a symmetrical fashion around the conductor. This keeps the stress on the conductor equal for the most part all around the GTO cable.

When a current-carrying conductor is installed in a nonmetallic wiring method, for example liquidtight flexible nonmetallic conduit, and a bonding conductor must be installed for bonding metal parts associated with the neon sign or neon installation, the magnetic lines of force (flux) will try to compress into the bonding conductor on that side of the conductor and remain expanded on the other sides. As the compressed magnetic field flows in this circuit, the current is flowing with this unbalanced magnetic flux from zero to maximum voltage and back to zero with every cycle of ac current flow. The stresses on the secondary GTO conductor are much greater on one side than the other and will continue this unbalanced stress condition which causes degradation of the high voltage secondary conductor insulation in time. By spacing this bonding conductor at the required intervals, the amount of unbalanced stresses imposed on the high voltage secondary conductors will be significantly reduced.

Sections 600.7 and 600.32(A)(4) both require a spacing of 38 mm (1½ in.) to be maintained between grounded metal parts or bonding conductors when

Flanged receptacle housing fastened to metal letters establishing bonding connection

Flexible metal conduit 30 m (100 ft.) maximum total accumulative length for bonding only
[Length limited by 600.32(J)]
[6 m (20 ft) metallic]
[15 m (50 ft) nonmetallic]

Neon Sign

NEC 600.7
NEC 600.32(J)

Figure 16-37. Bonding with Flexible Metal Conduit

Pull Box for taking up slack in GTO

14 AWG copper bonding jumper
required for bonding isolated
metal wiring methods

Branch Circuit

Figure 16-38. Bonding Metal Parts

Liquidtight Flexible
Nonmetallic Conduit

Primary
Branch
Circuit

1 1/2"

1 1/2"

600.7(C)
600.32(A)(4)

Transformer or
Power Supply

Maintain 38 mm (1 ½ in.) Minimum distance
between conduit and bonding conductor
and grounded metal parts other than at
points of termination to enclosures

Figure 16-39. Secondary Circuit Operating at 100 HZ or Less

Figure 16-40. Secondary Circuit Operating at Over 100 HZ

the secondary circuit operates at 100 Hz or less. When the secondary circuit operates at over 100 Hz, the spacing requirement increases to 45 mm (1 $^3/_4$ in). This conductor must not be smaller than 14 AWG.

"Metal parts of a building or structure shall not be permitted as a secondary return conductor or equipment grounding conductor."[58] Return secondary leads cannot be connected to the metal parts of a building or structure and used as a return for a mid-point return wired secondary circuit. This is a bad situation that can lead to fire and shock hazards.

[1] NFPA 70, *National Electrical Code* 2002, 250.106 (National Fire Protection Association, Quincy, MA, 2002), p. 70-111.

[2] NFPA 70, 250.100, p. 70-109.

[3] NFPA 70, 501.16(A), p.70-356.

[4] NFPA 70, 501.16(B), p 70-356.

[5] NFPA 70, 501.16(B), p. 70-356.

[6] NFPA 70, 547.9(C), p. 70-439.

[7] NFPA 70, 547.5(F), p. 70-438.

[8] NFPA 70, 250.112, p. 70-112.

[9] NFPA 70, 547.9(B)(1), p. 70-438.

[10] NFPA 70, 547.9(B)(1), p. 70-438.

[11] NFPA 70, 547.9(B)(1), p. 70-438 & 70-439.

[12] NFPA 70, 547.10(A), p.70-439.

[13] NFPA 70, 547.10(C), p.70-439.

[14] NFPA 70, 547.10(C), p.70-439.

[15] NFPA 70, 547.5(G), p. 70-438.

[16] NFPA 70, 517.11, FPN, p. 70-402.

[17] NFPA 70, 517.13(B), p. 70-402.

[18] NFPA 70, 517.13(B), p. 70-402.

[19] NFPA 99, *Standard for Health Care Facilities*.

[20] NFPA 99.

[21] NFPA 70, 517.14, p. 70-402.

[22] NFPA 70, 517.19(C), p. 70-404.

[23] NFPA 70, 517.19(D), p. 70-404.

[24] NFPA 70, 517.19(E), p. 70-404.

[25] NFPA 70, 517.19(F), FPN, p. 70-404.

[26] NFPA 70, 517.20, p. 70-404.

[27] NFPA 70, 517.20(A), Exception, p. 70-404.

[28] NFPA 70, 517.20(B), p. 70-404.

[29] NFPA 70, 517.21, p. 70-404.

[30] NFPA 70, 680.1, p. 70-523.

[31] NFPA 70, 680.23(F)(3), p. 70-529.

[32] NFPA 70, 680.22(A)(3), p. 70-527.

[33] NFPA 70, 680.22(A)(6), p. 70-527.

[34] NFPA 70, 680.26(B), p. 70-531.

[35] NFPA 70, 680.26(C), p. 70-531.

[36] NFPA 70, 680.24(A), p. 70-529.

[37] NFPA 70, 680.24(A)(2)(b), p. 530.

[38] NFPA 70, 680.6, p. 70-525.

[39] NFPA 70, 680.24(F), p. 70-530.

[40] NFPA 70, 680.23(F)(2), p. 70-529.

[41] NFPA 70, 680.23(B)(3), p. 70-529.

[42] NFPA 70, 680.21(A), p. 70-526.

[43] NFPA 70, 680.21(A)(4), p. 70-526.

[44] NFPA 70, 250.32(A)(2), p. 70-103.

[45] NFPA 70, 680.42, p. 70-533.

[46] NFPA 70, 680.43(A), p. 70-533 & 534.

[47] NFPA 70, 680.43(D), p. 70-534.

[48] NFPA 70, 680.43(E), p. 70-534.

[49] NFPA 70, 680.44, p. 70-535.

[50] NFPA 70, 680.70, p. 70-537.

[51] NFPA 70, 680.74, p. 70-537.

[52] NFPA 70, Article 100, p. 70-33.

[52] NFPA 70, 600.7(B), p. 70-478.

[54] NFPA 70, 600.7(C), p. 70-478.

[55] NFPA 70, 600.7(C), p. 70-478.

[56] NFPA 70, 600.7(D), p. 70-478.

[57] NFPA 70, 600.7(C), p. 70-478.

[58] NFPA 70, 600.32(A)(5), p. 70-481.

Chapter Sixteen: The questions included here were developed using material included in this chapter. The answers can be found by reviewing the text. It is also important that students make use of the 2002 *NEC*, where many of the answers can be found.

1. Protection from lightning discharges is provided by ____.
    a. building low-rise structures
    b. building structures of wood construction
    c. providing a low-resistance path to earth
    d. providing a high-resistance path to earth

2. It is required that ground terminals for lightning protection systems be ____.
    a. bonded to the electrical system grounding electrode system
    b. bonded to the building water pipe system
    c. isolated from the building electrical system by at least six feet
    d. isolated from the building grounding electrode system

3. Generally, in hazardous (classified) locations, locknuts installed on each side of an enclosure, or a locknut on the outside and a bushing on the inside ____ be used for bonding.
    a. shall
    b. can
    c. cannot
    d. may

4. Bonding locknuts or bonding bushings with bonding jumpers must be used in hazardous (classified) locations to ensure the integrity of the bond and its capability of carrying any ____ that may be imposed.
    a. unidentified currents
    b. neutral currents
    c. objectionable current
    d. fault current

5. Where flexible metal conduit is permitted and used in a hazardous (classified) location, it must have an internal or external bonding jumper installed to supplement the conduit. Where the jumper is installed on the outside, it is limited to ____ feet in length.
    a. 6
    b. 7
    c. 8
    d. 9

6. Where equipment is installed in an agricultural building that requires grounding, a ____ equipment grounding conductor must be installed to ground the equipment.
    a. aluminum
    b. copper
    c. copper-clad
    d. steel

7. Minimizing or elimination of stray voltages in an agricultural building is accomplished by bonding together the reinforcing rods and wire mesh in the concrete and then connecting these to the metal piping systems, stanchions and grounding electrode systems with a copper conductor not smaller than. ____ AWG, .
    a. 8
    b. 6
    c. 4
    d. 2

8. Which of the following statements is true about wiring in an agricultural building?
    a. An 8-ft. ground rod is required at each corner of the building.
    b. GFCI protection is requiured for all 125-volt, 1-phase, 15- and 20-ampere receptacles.
    c. Equipment grounding conductors installed underground where livestock is housed shall be insulated or covered copper.
    d. Line-to-neutral loads are not permitted.

9. A redundant means of grounding is required in ____ of health care facilities.
    a. limited-care facilities
    b. office areas
    c. all areas
    d. patient care areas

10. Where used in patient care areas, Type MI cable or Type ____ cable is required to have outer metal armor or sheath that is ____ as an acceptable grounding return path.
    a. MV - approved by special permission
    b. SE - permitted
    c. SNM - approved
    d. MC - identified

11. Where metal raceways or Type MC or MI cables are used for feeders supplying critical care areas in health care facilities, grounding of panelboards and switchboards must be assured. Acceptable methods include all of the following EXCEPT ____

    a.   grounding bushings with bonding jumpers
    b.   threaded bosses
    c.   bonding locknuts or bushings
    d.   standard locknuts

12. Equipment in the vicinity of swimming pools must be bonded together with a 8 AWG or larger copper conductor. This also includes metal raceways and cables, piping and fixed metal parts within ____ feet horizontally or ____ feet vertically of the pool, observation stands, towers, platforms or diving structures.

    a.   5 – 18
    b.   6 – 20
    c.   5 – 12
    d.   9 – 14

13. A common bonding grid at a pool is permitted to be the structural reinforcing steel of a concrete pool, the wall of a bolted or welded pool, or a ____ AWG or larger solid copper conductor.

    a.   8
    b.   6
    c.   4
    d.   2

14. A dry-niche lighting fixture is required to be provided with a provision for the drainage of water and a means for accommodating ____ equipment grounding conductor(s) for each conduit entry.

    a.   1
    b.   2
    c.   3
    d.   0 or none

15. All 125-volt receptacles located within ____ feet of the inside walls of a pool are required to be protected by a ground-fault circuit interrupter.

    a.   30
    b.   25
    c.   22
    d.   20

16. Ground-fault circuit interrupters used around pools are permitted to be installed in a ____ that contains circuits protected by other than ground-fault circuit interrupters.

    a.   cabinet
    b.   metal device box
    c.   panelboard
    d.   nonmetallic box

17. For other than storable pools, a flexible cord is not permitted to exceed ____ feet in length and must have a copper equipment grounding conductor not smaller than 12 AWG with a grounding-type attachment plug.

    a.   3
    b.   4
    c.   5
    d.   6

18. Ground-fault circuit interrupters used around pools are required to be self-contained units, circuit-breaker types, receptacle-types or other ____ types.

    a.   listed
    b.   acceptable
    c.   approved
    d.   labeled

19. A ground-fault circuit interrupter is required to be installed in the branch circuit supplying underwater lighting fixtures operating at more than ____ volts, so that there is no shock hazard during relamping.

    a.   6
    b.   10
    c.   12
    d.   15

20. The installation of the ground-fault circuit interrupter installed to protect an underwater lighting fixture must be such that there is no shock hazard with any likely fault-condition combination that involves a person in a conductive path from any ungrounded part of the ____ or the fixture to ground.

    a.   service
    b.   system
    c.   feeder
    d.   branch circuit

21. Where metal conduit is used to supply a ____ niche fixture, it is required to be of brass or other approved corrosion-resistant metal.
    a.   dry-
    b.   wet-
    c.   no-
    d.   metal-

22. A wet-niche fixture is required to be bonded to and secured to the forming shell by a positive locking device that assures a low-resistance contact and requires a ____ to remove the fixture from the forming shell.
    a.   tool
    b.   permit
    c.   fitting
    d.   helper

23. Transformers used for the supply of underwater fixtures, together with the transformer enclosure, are required to be ____ for the purpose.
    a.   suitable
    b.   listed
    c.   acceptable
    d.   approved

24. The transformer used to supply underwater lighting fixtures is required to be a ____ winding type having a grounded metal barrier between the primary and secondary windings.
    a.   three
    b.   isolated
    c.   one
    d.   four

25. Signs and ___ must be grounded.
    a.   all line-side equipment
    b.   electrode receptacle housings
    c.   metal equipment of outline lighting
    d.   electrical-discharge tubing

26. Where flexible nonmetallic conduit is used on the secondary of a sign transformer, the bonding conductor must be:
    a.   installed inside the conduit
    b.   not more than 6-feet long
    c.   installed separate and remote from the nonmetallic conduit
    d.   listed GTO cable

27. Listed flexible metal conduit that is used to enclose the secondary wiring of a sign transformer is permitted as the bonding means for total accumulative lengths not exceeding:
    a.   6 feet.
    b.   25 feet
    c.   50 feet
    d.   100 feet

Key Notes

Grounding and Bonding for Electronic Equipment

Article 645 covers electrical requirements for information technology equipment. Section 645.1 reads, "**Scope**. This article covers equipment, power-supply wiring, equipment interconnecting wiring, and grounding of information technology equipment and systems, including terminal units, in an information technology equipment room." Articles 725 and 800 apply to interconnection of information technology equipment located outside the computer room, such as for personal computers, workstations, and printers. A FPN to 645.1 reads, "For further information, see NFPA 75-1999, *Standard for the Protection of Electronic Computer/Data Processing Equipment.*"

OBJECTIVES

After studying this chapter, the reader will be able to understand

- Requirements for grounding of information technology equipment.
- Isolated equipment grounding conductors.
- Proper grounding methods for a data processing system.
- Reducing ground loop effects.
- Equalizing potential in a computer room.
- High frequency effects in grounding conductors.
- Surge arresters and TVSS usage.
- Harmonic currents.

As indicated, some grounding requirements for information technology equipment are contained in Article 645. Due to the rules on organization found in 90.3, the requirements for grounding in Article 250 apply to information technology equipment except as modified in Article 645.

Special conditions are imposed in 645.2 before the provisions of Article 645, including those for grounding, are permitted to apply to the information technology room under consideration. These requirements are as identified in (1) through (5) below:

(1) Disconnecting means complying with 645.10 are provided.

(2) A separate heating/ventilating/air-conditioning (HVAC) system is provided that is dedicated for information technology equipment use and is separated from other areas of occupancy. Any HVAC system that serves other occupancies shall be permitted to also serve the information technology equipment room if fire/smoke dampers are provided at the point of penetration of the room boundary. Such dampers shall operate on activation of smoke detectors and also by operation of the disconnecting means required by 645.10.

FPN: For further information, see NFPA 75-1999, *Standard for the Protection of Electronic Computer/Data Processing Equipment*, Chapter 8, 8-1, 8-1.1, 8-1.2 and 8-1.3.

(3) Listed information technology equipment is installed.

(4) The room is occupied only by those personnel needed for the maintenance and functional operation of the installed information technology equipment.

(5) The room is separated from other occupancies by fire-resistant-rated walls, floors, and ceilings with protected openings.

FPN: For further information on room construction requirements, see NFPA 75-1999, *Standard for the Protection of Electronic Computer/Data Processing Equipment*, Chapter 3.[1]

Rooms for information technology equipment that do not meet all these conditions must be wired and grounded in accordance with the general requirements of *NEC* Chapters 1-4. It should also be pointed out that the provisions of Article 645 apply only within the information technology equipment room and not to other spaces, such

as in general office areas, where all the features of 645.2 (1) through (5) are not provided.

Grounding requirements in 645.15 read, "All exposed non-current-carrying metal parts of an information technology system shall be grounded in accordance with Article 250 or shall be double insulated." This is a general reference to Article 250 where grounding requirements for all types of equipment are found. Few additional requirements or modifications of the general grounding requirements in Article 250 are found in Article 645.

Where signal reference structures are installed, they shall be bonded to the equipment grounding system provided for the information technology equipment.

In addition, 645.15 provides that, "Power systems derived within listed information technology equipment that supply information technology systems through receptacles or cable assemblies supplied as part of this equipment shall not be considered separately derived for the purpose of applying 250.20(D)." Section 250.20(D) provides the requirements for grounding of electrical systems such as for services provided by the electric utility as well as for systems that are separately derived. Without the modification that is provided in 645.15, electrical systems produced in power distribution units or uninterruptible power systems (UPS) that are commonly used for information technology equipment must be grounded as separately derived systems in accordance with 250.30.

Finally, 645.15 requires, "Where signal reference structures are installed, they shall be bonded to the equipment grounding system provided for the information technology equipment." This requirement will be explored later in this chapter.

Two fine print notes provide additional information. FPN No. 1 reads, "The bonding and grounding requirements in the product standards governing this listed equipment ensure that it complies with Article 250." FPN No. 2 states, "Where isolated grounding-type receptacles are used, see 250.146(D) and 406.2(D)." [2]

An additional reference to grounding of IT equipment is found in 250.6 Objectionable Current over Grounding Conductors. The general requirement is that grounding conductors be installed to prevent *objectionable current* from flowing over the grounding path(s). The *NEC* does not define or describe what objectionable current is but, certainly, any current flowing on grounding conductors that would in any way impair their performance of their intended function would be objectionable.

Section 250.6(B) outlines the following steps that can be taken to stop objectionable current:

(1) Discontinue one or more but not all of such grounding connections.

(2) Change the locations of the grounding connections.

(3) Interrupt the continuity of the conductor or conductive path interconnecting the grounding connection.

(4) Take other suitable remedial and approved action.[3]

Section 250.6(D), however, specifically applies to IT equipment or more broadly to all electronic equipment. It reads, "The provisions of this section shall not be considered as permitting electronic equipment from being operated on ac systems or branch circuits that are not grounded as required by this article. Currents that introduce noise or data errors in electronic equipment shall not be considered the objectionable currents addressed in this section."[4] This requirement was added after reports that some IT manufacturers or installers were not installing the equipment grounding conductors with the branch circuit to the equipment but were connecting IT equipment only to a local grounding point such as a driven ground rod.

Another general requirement, the importance of which cannot be overstated, is that the neutral or grounded system conductor not be grounded past the service disconnecting means [see 250.24(A)(5)]. Grounding of the neutral or system grounded conductor on the load side of the service creates parallel paths for neutral current over each path that is established such as over conduit, piping systems, cable shields and cable trays. An exception to this rule is that separately derived systems must be grounded, if the system produced can be grounded under the conditions given in 250.20 (see chapter twelve for additional information on grounding requirements for separately derived systems).

Basic Rule 250.24(A)(5)

1. Line-side of service only

2. No load-side connection

Grounded conductor

Service equipment

Grounding electrode

Load side

Equipment grounding conductor

Figure 17-1. System Grounding Connections

Purposes of Grounding Electrical Equipment

A broad, general rule is provided in 250.4(A)(2) on grounding of electrical equipment. This rule applies to all electrical equipment, including information technology equipment, as there is no modification of this rule in Article 645. The rule states, "Non-current-carrying conductive materials enclosing electrical conductors or equipment, or forming part of such equipment, shall be connected to earth so as to limit the voltage to ground on these materials."[5]

Section 250.4(A)(5) contains another important rule that also clearly applies to information technology equipment. It reads, "Electrical equipment and wiring and other electrically conductive materials likely to become energized shall be installed in a manner that creates a permanent, low-impedance circuit capable of safely carrying the maximum ground-fault current likely to be imposed on it from any point on the wiring system where a ground fault may occur to the electrical supply source."

The electrical supply source is either the transformer by the electric utility for the service or a separately derived electrical system. Electrical systems all follow the basic laws of physics: a complete circuit is required before current will flow, and current will always return to its source. The task then, is to provide the "permanent, low-impedance circuit capable of safely carrying the maximum ground-fault current likely to be imposed on it."[6]

This section goes on to require, "The earth shall not be used as the sole equipment grounding conductor or effective ground-fault current path."[7]

Safety Grounds and Clean Grounds

The *NEC* does not use the terms *safety ground* or *clean ground*. These terms are sometimes used in the information-technology industry to differentiate between the equipment grounding conductor required in the branch circuit to the equipment and the isolated equipment grounding conductor sometimes installed for IT equipment. The reference to the safety ground in IT jargon is to the equipment grounding conductor as defined in Article 100. This safety ground is often the conduit for the branch circuit and is sometimes referred to as the *dirty ground* as it is felt to carry unwanted electrical noise from a variety of sources. Sometimes unsafe actions are

Figure 17-2. Isolated Ground Circuits

taken to interrupt the equipment ground (safety or dirty ground) and install grounding means near the IT equipment. Such practice is not permitted by any safety code and can create shock or electrocution hazards.

As mentioned above, 645.15 requires that IT

Usually, the owner, design engineer, or equipment manufacturer specifies the circuit to have isolated equipment grounds.

equipment be grounded in accordance with Article 250. Section 250.86 applies generally and requires, "Except as permitted by 250.112(I), metal enclosures and raceways for other than service conductors shall be grounded." The three exceptions to this rule do not apply to IT equipment, nor does the rule in 250.112(I).

However, 250.110 gives conditions under which "exposed non-current-carrying metal parts of fixed equipment likely to become energized shall be grounded." Conditions from this section that might apply to IT equipment and require grounding include:

(1) Where within 2.5 m (8 ft) vertically or 1.5 m (5 ft) horizontally of ground or grounded metal objects and subject to contact by persons

(3) Where in electrical contact with metal

(5) Where supplied by a metal-clad, metal-sheathed, metal-raceway, or other wiring method that provides an equipment ground ...[8]

It is safe to say that metal enclosures for IT equipment must be grounded by an equipment grounding conductor. This equipment grounding conductor is required to either be the metal raceway acceptable in 250.118 as an equipment grounding conductor or to be an equipment grounding conductor contained within the wiring method to the IT equipment. Under no circumstances should the fault-current path to the source of power be interrupted and an equipment grounding connection be made to a local grounding electrode such as water pipe, building steel or driven electrodes. To do so violates all the principles of safety. This connection certainly will provide a high-impedance path and

will constitute ineffective grounding. Serious shock hazards can occur in this arrangement since the overcurrent device that supplies the equipment will not clear the fault as little current will flow in the high-impedance ground-fault path.

As mentioned, the term *clean ground* often refers to the isolated equipment grounding conductor sometimes specified for IT equipment or receptacle outlets by the equipment manufacturer or owner. This isolated equipment grounding conductor is permitted in 250.96(B). However, the section begins by stating, "Where required for the reduction of electrical noise (electromagnetic interference) on the grounding circuit...." The *NEC* does not address the issue of who makes the decision about what the circumstances are that would justify the use of isolated equipment grounding. Usually, the owner, design engineer, or equipment manufacturer specifies the circuit to have isolated equipment grounds.

Section 250.96(B) goes on to read, "an equipment enclosure supplied by a branch circuit shall be permitted to be isolated from a raceway containing circuits supplying only that equipment by one or more listed nonmetallic raceway fitting(s) located at the point of attachment of the raceway to the equipment enclosure. The metal raceway shall comply with provisions of this article and shall be supplemented by an internal insulated equipment grounding conductor installed in accordance with 250.146(D) to ground the equipment enclosure." Note that this section does not require the use of a metal raceway to supply the IT equipment. If PVC conduit were used, the special insulating connector between the metal race-

Figure 17-3. Isolated Equipment Grounding

Figure 17-4. Isolated Grounding Circuits 250.146(D), 408.20 Exception

Figure 17-5. Isolated Grounding Receptacles

way and the IT equipment would not be required. Of course, the insulated equipment grounding conductor must be installed inside the PVC conduit to ground the IT equipment.

The reference here to installing the insulated equipment grounding conductor in accordance with 250.146(D) permits the equipment grounding conductor to pass through one or more panelboards without being connected to equip-

ment grounding terminal bars in the panelboard. The insulated equipment grounding conductor must be connected to the first of the following points to provide the low-impedance path required:

1. Source of the separately derived system if it supplies the branch circuit

2. Building disconnecting means

3. Service equipment

Note that if a metal raceway is used, the race-

Figure 17-6. Standard Grounding Receptacle

Figure 17-7. Grounding System Components

way must be bonded as necessary to ensure an adequate return fault-current path even though an insulated equipment grounding conductor is installed through the raceway for grounding the IT equipment. This ensures that the raceway does not become a shock hazard if a line-to-ground fault to the raceway occurs.

Proper Grounding Methods for a Data Processing System

To comply with the safety requirements of Article 250, an equipment grounding conductor of a type recognized in 250.118 must be installed as either the wiring method itself such as a metal conduit or an equipment grounding conductor inside the conduit or tubing. The equipment grounding means must be connected to the metal

> Many computer data signals operate at 5 V or less, which makes the system more susceptible to acquiring electrical noise from power circuits if cables are run closely together.

enclosure for the IT equipment. The equipment grounding conductor, where it is a wire, is sized from Table 250.122 based on the rating of the overcurrent protective device ahead of the branch circuit. For example, a 10 AWG copper wire is permitted as the equipment grounding conductor in branch circuits having from 30-ampere through 60-ampere overcurrent protection. This will ensure the ground-fault return path will meet the rules in 250.4(A)(5) in being a "permanent, low-impedance circuit, capable of safely carrying the maximum ground-fault current likely to be imposed on it." This rule applies regardless of whether the branch circuits that supply the IT equipment originate in the electrical system provided from the electric utility service or in a dedicated power distribution unit (PDU) for the IT equipment.

As mentioned above, under no conditions should the ground-fault return path be interrupted and be connected to some other ground-ing means such as a local grounding electrode. Doing so can create a shock or electrocution hazard and aggravate the very problem such action is intended to solve.

Reducing Ground Loop Effects

The term *ground loops* is not defined in the *NEC*. It is used in the IT industry to refer to installations where ground current can flow on multiple paths, and it is felt this ground current or electrical noise travels in a circular fashion, hence the term ground loops. Connecting computers and terminals on different circuits that are grounded to building steel while the shielded communications cable between the computer and terminal completes the ground loop can cause this situation. Other connections through equipment grounding paths can add additional ground loops. Differences in ground potential will let noise current flow and intercouple with data signals. Many computer data signals operate at 5 V or less, which makes the system more susceptible to acquiring electrical noise from power circuits if cables are run closely together.

One method for solving this problem is to supply all equipment from the same power supply. The equipment grounding means of the branch-circuit wiring will serve to keep the ground potential the same. This solution may be impractical where the computer is located a long distance from peripherals and is supplied by different power supplies or different panelboards that, in turn, are supplied by different feeders.

Known solutions to the problem of ground loops include:

1. Single point grounding. This solution may not be suitable at very high frequencies or where equipment is located a long distance from another and is connected by shielded cables.

2. A *balun coil*, which is a coil of insulated wire on a nonmetallic core. This can be installed in the power branch circuit or the data circuit. Manufacturers sometimes build this into equipment. Baluns are not effective at all frequencies.

3. Modems which are normally used as interfaces with telephone circuits.

4. Fiber optic transmission over completely non-conducting paths.

5. Optical isolators.

6. Interface devices such as surge arresters and transient voltage surge suppressors.

Equalizing Potential in a Computer Room

For the purposes of grounding IT equipment, an *equipotential plane* is described as, "a mass (or masses) of conducting material that, when bonded together, provide a uniformly low im-

> Information technology equipment is connected to the signal reference grid to equalize potential between components.

pedance to current flow over a large range of frequencies."[9] The *signal reference structure* is defined as, "A system of conductive paths among interconnected equipment that reduces noise-induced voltages to levels that minimize improper operation. Common configurations include grids and planes."[10] Keeping all electrical equipment, including IT equipment, at the same potential is

necessary in an attempt to minimize disruptions in the operation of IT equipment.

Advantages of an equipotential plane include:
(1) Low impedance return path for radio frequency (RF) noise currents
(2) Containment of electromagnetic (noise) fields between their source (cables and similar) and the plane
(3) Increasing filtering effectiveness of contained electromagnetic fields, and
(4) Shielding of adjacent circuits or equipment
Equipotential plane structures can be established in several forms. Typical forms include:
(1) A conductive grid embedded in, or attached to, a concrete floor
(2) Metallic screen or metal strips under floor tile
(3) Ceiling grid above equipment, and
(4) Supporting grid of raised access flooring typical in computer rooms

Purpose of a Signal Reference Grid

The term *signal reference grid* refers to a structure such as a computer floor or a copper mesh grid installed under the computer floor. Information technology equipment is connected to the signal reference grid to equalize potential between components. All equipment in the

● 6 AWG or larger copper conductor bonded to raised floor framing members

Electrical Panelboard →

Equipment grounding conductor terminal bar →

Typical raised floor in information technology room

Figure 17-8. Static Grounding & Signal Reference Grids

room, including equipment that is wall-mounted, should be connected to the grid. The *NEC* is silent on this subject but, as with other bonding or grounding methods used for IT equipment, such bonding cannot substitute for the equipment grounding conductor required to be installed with the branch-circuit conductors. The signal reference grid can be thought of as overlaying the equipment grounding conductors (sometimes referred to as safety grounds) that are required.

The signal reference grid serves as a signal reference plane over a broad range of frequencies. They are correctly referred to as broadband grounding systems. An effort is being made to reduce, eliminate or control the tendency of conductors connected to computers to resonate at higher frequencies. A grid provides multiple parallel conducting paths between its metal parts. If one path is a high-impedance path because of full or partial resonance, other paths of different lengths will be able to provide a lower impedance path. A signal reference grid could be constructed of continuous sheet copper or aluminum, zinc-plated steel, or any number of pure or composite metals with good surface conductivity. However, this type of construction would not

only be expensive but difficult to install in a computer room where other services have been or will be installed (see photo 17-1).

Grids of copper or aluminum strips are sometimes installed under a computer-raised floor. This grid provides a satisfactory constant potential reference network over a broad range of frequencies from dc to higher than 30 MHz. Typically, these have been formed of 4 AWG copper or aluminum conductors which have been electrically joined at their intersections, or by copper straps approximately 0.254 mm (0.010 in.) thick by 76 mm to 102 mm (3 to 4 in.) wide, also joined at their intersections. These grids typically lie directly upon the

Photo 17-1. Signal Reference Grids

Computer units connected to a signal reference grid and to AC ground

(1) through (4) are typical computer system modules.

(5) is the "green wire" safety equipment grounding conductor.

(6) is the signal reference grounding grid for raised floor structure.

Figure 17-9. Grounding Using A Signal Reference Grid

① 2' x 2' high-frequency signal ground grid floor of computer room

② Signal reference (short as possible)

③ Run in power supply raceway individual grounds (green conductor)

④ Terminal block power supply panel

⑤ Service or system grounding electrode-60 hertz ground

Figure 17-10. Equipment Grounding Conductor

subfloor under the IT room raised floor. Cables and conduits under the floor would normally lie below the raised floor but above the grid.

As mentioned, the metal structure for the raised floor is sometimes used for the signal reference grid. Typically, the floor grid is 600 mm x 600 mm (2 ft x 2 ft) as this size fits with standard cellular raised floor systems. The two essential requirements are bolted-down stringers (the lateral supporting braces installed between supporting pedestals), and suitably plated (tin or zinc) members so that low-resistance pressure connections can be made. It is important that all metal components of the floor structure be in good electrical contact with each other. This may require using bolts and clamps along with a copper or aluminum bonding conductor.

Some stringerless supporting structures are marketed in which the removable floor tiles lock onto the supports by gravity but it has not been demonstrated that the contact has a low enough resistance and is free from intermittent contact as people move about or loads are moved across the floor. The bolted down horizontal stringers (braces) with pressure-type spring washers or springs in the assembly have been shown to be highly suitable for the purpose.

Experience has shown that it is unnecessary that lift-out floor panels make a low-resistance contact with the supporting network of stringers. The plastic or synthetic rubber cushions or molded edging upon which the panels rest have been found to be adequate to drain static electricity from the panel if any should accumulate. A resistance as high as 20,000 megohms under 20 percent relative humidity or higher will satisfy this requirement.

Once the signal reference grid has been established in the IT room, the various IT units can each be connected to the grid by a flat braided copper strap. The connection should be made from each IT unit, preferably at a point near where the equipment grounding conductor is connected, to the nearest intersection of the reference grid. The strap should be no longer than necessary and should have few bends and very little loop or sag to minimize the impedance at high frequency. In addition, the effectiveness of the reference grid is improved if it is solidly connected to the power supply for the IT equipment by a very short strap.

If the IT equipment manufacturer does not subscribe to the philosophy of connecting the IT equipment to the grid at multiple locations, the stray capacitance between the IT equipment and the raised floor reference grid will still help reduce voltage differences between grounded parts of the IT equipment.

High Frequency Effects in Grounding Conductors

Avoiding resonances at radio frequencies has become more important as the frequencies of digital signal circuits have increased beyond the 3 to 10 MHz clock repetition rate. Resonance occurs when the length of a conductor and the frequency of the alternating current are in tune. This is the principle of tuning a radio transmitting tower or antenna for maximum resonance and maximum radiation.

At frequencies slightly below and above resonance, the partial resonance still increases the impedance so it is ineffective as a constant potential conductor such as that needed as a good ground reference. Good engineering practice requires that that a conductor any longer than 1/20 of a wavelength cannot be counted upon to equalize voltages between its ends. This amounts to only 1.34 m (4.4 ft) at 10 MHz.

The significance of this is that unless conductors can be limited to less than 1.22 m to 1.52 m

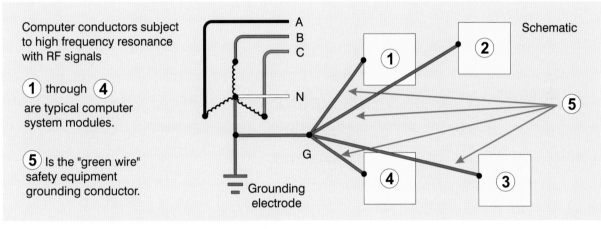

Figure 17-11. Multi-point Grounding Using Signal Reference Grid

Figure 17-12. Multi-point Grounding Using Signal Reference Grid

Computer conductors subject to high frequency resonance with RF signals

(1) through (4) are typical computer system modules.

(5) is the "green wire" safety equipment grounding conductor.

Figure 17-13. Equipment Grounding Conductors Connected Back to Source (PDU).

(4 to 5 ft) in length, conventional grounding techniques with single-point grounds may not be effective for signal and noise frequencies up to 10 MHz. At higher frequencies and longer length of conductors, other techniques are needed to avoid resonances.

The use of multipoint grounding and short conductors appear to be the simplest and most reliable methods for dealing with signals over 10 MHz in frequency. If bypass capacitors are installed instead of bonding straps, it is possible to make a conductor shield appear to be grounded at frequencies where the capacitors appear to have very low impedance. At low frequencies, they carry very little current so the same system may have the characteristics of a single-point ground at power frequencies and other low frequencies. A signal reference grid is an effective way to ground systems and equipment installed in computer rooms.

Establishing a Single-Point Ground Reference

Deciding whether to use single-point grounding or multipoint grounding typically depends on the frequency range of equipment. Single-point grounding is typically applied to analog circuits with signal frequencies up to 300 kHz, while digi-

tal circuits with signal frequencies in the MHz range should utilize multipoint grounding.

The concept of single-point grounding is similar to the concept of equipotential grounding used in swimming pools. The intent is to maintain potential (voltage) differences among various IT equipment as low as possible. It should be noted that single-point grounding includes a set of grounding conductors that are in addition to the required equipment grounding conductors (sometimes referred to as "safety grounds"). Single-point grounding should never be installed instead of the required equipment grounding conductors.

Single-point-grounding conductors are often connected from all equipment to a supplementary (enhanced) grounding electrode in the vicinity of IT equipment. However, a bonding conductor must be installed from this supplementary grounding electrode to the grounding electrode system for the electrical system. The supplementary grounding electrode must never be isolated from the electrical system grounding electrode or be permitted to serve in place of the equipment grounding conductors (safety grounds). This clearly violates the installation instructions for equipment that is listed by a qualified electrical testing laboratory.

The term *surge arrester* is defined in 280.2 as "A protective device for limiting surge voltages by discharging or bypassing surge current, and it also prevents continued flow of follow current while remaining capable of repeating these functions." Rules for installing surge arrestors can be found in Article 280. They are permitted to be installed either at the service equipment or on the load side of the service equipment. Surge arresters installed on circuits of less than 1000 volts must be listed for the purpose. Surge arresters must be made inaccessible to other than qualified persons unless listed for installation in accessible locations. Special rules for installing surge arresters on circuits of 1 kV and over can be found in Article 280.

Surge arresters are of several basic types with different characteristics serving different functions in different ways. When used in combination in a coordinated manner, they provide protection by limiting overvoltage, reducing stress and unnecessary failures in components and reducing component failures in equipment caused by surges.

Two types of surge arresters become conductive when their threshold voltage is exceeded. The first type becomes and stays conductive until the current that flows is reduced to zero such as when an alternating current passes through zero in changing direction of flow. The second type becomes conductive when its voltage threshold is exceeded but returns to its open circuit state as soon as the voltage drops below its threshold.

Surge arresters are of several basic types with different characteristics serving different functions in different ways

Pellet type and gas discharge type arresters are typical of the first type of surge arresters. They are usually capable of handling very high currents because the voltage drop through them becomes very low when they are conducting. However, if a weak, short duration impulse of a millisecond triggers the arrester, the shunting action would essentially short circuit the line for at least half of a cycle or possibly longer. The transient caused by shorting the power conductor may be greater than the event that triggered it.

The second type, typical of nonlinear insulat-

Photo 17-2. Various Transient Voltage Surge Suppressors and Associated Markings

ing material, becomes conductive when voltage gradients exceed their threshold values. Some semiconductors in an avalanche mode exhibit similar characteristics not unlike Zener diodes. Metal oxide varistors or MOVs are marketed and widely used for arresting surges. These are constructed of an amorphous oxide material and exhibit these characteristics. Other types of surge arrestors of a silicon solid-state devices are marketed which perform similar functions.

Surge arresters divert surge currents rather than absorb them. They are connected from ungrounded terminals in service equipment or panelboard to the ground terminal. Other common connection points include power conditioning equipment, the line and load side of transformers, motor generators and UPS systems.

There are risks associated with these devices and they must be installed in compliance with manufacturer's installation instructions. Some devices may discharge flaming materials when they fail and must be mounted in suitable enclosures. The devices should be connected to the protected equipment with short conductors not more than .304 mm (1 ft) or so in length.

Some depth of technical understanding of surge voltages, the nature of traveling pulse waves, their reflection and diversion may be needed for proper selection of surge arresters and the locations where they are most likely needed and can be most effective. In general, they are most likely needed wherever current in an inductive circuit element can be interrupted. They also might be useful where long cable leads interconnect widely separated systems as in different buildings or parts of a system within a very large building. Surge arresters are often installed at entrances of power and communications cables into a building to divert lightning and power switching surges before they reach IT equipment, power conditioning equipment or air-conditioning equipment.

TVSS Usage

Transient Voltage Surge Suppressors are covered in Article 285. The term *transient voltage surge suppressor (TVSS)* is defined in 285.2 as "A protective device for limiting transient voltages by diverting or limiting surge current; it also prevents continued flow of follow current while remaining capable of repeating these functions." Specific installation requirements similar to those for

surge arresters can be found in Article 285. Unlike surge arresters, TVSS devices are required to be listed and do not have a rating in excess of 600 volts. They are not permitted on ungrounded systems. According to 285.4, "where used at a point on a circuit, the TVSS shall be connected to each ungrounded conductor." For other than receptacle-type devices, TVSS devices must be marked with their short-circuit current rating and are not permitted to be installed at any point on the electrical system where the available fault current is in excess of that rating (see photo 17-2).

Harmonic Currents

Harmonics in the power source voltage to IT equipment can interfere with the proper operation of some IT equipment internal regulators. It can also create extra losses, excessive ground leakage current in line filters, and couple unwanted signals into low signal level conductors.

Harmonic currents are produced by equipment classed as being a nonlinear load. The term *nonlinear load* is defined in Article 100 as, "A load where the wave shape of the steady-state current does not follow the wave shape of the applied voltage." A FPN following the definition indicates that

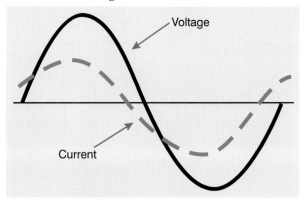

Figure 17-14. Capacitive Linear Load

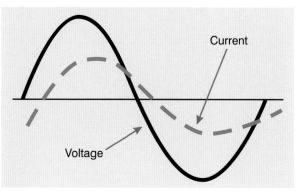

Figure 17-15. Inductive Linear Load

"Electronic equipment, electronic/electric-discharge lighting, adjustable-speed drive systems, and similar equipment may be nonlinear loads." The power supplies in information technology equipment often are switching mode power supplies and are nonlinear loads. This nonlinear load can cause several problems in the electrical supply system including overloading neutral conductors and creating excessive heat in transformers.

Harmonic currents appear as reflected waveforms that distort the basic sine wave produced and provided by the electric utility. These reflected waves are multiples of the fundamental root frequency (60 Hz) sine wave. The odd multiples, such as the 3rd (180 Hz), 5th (300 Hz), 7th (420 Hz) add to the distortion of the sine wave and will appear as a single distorted waveform on an oscilloscope.

When the load current of IT units is distorted and contains harmonics, they interact with the power source impedance to create voltage drops

Figure 17-16. Nonlinear Load

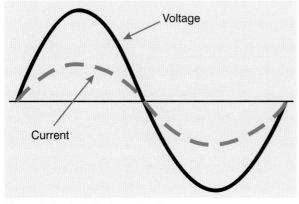

Figure 17-17. Resistive Linear Load

at the harmonic currents that may be present. Line voltage distortion from this cause is the result of IT load characteristic interacting with power source impedance. It is not correctable at the

Harmonic currents appear as reflected waveforms that distort the basic sine wave produced and provided by the electric utility.

power source except to the extent that the source impedance can be reduced at each harmonic frequency by filters. Capacitors in parallel with the load may help as well but may cause unwanted resonances at specific frequencies.

Engineers will often specify *K-rated* transformers for IT installations. These transformers are designed to handle, without failure, the extra heat produced in the windings by the harmonic currents. K-rated transformers are available with ratings from K-4 through K-20. In addition, design engineers often specify oversize neutrals to handle the extra nonlinear load without excessive voltage drop or heating.

---

[1] NFPA 70, *National Electrical Code* 2002, 645.2 (National Fire Protection Association, Quincy, MA, 2002), p. 70-509.

[2] NFPA 70, 645.15, p. 511.

3 NFPA 70, 250.6(B), p. 70-98.

[3] NFPA 70, 250.6(D), p. 98.

[4] NFPA 70, 250.4(A)(2), p. 97.

[5] NFPA 70, 250.4(A)(5), p.97.

[6] NFPA 70, 250.4(A)(5), p.97.

[7] NFPA 70, 250.110, p. 111.

[8] IEEE Std 1100-1999, *IEEE Recommended Practice for Powering and Grounding Electronic Equipment.*

For more detailed information on Grounding and Bonding requirements for Information Technology Equipment see *FIPS 94 Guidelines on Electrical Power for ADP Installations and* FIPS Federal Information Processing Standards Publication

[9] IEEE Std 1100-1999, *IEEE Recommended Practice for Powering and Grounding Electronic Equipment.*

[10] Burke, Thomas M., "Listing Requirements for ITE Installed in Computer Rooms," *IAEI News*, Volume 73, Number 1, January/February 2001, p. 11.

[11] Jones, Robert A., "Disconnecting Means—Requirements as Stated in 645.20 & 11," *IAEI News*, Volume 73, Number 1, January/February 2001, p. 22.

[12] Johnston, Michael J., "Installations and Inspections of Information Technology Equipment," *IAEI News*, Volume 73, Number 1, January/February 2001, p. 28.

Chapter Seventeen: The questions included here were developed using material included in this chapter. The answers can be found by reviewing the text. It is also important that students make use of the 2002 *NEC*, where many of the answers can be found.

1. Which of the following Articles in the NEC covers electrical requirements for information technology equipment specifically?
   a.   Article 250
   b.   Article 645
   c.   Article 725
   d.   Article 800

2. The grounding requirements for information technology equipment in Article 250 apply
   a.   To the entire installation
   b.   Unless amended in Article 90
   c.   Unless amended in Article 250
   d.   Unless amended in Article 645

3. The wiring methods permitted in Article 645 for information technology equipment are permitted to be used
   a.   Only where all the conditions of 645.2 are complied with
   b.   For computer room equipment located in dedicated spaces
   c.   For interconnection of peripheral equipment in open office spaces
   d.   Provided a disconnecting means is located at exit doors

4. Power systems derived within listed information technology equipment that supply information technology systems
   a.   Are considered separately derived systems and must be grounded
   b.   Are considered separately derived systems but are not required to be grounded
   c.   Are not considered separately derived systems
   d.   Are considered uninterruptible power Systems (UPS)

5. Currents that introduce noise or data errors in electronic equipment
   a.   Are considered a design flaw and are corrected as desired
   b.   Are considered an unavoidable event in computer operation
   c.   Are not considered objectionable currents
   d.   Are not considered in design of systems

6. Which of the following sections in Article 250 generally requires that the neutral not be grounded past the service?
   a.   250.20
   b.   250.24(A)(5)
   c.   250.30(B)(2)
   d.   250.66

7. Which of the following statements most accurately describes effective grounding?
   a.   permanent
   b.   low-impedance
   c.   adequate current-carrying
   d.   all of the above

8. The terms *safety ground* and *clean ground*
   a.   are defined in NFPA-70
   b.   are defined in NFPA 75
   c.   are IT jargon
   d.   all of the above

9. Grounding IT equipment to *local* grounding means such as a ground rod rather than using the *safety* ground
   a.   is never permitted
   b.   is a design choice
   c.   is permitted to solve *dirty ground* problems
   d.   is not likely to create a shock or electrocution risk

10. Which of the following statements is true regarding isolated grounding receptacles or equipment?
   a.   They are permitted to be installed only where necessary.
   b.   An insulated equipment grounding conductor is required.
   c.   The equipment grounding conductor can pass through one or more panelboards.
   d.   All of the above.

11. The insulated equipment grounding conductor for isolated grounding receptacles or equipment is permitted to terminate at which of the following locations?
   a. Source of the separately derived system if it supplies the branch circuit.
   b. The computer room disconnecting means.
   c. The local grounding electrode near the computer room.
   d. None of the above.

12. To reduce *ground loops* which of the following actions is sometimes taken?
   a. Single-point grounding.
   b. Insert a balun coil.
   c. Fiber optic cables.
   d. All of the above.

13. Advantages of establishing an equipotential plane in IT equipment rooms include which of the following?
   a. Low-impedance return path for radio frequency (RF) noise currents.
   b. Containment of electromagnetic (noise) fields between their source (cables and similar) and the plane.
   c. Increasing filtering effectiveness of contained electromagnetic fields.
   d. All of the above.

14. A *signal reference grid* can be thought of or is permitted to
   a. Replace the equipment grounding conductor of the branch circuit.
   b. Overlay the equipment grounding conductor of the branch circuit.
   c. Be used in addition to the equipment grounding conductor of the branch circuit.
   d. All of the above.

15. Signal reference grids are typically constructed of which of the following materials or in which method?
   a. Copper or aluminum strips.
   b. 4 AWG copper or aluminum conductors.
   c. The metal structure for the raised floor.
   d. All of the above.

16. Which of the following statements about high frequencies in computer room circuits is or are true?
   a. Digital signal circuits have increased beyond the 8 to 12 MHz clock repetition rate.
   b. Resonance occurs when the length of a conductor and the frequency of the alternating current are in tune.
   c. Good engineering practice requires that that a conductor any longer than $1/6$ h of a wavelength cannot be counted upon to equalize voltages between its ends.
   d. All of the above.

17. Which of the following statements about single-point grounding in computer rooms is or are true?
   a. The intent is to maintain potential (voltage) differences among different IT equipment as low as possible.
   b. Single-point grounding is typically applied to digital circuits with signal frequencies up to 300 kHz
   c. Single-point grounding includes a set of grounding conductors that replace the equipment grounding conductors.
   d. Supplementary grounding electrodes are permitted to be isolated from the electrical system grounding electrode.

18. Which of the following statements about surge arrestors and transient voltage surge suppressors is NOT true?
   a. Surge arresters protect equipment by absorbing excessive voltages.
   b. Surge arresters installed on circuits of less than 1000 volts are required to be listed for the purpose.
   c. TVSS devices are required to be listed and do not have a rating in excess of 600 volts.
   d. Two types of surge arrestors become conductive when their threshold voltage is exceeded.

19. Which of the following statements about harmonic currents and nonlinear loads is NOT true?

   a. Nonlinear loads can cause several problems in the electrical supply system including overloading neutral conductors and creating excessive heat in transformers.

   b. Harmonic currents appear as reflected waveforms that distort the basic sine wave produced and provided by the electric utility.

   c. The even multiples of the sine wave, such as the 2nd (120 Hz), 4th (240 Hz), 6th (360 Hz) are the most troublesome harmonic currents.

   d. Engineers often specify oversize neutrals to handle the extra nonlinear load without excessive voltage drop or heating.

Key Notes

CHAPTER EIGHTEEN

Low-Voltage and Intersystem Grounding and Bonding

Grounding requirements for low-voltage circuits and systems less than 50 volts are covered in 250.20(A). Typical low-voltage systems falling into this category can include motor control circuits, Class 1, 2, or 3 circuits, and remote control or signaling circuits, such as low-voltage circuits for electric door release equipment or doorbell circuits. The *Code* requires these systems to be grounded under any of the following conditions:

"(1)Where supplied by transformers, if the transformer supply system exceeds 150 volts to ground."[1] A low-voltage ac system must be grounded if it is supplied by a transformer secondary that is supplied on the primary side with a voltage exceeding 150 volts from the ungrounded conductors to ground. The primary circuits for these transformers are from a grounded system, so there will be measurable voltages from any of the phase (ungrounded) conductors to earth ground or grounded metal parts.

"(2)Where supplied by transformers, if the transformer supply system is ungrounded."[2] A secondary system must also be grounded where the transformers for these circuits or systems are supplied on the primary side by an ungrounded system, such as by an ungrounded 460-volt or 240-volt power system, which is not required to be grounded by 250.20(B).

"(3) Where installed as overhead conductors outside of buildings."[3] When the conductors are installed in exterior locations as overhead conductors outside of buildings, the previously mentioned low-voltage systems or circuits must be grounded. These would include, but are not limited to, signal circuits, fire alarm, and other control circuits or systems.

Where low-voltage circuits or systems do not fall into the requirements of 250.20(A) as outlined above, grounding is optional. In other words, these types of systems or circuits could be operated ungrounded with no conductor of the system or circuit supplied by the secondary, if the transformer is intentionally grounded.

Low-voltage Motor Control Circuits

Low-voltage motor control circuits are required to be grounded if any of the conditions in 250.20(A)(1) or 250.20(A)(2) exist.

Motor controllers and combination motor control units are often supplied with control-circuit transformers that are factory grounded on the secondary of the low-voltage side by the manufacturer to ensure compliance with this section of the *Code*. If the motor control circuits are field installed and wired, then grounding may or may not be required, depending on the voltage or characteristics of the circuit on the primary side of the control circuit transformer. The determining factors for grounding a low-voltage motor control circuit on the secondary of the transformer are

either: (a) if it is supplied on the primary by a circuit derived from an ungrounded system, or (b) the primary is supplied by a circuit that exceeds 150 volts between any of the primary conductors to ground. The requirements for motor control circuits are located in Part VI of Article 430.

The *motor control circuit* is defined as the circuit of a control apparatus or system that carries the electric signals directing the performance of the controller but does not carry the main power current of the motor itself. The requirements for motor control circuits include overcurrent protection, physical protection, disconnecting means, and control circuit transformers located in enclosures. Section 430.72(C) includes the overcurrent protection requirements for the control circuit transformers. Motor control circuit transformers supplying "a Class 1 power-limited circuit, Class 2, or Class 3 remote-control circuit" must comply with the requirements of Article 725.

Other Signaling Circuits

Low-voltage signaling circuits are required to be grounded circuits or systems if either 250.20(A)(1) or (2) applies. For example, the circuit is supplied by a transformer that has a primary voltage of 480-volts single-phase and steps the voltage down for the signal circuit to 24-volts AC for a horn or bell annunciation.

Intrinsically Safe Systems and Circuits

Branch circuits that include an equipment grounding conductor as covered in 250.118 must supply intrinsically safe systems. The equipment grounding conductor is for grounding the metal enclosure(s) and other metal parts and equipment of the system.

Article 504 includes the requirements for intrinsically safe systems, with 504.50 and 504.60 including the grounding and bonding requirements for these systems.

Section 504.50(A) requires "Intrinsically safe apparatus, associated apparatus, cable shields, enclosures, and raceways, if of metal, shall be grounded." A control drawing is required for these systems to provide specific information and instructions. These control drawings include grounding and bonding information critical to the integrity of the intrinsically safe system or circuit(s). Supplementary bonding to the grounding electrode may be needed for some associated apparatus (zener diode barriers, for example) if specified in the control drawing. Additional information relative to these systems can be found in ANSI/ISA RP 12.6-1995, *Wiring Practices for Hazardous (Classified) Locations Instrumentation Part 1: Intrinsic Safety.*

It is not uncommon for the required control drawing(s) to specify a grounding electrode conductor connection for these systems in addition to connection to an equipment grounding conductor. Usually, terminals for both of these conductors are located within the system enclosure or control panel. If a grounding electrode is required, then one of the electrodes in 250.50 must be used. The choice of which grounding electrode is required is governed by the same means specified in 250.30(A)(4) for separately derived systems The electrode must be "as close as practicable and preferably in the same area as the grounding electrode conductor connection to the system." The grounding electrode shall be the nearest one of the following:

(1) An effectively grounded structural metal member of the structure

(2) An effectively grounded metal water pipe within 1.5 m (5 ft) from the point of entrance into the building

(3) Other electrodes as specified by 250.52 where

the electrodes specified by 250.30(A)(4)(1) or (A)(4)(2) are not available[4]

The size of this grounding conductor is specified by the manufacturer's control drawing. "Where shielded conductors or cables are used, shields shall be grounded," unless the shield conductor is itself part of the intrinsically safe circuit [see 504.50(C) and Exception]. The bonding requirements for metallic enclosures and raceways enclosing intrinsically safe systems or circuits in hazardous locations must be in accordance with any of the methods that are suitable for the bonding on the line side of a service.

Section 250.100 indicates that "Regardless of the voltage of the electrical system, the electrical continuity of non-current-carrying metal parts of equipment, raceways, and other enclosures in any hazardous (classified) location as defined in Article 500 shall be ensured by any of the methods specified for services in 250.92(B) that are approved for the wiring method used."

Sections 501.16, 502.16, and 503.16 also provide information relative to restrictive bonding requirements in hazardous locations. Bonding must be extended and maintained for all intervening raceways and enclosures from the hazardous location all the way to the applicable ser-

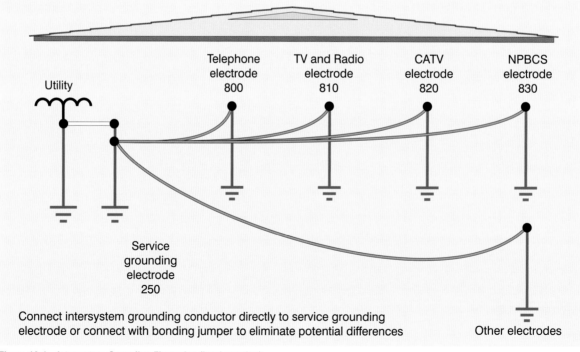

Connect intersystem grounding conductor directly to service grounding electrode or connect with bonding jumper to eliminate potential differences

Figure 18-1. Intersystem Grounding Electrodes (bond together)

vice or separately derived system (see chapter sixteen for more detailed information on bonding requirements in hazardous locations).

The Purpose of Intersystem Grounding and Bonding

Intersystem grounding and bonding requirements serve a few important purposes. The systems and circuits covered in *NEC* Chapter 8 must be grounded to earth and bonded to the electrical power distribution systems for the building or structure. By providing the connection to a grounding electrode, the systems and circuits are afforded reasonable protection from spike and surge currents and also brief elevated potentials due to lightning strikes. Bonding the electrodes of the systems listed in Chapter 8 to the grounding electrode system for the electrical system(s) of the building or structure limits the potential differences during normal operation and during surge or spike events and the effects of lightning discharges into the earth at close proximities. Bonding the electrodes of the two systems together limits the potential differences and shock hazards that could result from isolated earthing connections.

The *NEC* is specific in Articles 800, 810, 820, and 830 about grounding electrodes to be used

with these systems. Each article requires an effective bonding connection to the electrical power grounding electrode(s) system with a minimum 6 AWG copper conductor. This means that where items such as satellite dishes, TV antennas, radio antennas, and the like are installed at any occupancy, electrodes are required and the electrode must be the same electrode used by the electrical service or system. All grounding electrodes that are installed must be bonded to the building grounding electrode system

The grounding and bonding for these systems, although referred to as *intersystem grounding and bonding*, is usually not the grounding of the particular system itself. Part II of Article 250 is specifically titled Circuit and System Grounding. Section 250.20 specifies systems required to be grounded, systems permitted to be grounded, and systems not permitted to be grounded.

The requirements for safety of communications systems, antenna systems, CATV and radio systems, and network-powered broadband communications systems are located in *NEC* Chapter 8 entitled Communications Systems. Section 250.94 includes the requirements for intersystem bonding and grounding. The systems that the *Code* points to here are low-voltage systems and frequency signals or communications circuits and

Figure 18-2. Bonding for Other Systems 250.94

systems. The systems here are not normally grounded systems as covered in 250.20, where one conductor of the system or service is intentionally grounded. The grounding and bonding requirements for the systems covered provide safety from shock hazards and fire hazards and, in addition, provide important equipment and system/circuit protection from damaging surges or dips in the normal electrical supply. The following definitions are applicable to the installations of these voice, data, and video (VDV) systems.

Grounding Conductor. A conductor used to connect equipment or the grounded circuit of a wiring system to a grounding electrode or electrodes.[5]

Bonding (Bonded). The permanent joining of

Figure 18-3. Intersystem Bonding

Figure 18-4. Communications Systems Grounding and Bonding

metallic parts to form an electrically conductive path that ensures electrical continuity and the capacity to conduct safely any current likely to be imposed.[6]

Bonding Jumper. A reliable conductor to ensure the required electrical conductivity between metal parts required to be electrically connected.[7] Article 800, titled Communications Circuits, covers telephone, telegraph (except radio), outside wiring for fire alarm and burglar alarm, and similar central station systems; and telephone systems not connected to a central station system but using similar types of equipment, methods of installation, and maintenance for such circuits and systems. Grounding and bonding requirements for communications systems are found in Part IV and specifically in 800.40 and 800.41.

(1) Insulation. The grounding conductor shall be insulated and shall be listed as suitable for the purpose.

(2) Material. The grounding conductor shall be copper or other corrosion-resistant conductive material, stranded or solid.

(3) Size. The grounding conductor shall not be smaller than 14 AWG.

(4) Length. The primary protector grounding conductor shall be as short as practicable. In one- and two-family dwellings, the primary protector grounding conductor shall be as short as practicable, not to exceed 6.0 m (20 ft) in length.[8]

This grounding also serves as a path of discharge for lightning strikes and other events that work to raise potentials on these systems. By keeping the length of this grounding conductor as short as practicable, the effectiveness of the grounding conductor to provide a low-impedance path to ground to dissipate lightning and other surge events is established.

The Exception 800.40(A)(4) reads *"In one- and two-family dwellings where it is not practicable to achieve an overall maximum primary protector grounding conductor length of 6.0 m (20 ft), a separate communi-*

Figure 18-5. **Length of Grounding Conductor for Cable and Primary Protector (Dwelling Units)**

Figure 18-6. **Length of Grounding Conductor for Cable and Primary Protector (Dwelling Units)**

cations ground rod meeting the minimum dimensional criteria of 800.40(B)(2)(2) shall be driven, the primary protector shall be grounded to the communications ground rod in accordance with 800.40(C), and the communications ground rod bonded to the power grounding electrode system in accordance with 800.4(D)." In this case, the electrode for the communications system is to be installed and bonded to the service grounding electrode system with a bonding jumper sized not smaller than 6 AWG copper (see figures 18-5 and 18-6).

Section 800.40(A)(5) also requires that "The grounding conductor shall be run to the grounding electrode in as straight a line as practicable." If the grounding conductor is installed in a metal raceway, both ends of the grounding conductor must be bonded to the grounding conductor.

Electrodes

For communications systems, the goal is to establish a connection to a grounding electrode. The *NEC* specifies several acceptable points on an existing building or structure grounding electrode system where this connection can be made. This preferred method eliminates distances and potential differences between the earthing circuits of the two systems. Section 800.40(B) requires the grounding conductor be connected "to the nearest accessible location on the following:

(1) The building or structure grounding electrode system as covered in 250.50
(2) The grounded interior metal water piping system, within 1.5 m (5 ft) from its point of entrance to the building, as covered in 250.52
(3) The power service accessible means external to enclosures as covered in 250.94
(4) The metallic power service raceway
(5) The service equipment enclosure
(6) The grounding electrode conductor or the grounding electrode conductor metal enclosure
(7) The grounding conductor or the grounding electrode of a building or structure disconnecting means that is grounded to an electrode as covered in 250.32.

For purposes of this section, the mobile home service equipment or the mobile home disconnecting means, as described in 800.30(B), shall be considered accessible.[9]

In buildings or structures that have no grounding electrode or grounding electrode system, an electrode is required to be installed. The connections of grounding conductors to grounding electrodes shall comply with the provisions in 250.70. "Connectors, clamps, fittings, or lugs used to attach grounding conductor and bonding jumpers to grounding electrodes or to each other that are to be concrete-encased or buried in the earth shall be suitable for its application."[10]

Intersystem Grounding and Bonding at Mobile Homes

The grounding for communications systems on a mobile home generally follows the same requirements as for permanent buildings or structures. Electrically, the objectives are similar. The goal is to establish a grounding or earthing connection to eliminate differences of potential, and to

provide a path into the earth to dissipate lightning strikes.

"Where there is no mobile home service equipment located in sight from, and not more than 9.0 m (30 ft.) from, the exterior of the mobile home it serves, or there is no mobile home disconnecting means grounded in accordance with 250.32 and located within sight from and not more than 9.0 m (30 ft.) from, the exterior wall of the mobile home it serves, the primary protector ground shall be in accordance with 800.40(B)(2)."[11] The bonding requirements are a bit more specific for the mobile home applications.

In buildings or structures that have no grounding electrode or grounding electrode system, an electrode is required to be installed.

"**(B) Bonding.** The primary protector grounding terminal or grounding electrode shall be bonded to the metal frame or available grounding terminal of the mobile home with a copper conductor not smaller than 12 AWG under any of the following conditions:

(1) Where there is no mobile home service equipment or disconnecting means as in 800.41(A)
(2) Where the mobile home is supplied by cord and plug"[12]

Article 810 is titled Radio and Television Equipment and "covers antenna systems for radio and television receiving equipment, amateur radio transmitting and receiving equipment, and certain features of transmitter safety. This article covers antennas such as multi-element, vertical rod, and dish, and also covers the wiring and cabling that connects them to equipment. This article does not cover equipment and antennas used for coupling carrier current to power line conductors."

Article 820 is titled Community Antenna Television and Radio Distribution Systems and "covers coaxial cable distribution of radio frequency signals typically employed in community antenna television (CATV) systems."

Article 830 is titled Network-Powered Broadband Communications Systems and includes the minimum safety requirements for these "systems that provide any combination of voice, audio, video, data, and interactive services through a network interface unit."

Each of these articles includes the grounding and bonding requirements for these particular systems although the system itself is usually ungrounded.

As indicated by 90.3, *NEC* Chapter 8 is independent of the other chapters and stands alone unless specifically referenced from therein. The term *grounding conductor* as used in Chapter 8 in any of these articles is referring to the conductor that is connected to an electrode that functions similar to a grounding electrode conductor. The *Code* permits the minimum sizes of these grounding or earthing conductors to be smaller than normally is allowed by Chapter 2 in Article 250. The communications system grounding conductor is permitted to be as small as a 14 AWG. However, the bonding conductor required to connect the grounding electrode for the power distribution system to the grounding electrode for the communication system is required to be a minimum 6 AWG.

Cable Grounding Requirements

The metallic sheaths of communications cables entering buildings shall be grounded as close as practicable to the point of entrance or shall be interrupted by an insulating joint or equivalent device located as close as practicable to the point of entrance. If the metallic shields of the cables are grounded, they are required to be grounded by a conductor that is insulated and suitable for the purpose. This conductor is required to be copper or other corrosion resistant conductor and can be solid or stranded not smaller than 14 AWG. It is important that the routing and orientation of the conductor be installed so that it will be in as straight a line as practicable. When functioning to dissipate the effects of lightning, it is important that the conductor be installed with straight lines and gradual bends. The grounding conductor is required to be installed in a manner so as to be protected from physical damage. This is often accomplished by installing it in a raceway. If the grounding conductor is installed in a metal raceway, both ends of the raceway shall be bonded to the grounding conductor or the same terminal or electrode to which the grounding conductor is connected. The grounding electrode to be used shall be electrode or point of grounding connection as indicated in 800.40(B)(1):

"**(B) Electrode.** The grounding conductor shall be connected in accordance with 800.40(B)(1) and (B)(2):

(1) In Buildings or Structures with Grounding Means. To the nearest accessible location on the following:

(1) The building or structure grounding electrode system as covered in 250.50

(2) The grounded interior metal water piping system, within 1.5 m (5 ft) from its point of entrance to the building, as covered in 250.52

(3) The power service accessible means external to enclosures as covered in 250.94

(4) The metallic power service raceway

(5) The service equipment enclosure

(6) The grounding electrode conductor or the grounding electrode conductor metal enclosure

(7) The grounding conductor or the grounding electrode of a building or structure disconnecting means that is grounded to an electrode as covered in 250.32."[13]

Differences of Potential

Bonding together all separate electrodes will limit potential differences between them and between their associated wiring systems. Grounding electrodes of different systems or circuits that are not bonded together can present hazards for property and persons. It is important for eliminating shock hazards to bond these intersystem electrodes to the power distribution system grounding electrode system.

Bonding Electrodes of Different Systems

Section 800.40(D) states that "A bonding jumper not smaller than 6 AWG copper or equivalent shall be connected between the communications grounding electrode and power grounding electrode system at the building or structure served where separate electrodes are used. Bonding together of all separate electrodes shall be permitted."

Radio and TV Antennas

Grounding and Discharge Equipment

Masts and metal structures supporting antennas shall be grounded in accordance with 810.21. Where required, each conductor of a lead-in from an outdoor antenna shall be provided with a listed antenna discharge unit. The antenna discharge unit shall be grounded in a manner specified by 810.21.

Grounding Conductor Installations

The grounding conductor for a radio or TV antenna system is required to be connected to an electrode or other suitable grounding connection point as specified by 810.21(F). These requirements are the same as the requirements for communications systems in Article 800. The size of the grounding conductors is a bit different, however, than the size required in Article 800.

(A) **Material.** The grounding conductor shall be of copper, aluminum, copper-clad steel, bronze, or similar corrsion-resistant material. Aluminum or copper-clad aluminum grounding conductors shall not be used where in direct contact with masonry or the earth or where subject to corrosive conditions. Where

used outside, aluminum or copper-clad aluminum shall not be installed within 450 mm (18 in.) of the earth.

(C) **Supports.** The grounding conductors shall be securely fastened in place and shall be permitted to be directly attached to the surface wired over without the use of insulating supports.

(D) **Mechanical Protection.** The grounding conductor shall be protected where exposed to physical damage, or the size of the grounding conductors shall be increased proportionately to compensate for the lack of protection.

(E) **Run in Straight Line.** The grounding conductor for an antenna mast or antenna discharge unit shall be run in as straight a line as practicable from the mast or discharge unit to the grounding electrode.

(G) **Inside or Outside Building.** The grounding conductor shall be permitted to run either inside or outside the building.

(H) **Size.** The grounding conductor shall not be smaller than 10 AWG copper, 8 AWG aluminum, or 17 AWG copper-clad steel or bronze.

(I) **Common Ground.** A single grounding conductor shall be permitted for both protective and operating purposes.[14]

Electrode

The *Code* specifies several acceptable connection points on an existing building or structure grounding electrode system. This would be the preferred method, and eliminates distances and potential differences between the earthing circuits of the two systems. Section 810.21(F) requires the grounding conductor to be connected to the nearest accessible location on the following:

"a. The building or structure grounding electrode system as covered in 250.50

b. The grounded interior metal water piping systems, within 1.52 m (5 ft) from its point of entrance to the building, as covered in 250.52

c. The power service accessible means external to the building, as covered in 250.94

d. The metallic power service raceway

Radio

TV

Radio and Television grounding conductors per 810.21

Connection to be made within 1.5 m (5 ft) from the point of entrance of the water pipe

Ground rod

Effectively grounded water pipe

Figure 18-7. Interior Metal Water Piping as an Electrode

e. The service equipment enclosure, or

f. The grounding electrode conductor or the grounding electrode conductor metal enclosures..."[15]

If the building or structure has no grounding electrode or grounding electrode system, any of the electrodes in 250.52 can be established. It is required to install an electrode if any of the electrodes in 250.50 are not available. According to 810.21(G) and (H), "the grounding conductor is permitted to run either inside or outside the building," and "shall not be smaller than 10 AWG copper, 8 AWG aluminum, or 17 AWG copper-clad steel or bronze." The connection of the grounding conductor shall be made to grounding electrode and serves both as an operational grounding connection and as a protective earthing connection for lightning and other events affecting the system and potential differences

The grounding electrode for a radio or television antenna is required to be bonded to the electrode system for the building. "A bonding jumper not smaller than 6 AWG copper or equivalent shall be connected between the radio and television equipment grounding electrode and the power grounding electrode system at the building or structure served where separate electrodes are used"[16] (see figure 18-8).

Amateur Transmission and Receiving Stations

The installation of any lead-in antenna conductors shall be in such a manner that affords a degree of protection against accidental contact by personnel.

Section 810.56 requires that "lead-in conductors to radio transmitters shall be located or installed so as to make accidental contact with them difficult." Where transmitting or receiving stations are installed, "each conductor of a lead-in for outdoor antennas shall be provided with an antenna discharge unit or other suitable means that will drain static charges from the antenna system.

"Exception No. 1: Where protected by a continuous metallic shield that is permanently and effectively grounded.

Radio and TV antenna system grounding connection to metal water pipe electrode required to be made within 1.5 m (5 ft) from the point of entrance of the water pipe

Figure 18-8. Interior Metal Water Piping as an Electrode

Exception No. 2: Where the antenna is permanently and effectively grounded."[17]

There are two grounding conductors required for this specific equipment: an operating grounding conductor and a protective grounding conductor. These grounding conductors for amateur transmitting and receiving stations are required to be installed in accordance with 810.58(A), (B), or (C) as follows:

(A) Other Sections. All grounding conductors for amateur transmitting and receiving stations shall comply with 810.21(A) through (J).

(B) Size of Protective Grounding Conductor. The protective grounding conductor for transmitting stations shall be as large as the lead-in but not smaller than 10 AWG copper, bronze, or copper-clad steel.

(C) Size of Operating Grounding Conductor. The operating grounding conductor for transmitting stations shall not be less than 14 AWG copper or its equivalent.[18]

Community Antenna Television and Radio Distribution (CATV) Systems

CATV systems are usually wired with conductors that are coaxial and have shields inherent to the cable assembly. Where coaxial cable is exposed to lightning or to accidental contact with lightning arrester conductors or power conductors operating at a voltage of over 300 volts to ground, "the outer conductive shield of the coaxial cable shall be grounded at the building premises as close to the point of cable entrance or attachment, as practicable."[19] Where these shields are required to be grounded, selecting a grounding location to achieve the shortest practicable grounding conductor will help limit potential differences between CATV and other metallic systems. "Where the outer conductive shield of a coaxial cable is grounded, no other protective devices shall be required."[20]

Where the shielded cables are grounded, they shall be grounded in a manner specified in 820.40(A) through (D) as follows:

(A) Grounding Conductor.

(1) Insulation. The grounding conductor shall be insulated and shall be listed as suitable for the purpose.

(2) Material. The grounding conductor shall be copper or other corrosion-resistant conductive material, stranded or solid.

(3) Size. The grounding conductor shall not be smaller than 14 AWG. It shall have a current-

Figure 18-9. Length of Grounding Conductor for Coaxial Cable

carrying capacity approximately equal to that of the outer conductor of the coaxial cable...

(4) Length. The grounding conductor shall be as short as practicable. In one- and two-family dwellings, the grounding conductor shall be as short as practicable, not to exceed 6.0 m (20 ft) in length." [*Editor's note:* An exception permits the grounding conductor to be connected to a separately installed electrode where the maximum distance specified cannot be achieved. In this case the separate electrode is to be installed and the electrode for the electrical service and the CATV electrode are required to be bonded together with a minimum 6 AWG copper bonding jumper as specified in 820.40(D) (see figure 18-9).

(5) Run in Straight Line. The grounding conductor shall be run to the grounding electrode in as straight a line as practicable.

(6) Physical Protection. Where subject to physical damage, the grounding conductor shall be adequately protected. Where the grounding conductor is run in a metal raceway, both ends of the raceway shall be bonded to the grounding conductor or the same terminal or electrode to which the grounding conductor is connected.

(B) Electrode. The grounding conductor shall be connected in accordance with 820.40(B)(1) and (B)(2).

(1) In Buildings or Structures with Grounding Means. To the nearest accessible location on the following:

(1) The building or structure grounding electrode system as covered in 250.50;
(2) The grounded interior metal water piping system, within 1.5 m (5 ft) from its point of entrance to the building, as covered in 250.52;
(3) The power service accessible means external to enclosures as covered in 250.94;
(4) The metallic power service raceway;
(5) The service equipment enclosure;
(6) The grounding electrode conductor or the grounding electrode conductor metal enclosure; or
(7) To the grounding conductor or to the grounding electrode of a building or structure disconnecting means that is grounded to an electrode as covered in 250.32.

(2) In Buildings or Structures Without Grounding Means. If the building or structure served has no grounding means, as described in 820.40(B)(1):

Figure 18-10. Interior Metal Water Piping as an Electrode

(1) To any one of the individual electrodes described in 250.52 (A)(1), (2), (3), (4); or,

(2) If the building or structure served has no grounding means as described in 820.40(B)(1) or (B)(2)(1), to an effectively grounded metal structure or to any one of the individual electrodes described in 250.52(A)(5), (6), and (7).[21]

Where the CATV system and or shielding of coaxial cables are connected to grounding electrodes, they are required to be connected in accordance with 820.40(C) and (D). The "connections to grounding electrodes must also comply with applicable requirements of 250.70."[22]

The electrode(s) of the CATV system must be bonded to the electrical service or system grounding electrode by means of a bonding jumper. Section 820.40(D) requires that "A bonding jumper not smaller than 6 AWG copper or equivalent shall be connected between the antenna systems grounding electrode and the power grounding electrode system at the building or structure served where separate electrodes are used." Selecting a grounding location to achieve the shortest practicable grounding conductor will help limit potential differences between CATV and other metallic systems.

Network-Powered Broadband Communications Systems

A typical basic network-powered broadband

> Selecting a grounding location to achieve the shortest practicable grounding conductor will help limit potential differences between CATV and other metallic systems.

communications system configuration "includes a cable supplying power and broadband signal to a network interface unit that converts the broadband signal to the component signals. Typical cables are coaxial cable with both broadband

Figure 18-11. Cable Network Interface Unit and Primary Protector Grounding

signal and power on the center conductor, composite metallic cable with a coaxial member for the broadband signal and a twisted pair for power, and composite optical fiber cable with a pair of conductors for power. Larger systems may also include network components such as amplifiers that require network power."[23]

"Network interface units containing protectors, NIUs with metallic enclosures, primary protectors, and the metallic members of the network-powered broadband communications cable that are intended to be grounded shall be grounded as specified in 830.40(A) through (D)."[24] These requirements are similar to those in Articles 800, 810, and 820. The purpose of these grounding and bonding connections is also the same.

The size of the grounding conductor for the NPBCS is required to be not "smaller than 14 AWG" copper "and shall have a current-carrying capacity approximately equal to that of the grounded metallic member(s) and protected conductor(s) of the network-powered broadband communications cable. The grounding conductor shall not be required to exceed 6 AWG."[25] "Connectors, clamps, fittings, or lugs used to attach grounding conductors and bonding jumpers to grounding electrodes or to each other that are to be concrete encased or buried in the earth shall be suitable for its application."[26] The grounding clamps and connectors suitable for use in accordance with 250.64 are suitable as well as listed clamps for use specifically with these low-voltage/current systems.

(B) Electrode. The grounding conductor shall be connected as follows.

(1) In Buildings or Structures with Grounding Means. To the nearest accessible location on the following:

(1) The building or structure grounding electrode system as covered in 250.50;

(2) The grounded interior metal water piping system, within 1.5 m (5 ft) from its point of entrance to the building, as covered in 250.52;

(3) The power service accessible means external to enclosures as covered in 250.94;

(4) The metallic power service raceway;

(5) The service equipment enclosure;

(6) The grounding electrode conductor or the grounding electrode conductor metal enclosure; or

(7) To the grounding conductor or to the grounding electrode of a building or struc-

Figure 18-12. Cable Network Interface Unit and Primary Protector Grounding

Where length exceeds 6.0 m (20 ft), ground rod permitted for primary protector

Grounding electrode conductor

AUDIO
DATA
VOICE
VIDEO

NIU

Network interface unit

Ground rod for primary protector permitted but required to be bonded to power system grounding electrode ture disconnecting means that is grounded to an electrode as covered in 250.32.

(2) In Buildings or Structures Without Grounding Means. If the building or structure served has no grounding means, as described in (B)(1):

(1) To any one of the individual electrodes described in 250.52 (A)(1), (2), (3), (4); or

(2) If the building or structure served has no grounding means as described in 820.40(B)(1) or (B)(2)(1), to an effectively grounded metal structure or to a ground rod or pipe not less than 1.5 m (5 ft) in length and 12.7 mm ($^1/_2$ in.) in diameter, driven, where practicable, into permanently damp earth and separated from lightning conductors as covered in 800.13 and at least 1.8 m (6 ft) from electrodes of other systems. Steam or hot water pipes

Grounding clamps are available that are suitable for these grounding connections, but these grounding clamps are not permitted to be used for power distribution system grounding electrode connections unless so listed.

or lightning-rod conductors shall not be employed as electrodes for protectors, NIUs, with integral protection, grounded metallic members, NIUs with metallic enclosures, and other equipment.[27]

Mobile home service equipment or mobile home disconnecting means are considered accessible points for connection of the grounding conductors for network-powered broadband communications systems.

Lightning and Other Hazards

"On network-powered broadband communications conductors not exposed to lightning or accidental contact with power conductors, providing primary electrical protection in accordance with [830.30(A)] will help protect against other hazards, such as ground potential rise caused by power fault currents, and above-normal voltages

induced by fault currents on power circuits in proximity to the network-powered broadband communications conductors.

FPN No. 2: Network-powered broadband communications circuits are considered to have a lightning exposure unless one or more of the following conditions exist:

(1) Circuits in large metropolitan areas where buildings are close together and sufficiently high to intercept lightning.

(2) Areas having an average of five or fewer thunderstorm days per year and earth resistivity of less than 100 ohm-meters. Such areas are found along the Pacific coast.[28]

Bonding of Electrode Systems

It is also required to bond the grounding electrode for a network-powered broadband communications system to the power system grounding electrode or electrode system for the building or structure. "A bonding jumper not smaller than 6 AWG copper or equivalent must be connected between the network-powered broadband communications system grounding electrode and the power grounding electrode system at the building or structure served where separate electrodes are used."[29]

The grounding and bonding requirements for the systems covered in *NEC* Chapter 8 are important for protection against electric shock and fire, as well as providing a level of protection from surges, spikes, dips, and lightning strikes. It is important to bond grounding electrodes of different systems together for safety and to comply with the minimum *NEC* rules. Effective grounding and bonding for these systems includes a connection of an equipment grounding conductor with the supply circuit for grounding equipment.

Section 250.94 covers the requirements for intersystem grounding and bonding. This section requires "an accessible means external to enclosures for connecting intersystem bonding and grounding conductors shall be provided at the service equipment and at the disconnecting means for any additional buildings or structures by at least one of the following means:

(1) Exposed nonflexible metallic raceways

(2) Exposed grounding electrode conductor

(3) Approved means for the external connection of a copper or other corrosion-resistant bonding or grounding conductor to the grounded raceway or equipment."[30]

Grounding clamps are available that are suitable for these grounding connections, but these grounding clamps are not permitted to be used for power distribution system grounding electrode connections unless so listed. An example would be the perforated metal-strap type grounding clamp used to connect grounding conductors of these systems

Failure to do so can result in shock hazards, fire hazards, and damage to electronic components of these systems.

to the nonflexible metallic service raceways.

The authority having jurisdiction would have the responsibility of approval of any provided external connecting or bonding means. The fine print note indicates that "6 AWG copper conductor with one end bonded to the grounded nonflexible metallic raceway or equipment and with 152 mm (6 in.) or more of the other end made accessible on the outside wall is an example of the approved means covered in 250.94(3)."[31]

Summary

The grounding and bonding requirements for voice, data, and video systems covered by *NEC* Chapter 8 are minimum requirements for safety. The *Code* includes grounding and bonding requirements for systems, circuits, and enclosures in Chapter 2, Wiring and Protection. Chapter 8 is independent from the requirements of Chapter 2, unless specifically referenced from the particular article in Chapter 8. There are requirements specified in the articles of Chapter 8 and specific requirements for grounding and bonding required for safety for these special systems. Proper grounding and bonding is essential for safety and operation of these systems. One of the most important requirements of the *Code* is to bond the electrodes or electrodes of different systems together. Failure to do so can result in shock hazards, fire hazards, and damage to electronic components of these systems. The *Code* includes the minimum re-

quirements for personnel and building safety, this simply means that one must do at least that much to comply. It is not uncommon to see the minimum requirements of the *NEC* be exceeded relative to the grounding and bonding for these systems, but the minimum requirements must be met. For information on enhanced grounding electrodes and grounding and bonding for sensitive electronic equipment, see chapter seventeen of this text.

See 250.60 for use of air terminals (lightning rods), and NFPA 780, *Lightning Protection Systems*.

For more information relative to the applicable requirements for a listed secondary protector refer to UL 497A-1996, *the Standard for Secondary Protectors for Communications Circuits*.

For more information relative to determining acceptable installation practices for telecommunications systems, circuits, and equipment refer to nationally recognized standards such as ANSI/EIA/TIA 568-A-1995, *Commercial Building Telecommunications Cabling Standard*; ANSI/EIA/TIA 569-1990, *Commercial Building Standard for Telecommunications Pathways and Spaces*; and ANSI/EIA/TIA 570-1991, *Residential and Light Commercial Telecommunications Wiring Standard*.

[1] NFPA 70, *National Electrical Code* 2002, 250.20(A), (National Fire Protection Association, Quincy, MA, 2002), p. 70-98.
[2] NFPA 70, 250.20(A), p. 70-98.
[3] NFPA 70, 250.20(A), p. 70-98.
[4] NFPA 70, 250.30(A)(4), p. 70-102.
[5] NFPA 70, Article 100, p. 70-35.
[6] NFPA 70, Article 100, p. 70-33.
[7] NFPA 70, Article 100, p. 70-33.
[8] NFPA 70, 800.40(A), p. 70-592.
[9] NFPA 70, 800.40(B)(1), p. 70-592 & 70-593.
[10] NFPA 70, 800.40(C), p. 70-593.
[11] NFPA 70, 800.41(A), p. 70-593.
[12] NFPA 70, 800.41(B), p. 70-593.
[13] NFPA 70, 800.40(B)(1), p. 70-592 & 70-593.
[14] NFPA 70, 810.21, p. 70-599 & 70-600.
[15] NFPA 70, 810.21(F), p. 70-559 & 70-600.
[16] NFPA 70, 810.21(J), p. 70-600.
[17] NFPA 70, 810.57, p. 70-600.
[18] NFPA 70, 810.58, p. 70-601.
[19] NFPA 70, 820.33, p. 70-603.
[20] NFPA 70, 820.33(A), p. 70-603.
[21] NFPA 70, 820.40, p. 70-592 & 70-593.
[22] NFPA 70, 820.40(C), p. 70-604.
[23] NFPA 70, 830.1 FPN, p. 70-607.
[24] NFPA 70, 830.40, p. 70-613.
[25] NFPA 70, 830.40(A)(3), p. 70-613.
[26] NFPA 70, 830.40(C), p. 70-613
[27] NFPA 70, 830.40, p. 70-613.
[28] NFPA 70, 830.30 FPN No. 1 and No. 2, p. 70-612.
[29] NFPA 70, 810.21(H), p. 70-600.
[30] NFPA 70, 250.94, p. 70-109.
[31] NFPA 70, 250.94, p. 70-109.

Chapter Eighteen: The questions included here were developed using material included in this chapter. The answers can be found by reviewing the text. It is also important that students make use of the 2002 *NEC*, where many of the answers can be found.

1. The conductor used to connect the equipment or grounded circuit conductor of a wiring system to a grounding electrode is a _____.
   a.   equipment grounding conductor
   b.   grounding electrode conductor
   c.   neutral conductor
   d.   grounding conductor

2. An external, accessible means of connecting intersystem bonding and grounding conductor shall be provided at _____.
   a.   a separately derived system
   b.   the service equipment or disconnecting means for additional buildings
   c.   the telephone equipment mounting board
   d.   the point of entry of a cable to the building

3. The minimum size grounding conductor for a receiving station shall be not less than_____.
   a.   12 AWG copper
   b.   10 AWG copper
   c.   6 AWG aluminum
   d.   6 AWG copper

4. For a one-family dwelling a grounding conductor for a coaxial cable shall be as short as practicable and not exceed _____.
   a.   3.0 m (10 ft)
   b.   9.0 m (30 ft)
   c.   6.0 m (20 ft)
   d.   1.5 m (5 ft)

5. The shield of a coaxial cable shall be grounded with a grounding conductor not less than _____.
   a.   4 AWG copper
   b.   6 AWG copper
   c.   14 AWG copper or other corrosion resistant material
   d.   12 AWG copper

6. Where there is a grounding electrode installed for a coaxial cable antenna system that is greater than 6.0 m (20 ft) from the power grounding electrode system, there shall be a _____ bonding jumper installed between the two electrodes
   a.   4 AWG copper
   b.   8 AWG copper
   c.   6 AWG copper
   d.   12 AWG copper

7. A grounding conductor for a coaxial cable system shall be permitted to be connected to which of the following:
   a.   the building or structure grounding electrode system
   b.   the metallic power service raceway
   c.   the service equipment enclosure
   d.   all of the above

8. The minimum size bonding conductor connected to a primary protector grounding terminal for a network-powered broadband communications system at a mobile home that is cord- and plug-connected shall be not less than _____.
   a.   2 AWG copper
   b.   4 AWG aluminum
   c.   12 AWG copper
   d.   10 AWG bronze

9. Where a grounding conductor for a communications system is installed in a metal raceway, the raceway shall be _____
   a.   continuous
   b.   bonded to both ends of the grounding conductor or to the same terminal to which the grounding conductor is connected.
   c.   rigid metal conduit
   d.   electrical metallic tubing

10. Where a grounding conductor for network-powered broadband communications systems is connected to concrete-encased or buried electrodes the connections shall be_____.
   a.   accessible
   b.   stainless steel
   c.   suitable for its application
   d.   irreversible compression-type

11. Grounding conductors for network-powered broadband communications systems shall be spaced at least _____ from lightning conductors where practicable.
   a.   6 m (20 ft)
   b.   1.8 m (6 ft)
   c.   15 m (50 ft)
   d.   900 mm (3 ft)

12. A 24-volt ac system is required to be grounded where _____.
   a.   supplied by a transformer with a supply system of 120 volts
   b.   where installed as overhead conductors outside of buildings
   c.   where the supply system is grounded
   d.   where ground detectors are not installed

13. Where an intrinsically safe system is required to be connected to a grounding electrode, the electrode shall be which of the following:
   a.   a ground rod
   b.   a pipe electrode
   c.   the electrodes specified by 250.52(a)(1),(2),(3), and (4) where available
   d.   a plate electrode

14. Where shields are used with intrinsically safe systems, and the shield is part of the intrinsically safe circuit, the shield is required to _____.
   a.   not be grounded
   b.   be grounded
   c.   be isolated
   d.   be insulated

15. At a mobile home that is supplied by a cord-and plug-connection, the primary protector bonding conductor for a communications system is required to be bonded to the_____.
   a.   water pipe at the sink location
   b.   metal frame or available grounding terminal of the mobile home
   c.   a ground rod
   d.   the panelboard enclosure in the mobile home

1000 Volts and Over

OBJECTIVES

After studying this chapter, the reader will be able to understand

- Grounding rules for high voltage (1 kV and over).
- Use of surge arrestors.
- Grounding of outdoor industrial substations.

For grounding of systems or circuits of 1 kV and over, the *NEC* indicates that where such high-voltage systems are grounded, they shall comply with Article 250 plus other rules in Part X of Article 250 which supplement or modify other rules in Article 250.

Most medium-voltage systems (2.4 to 15 kV) are resistance grounded while others are high-resistance grounded. The only difference in the various methods of resistance grounding is the amount of ground-fault current that will flow due to the varying resistance.

Most high-voltage systems (above 15 kV) are effectively grounded utilizing surge arresters because the transmission lines are open and commonly subjected to lightning surges and transient overvoltages (see figure 19-1).

Systems Rated 2400 Volts to 13,800 Volts

Grounding may be solid or low-resistance depending on the available ground-fault current and the size of the system. Reactance grounding in this voltage range is preferred if the circuits are overhead and thus subject to lightning exposure, if they serve rotating machinery, and if excessive ground-fault current will not develop. It is common in industrial systems to ground the neutral of systems rated 2400 volts and above through a resistor.

Owing to the higher voltage of these systems, compared to systems of 600 volts and less, the ground-fault current levels that will flow in the grounding electrode conductor and grounding electrode is sufficiently high that a current transformer may be placed in the grounding electrode conductor to give positive indication that a ground fault exists or may be used to clear the circuit in the event the ground fault persists over a predetermined time. Compare this with a system of 600 volts or less. There, the current transformer cannot be placed satisfactorily in the grounding electrode conductor, for little ground-fault current will flow in that conductor.

Systems Rated 15,000 Volts or More

Such systems are nearly always outdoors and have no rotating equipment served by them. Usually they are solidly grounded. Such solid grounding permits the use of grounded-neutral type surge arresters, which cost less and provide better protection from overvoltage.

In systems over 600 volts, there will be different voltage levels for various utilization purposes. Grounding must be resorted to at each voltage level as previously described for systems of 600 volts or less.

Grounding Outdoor Industrial Substations

The grounding of outdoor industrial substations involves grounding the fence and other supporting structures, all equipment and conductor enclosures and the grounding of the neutral.

A ground bus or grid should be established extending about 3 feet outside the periphery of the fence. The bus should be connected to many ground rods around its periphery and, in addition, should be connected to a metallic underground water piping system where available. The ground bus should have a minimum size of 4/0 AWG and be approximately 25 percent of the

Figure 19-1. Surge Arresters

capacity of the system. Cable may be sized on the basis of the capacity of bare conductor in free air and copper bus on the basis of 1200 amperes per square inch. Connections from the fence and from all equipment within the fence,

Figure 19-2. Methods of Grounding Over 600 Volts

Effectively grounded Grounded through surge arresters.

Reactance grounded. Grounded through an inductor.

Low resistance grounded. Grounded through a resistor. Typically allowing 50 to 600 amperes of ground fault current.

High-Resistance grounded. Grounded through high resistance. Typical designs usually provide for a maximum of 10 amperes ground-fault current.

Figure 19-3. Use of Grounding Transformers

that is, transformer cases, steel structures, switchgear including operating mechanisms for gang-operated disconnects, and so forth, should be not less than 4/0 AWG nor less than 25 percent of the capacity of the secondary conductors. To be sure of having a good, permanent neutral ground, it is best to connect the neutral to two points on the ground bus.

A grounding bus and connections are effective only if the mechanical construction is sound and as permanent as possible. No connection should be soldered. It is preferable to properly braze or weld all connections and to protect all cable from mechanical injury. It is common to use exothermic welding of these connections. In addition, compression connections using special tools and connectors is a common industry practice as well. If a metallic enclosure is used for the mechanical protection of the conductor, the enclosure should be connected in parallel with the cable being protected from end to end of the enclosure.

There is some difference of opinion as to whether the fence ground should be separated from the ground bus used for the station. If the fence is connected to the station bus and a fault occurs, the fence will be elevated above earth potential by the IZ drop. The potential gradient in the earth near the fence drops off very rapidly within the first few feet. Anyone contacting the fence may thus be subjected to a hazardous step-voltage under fault conditions.

If the fence ground was isolated from the station ground, then the fence will be elevated from the station ground, and under fault conditions anyone contacting the fence and the equipment will be subjected to a hazard.

In most industrial substations, the fence and equipment are close. In such cases, it is best to connect the fence ground and the station ground bus. The shock hazard when using a common ground for fence and equipment can be kept to a minimum by having the ground connection resistance as low as possible (see figure 19-2).

Derived Neutral Systems
Section 250.182 permits "A system neutral derived from grounding transformer shall be permitted to be used

Figure 19-4. 250.184(B) Multiple Grounding

Figure 19-5. Impedance Grounded System

Figure 19-6. Grounding Using Zig-Zag Transformers

for grounding high-voltage systems." These grounding transformers include zigzag, wye-delta, or T-connected types. Ground-fault protection may be provided, and positive tripping can be accomplished with low magnitudes of ground-fault current (see figure 19-4).

Solidly Grounded Neutral Systems

Section 250.184 also permits solidly grounded neutral systems. Multiple grounding is permitted at more than one point for services, direct buried portions of feeders employing a bare copper neutral and for overhead portions of a system. The neutral conductor of solidly grounded systems is generally required to have an insula-

tion level of not less than 600 volts. Bare copper conductors are permitted (see figure 19-4).

"Grounding shall be permitted at one or more of the following locations:

(1) Transformers supplying conductors to a building or other structure

(2) Underground circuits where the neutral is exposed

(3) Overhead circuits installed outdoors[1]

Multigrounded Neutral Conductor.

Where multiple grounded neutral systems are used the following requirements apply:

(1) The multigrounded neutral conductor

New Section 250.184(D) "Multi-Grounded Neutral Conductor" has been added.

New section includes provisions for sizing the multi-grounded neutral, locations of grounding connections, and maximum distance between grounding connections.

Neutral conductor shall be sized not less than 33-1/3 percent of the phase conductors, and the exception allows a neutral no smaller than 20 percent of the phase conductors under engineering supervision in industrial and commercial where it is assured that qualified staff are available to service the installation.

Figure 19-7. 250.6. Multi-Grounded Neutral Conductor

shall be of sufficient ampacity for the load imposed on the conductor but not less than 33 ½ percent of the ampacity of the phase conductors.

(2) The multigrounded neutral conductor shall be grounded at each transformer and at other additional locations by connections to a made or existing electrode.

(3) At least one grounding electrode shall be installed and connected to the multigrounded neutral circuit conductor every 400 m (1300 ft).

(4) The maximum distance between any two adjacent electrodes shall not be more than 400 m (1300 ft).

(5) In a multigrounded shielded cable system, the shielding shall be grounded at each cable join that is exposed to personnel contact.[2]

Impedance Grounded Neutral Systems

Section 250.186 permits impedance grounded neutral systems. Impedance grounding of these systems can be accomplished by reactance grounding, resistance grounding or high-resistance grounding.

Grounding of Systems Supplying Portable or Mobile Equipment

Special requirements apply to electrical systems that supply portable or mobile high-voltage equipment, other than substations installed on a temporary basis. The requirements are found in 250.188.

(A) Portable or Mobile Equipment. Portable or mobile high-voltage equipment shall be supplied from a system having its neutral grounded through an impedance. Where a delta-connected high-voltage system is used to supply portable or mobile equipment, a system neutral shall be derived.

(B) Exposed Non-Current-Carrying Metal Parts. Exposed non-current-carrying metal parts of portable or mobile equipment shall be connected by an equipment grounding conductor to the point at which the system neutral impedance is grounded.

(C) Ground-Fault Current. The voltage developed between the portable or mobile equipment frame and ground by the flow of maximum ground-fault current shall not exceed 100 volts. [*Editor's note:* This somewhat limits the shock hazard.]

(D) Ground-Fault Detection and Relaying. Ground-fault detection and relaying shall be provided to automatically de-energize any high-voltage system component that has developed a ground fault. The continuity of the equipment grounding conductor shall be continuously monitored so as to de-energize automatically the high-voltage circuit to the portable or mobile equipment upon loss of continuity of the equipment grounding conductor.

(E) Isolation. The grounding electrode to which the portable or mobile equipment system neutral impedance is connected shall be isolated from and separated in the ground at least 6.0 m (20 ft) from any other system or equipment grounding electrode, and there shall be no direct connection between the grounding electrodes, such as buried pipe and fence, and so forth.

(F) Trailing Cable and Couplers. High-voltage trailing cable and couplers for interconnection of portable or mobile equipment shall meet the requirements of Part III of Article 400 for cables and 490.55 for couplers.[3]

Grounding of High-Voltage Equipment

Section 250.190 generally requires grounding of all non-current-carrying metal parts of fixed, portable or mobile equipment of high-voltage equipment including fences, housings, enclosures and supporting structures.

The exception to this section excludes from the grounding requirement equipment that is isolated from ground and located so as to prevent any person who can make contact with the ground from contacting such metal parts when the equipment is energized.

Surge Arresters

Where surge arresters are used on secondary services of less than 1000 volts, it is required that the connections to the service conductor and to the grounding conductor be as short as practicable.

Line and ground connecting conductors shall not be smaller than 14 AWG copper or 12 AWG aluminum. The arrestor grounding conductor shall be connected to one of the following:

(1) Grounded service conductor
(2) Grounding electrode conductor
(3) Grounding electrode for the service
(4) Equipment grounding terminal in the service equipment.[4]

The use of method (3) is practical only on an ungrounded system.

If a surge arrester protects the primary of a transformer which supplies a secondary distribution system, the grounding conductor may be interconnected to the secondary neutral, provided that in addition to the direct grounding connection at the arrester, the grounded conductor of the secondary has elsewhere a grounding connection to a continuous metallic underground water piping system. However, where there are not less than four secondary connections to any underground metallic water piping system per mile, the direct grounding connection at the arrester may be omitted.

The grounding electrode conductor of a lightning arrester also may be connected to the secondary neutral. This is true if, in addition to the direct grounding connection at the arrester, the grounding conductor of the secondary system is part of a multi-grounded neutral system of which the primary neutral has at least four grounding connections in each mile of line, as well as a ground at each service.

Where the secondary is not grounded to a metallic water system but uses other available electrodes, an interconnection between surge arrester grounding conductor and the secondary neutral shall be made through a spark gap. The spark gap must have a breakdown voltage of at least twice the primary circuit voltage but not necessarily more than 10 kV. There shall be at least one other ground on the grounded conductor of

Figure 19-8. Stress Cones 310.6

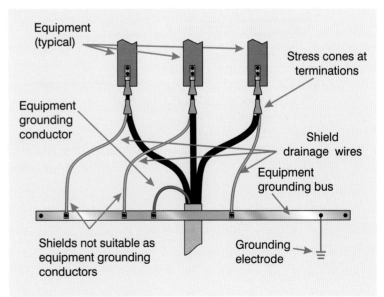

Figure 19-9. Stress Cones 310.6

the secondary not less than 6.0 m (20 ft) away from the surge arrester grounding electrode.

No other connection between a surge arrester ground and a secondary neutral is allowed by the *Code* except by special permission.

Stress Cones and Cable Shielding

Medium and high-voltage cables above 2000 volts generally are required to be shielded. This shielding is usually in the form of either a conductive tape or stranded shield conductors. The purpose of the shielding is to confine voltage stresses to the insulation and bleed off to ground capacitance at the termination points. Section 310.6 addresses shielding requirements for insulated conductors operating at over 2000 volts. These conductors are required to have an ozone resistant insulation and be shielded. These shields are required to be grounded through an effective grounding path meeting the requirements of 250.4(A)(5) or 250.4(B)(4) (see figures 19-7 and 19-8).

Exception: Nonshielded insulated conductors listed by a qualified testing laboratory shall be permitted for use up to 8000 volts under the following conditions:

(a) Conductors shall have insulation resistant to electric discharge and surface tracking, or the insulated conductor(s) shall be covered with a ma-terial resistant to ozone, electric discharge, and surface tracking.

(b) Where used in wet locations, the insulated conductor(s) shall have an overall nonmetallic jacket or a continuous metallic sheath.

(c) Where operating at 5001 to 8000 volts, the insulated conductor(s) shall have a nonmetallic jacket over the insulation. This insulation shall have a specific inductive capacity not greater than 3.6, and the jacket shall have a specific inductive capacity not greater than 10 and not less than 6.

(d) Insulation and jacket thickness shall be in accordance with Table 310.63.[5]

Article 490 includes grounding requirements for equipment over 600 volts, nominal. Generally the frames of switchgear and control assemblies are required to be grounded. "Devices with metal cases or frames, such as instruments, relays, meters, and instrument and control transformers, located in or on switchgear and control, shall have the frame or case grounded."[6]

[1] NFPA 70, *National Electrical Code* 2002, 250.184(B), (National Fire Protection Association, Quincy, MA, 2002), p. 70-119.
[2] NFPA 70, 250.184(D), p. 70-119.
[3] NFPA 70, 250.188, p. 70-120.
[4] NFPA 70, 280.21, p. 70-121.
[5] NFPA 70, 310.6, p. 70-135.
[6] NFPA 70, 490.37, p. 70-336.

Figure 19-10. Stress Cones 310.6

Chapter Nineteen: The questions included here were developed using material included in this chapter. The answers can be found by reviewing the text. It is also important that students make use of the 2002 *NEC*, where many of the answers can be found.

1. Most medium-voltage systems ____ to ____ are resistance grounded while others are high-resistance grounded.
   a.   5.4 - 20 kV
   b.   2.4 - 15 kV
   c.   3.3 - 17 kV
   d.   4.6 - 18 kV

2. Most high-voltage systems ____ are effectively grounded utilizing surge arresters because the transmission lines are open and commonly subjected to lightning surges and transient overvoltages.
   a.   above 15 kV
   b.   above 12 kV
   c.   above 10 kV
   d.   above 11 kV

3. It is common in industrial systems to ground the neutral of systems rated ____ volts and above through a resistor.
   a.   2000
   b.   2400
   c.   1800
   d.   1380

4. Systems rated ____ volts or more are nearly always outdoors and have no rotating equipment served by them.
   a.   10,000
   b.   12,000
   c.   15,000
   d.   13,000

5. In systems over ____ volts, there will be different voltage levels for various utilization purposes.
   a.   1000
   b.   2000
   c.   600
   d.   1500

6. The grounding of outdoor industrial substations is concerned with grounding the fence and other supporting structures, all equipment and conductor enclosures and the grounding of the neutral. A ground bus or grid should be established extending about ____ feet outside the periphery of the fence.
   a.   3
   b.   4
   c.   6
   d.   8

7. Cable can be sized on the basis of the capacity of bare conductor in free air and copper bus on the basis of ____ amperes per square inch.
   a.   1250
   b.   1000
   c.   1200
   d.   1300

8. Connections from the fence and from all equipment within the fence, that is, transformer cases, steel structures, switchgear including operating mechanisms for gang-operated disconnects, and so forth, should be not less than No. __ nor less than __ percent of the capacity of the secondary conductors.
   a.   3/0 – 50
   b.   4/0 – 25
   c.   2/0 – 40
   d.   1/0 – 30

9. To be sure of having a good, permanent neutral ground, it is best to connect the neutral to ____ points on the ground bus.
   a.   four
   b.   three
   c.   two
   d.   five

10. Impedance grounded neutral systems are permitted and can be accomplished by all but one of the following methods:
   a.   reactance grounding
   b.   resistance grounding
   c.   high-resistance grounding
   d.   low-resistance grounding

11. Without exception, where surge arresters are installed either indoors or outdoors they are required to be _____ to unqualified persons.
   a.   inaccessible
   b.   accessible
   c.   readily accessible
   d.   available

12. Where there are not less than four secondary connections to any underground metallic water piping system per each _____, the direct grounding connection at the arrester may be omitted.
   a. $^1/_4$ mile
   b.   mile
   c. $^1/_2$ mile
   d. $^3/_4$ mile

Key Notes

Tables

OBJECTIVES

After studying this chapter, the reader will
be able to understand

- Application of tables to determine
  proper types and sizing of grounding electrode conductor.

- How to determine where voltage
  drop will affect sizes.

- Determine minimum size of
  equipment grounding conductors.

- Dimensions and lengths of conduits
  and other raceways.

Included in this chapter are several tables that are vital or useful in studying the subject of grounding and correctly applying the *NEC* rules in that regard. In addition, several tables are included to assist the installer or inspector of electrical systems in properly designing, installing and maintaining these systems. By carefully utilizing the information in these tables, the electrical system will benefit from additional safety and should serve the owner for years to come.

The following tables are included:

Table 250.66
Grounding Electrode Conductor for AC Systems

Size of Largest Ungrounded Service-Entrance Conductor or Equivalent Area for (AWG/kcmil) Conductors[a]		Size of Grounding Electrode Parallel Conductor (AWG/kcmil)	
Copper	Aluminum or Copper-Clad Aluminum	Copper	Aluminum or Copper-Clad Aluminum[b]
2 or smaller	1/0 or smaller	8	6
1 or 1/0	2/0 or 3/0	6	4
2/0 or 3/0	4/0 or 250	4	2
Over 3/0 thru 350	Over 250 thru 500	2	1/0
Over 350 thru 600	Over 500 thru 900	1/0	3/0
Over 600 thru 1100	Over 900 thru 1750	2/0	4/0
Over 1100	Over 1750	3/0	250 kcmil

Notes:

1. Where multiple sets of service-entrance conductors are used as permitted in Section 230.40, Exception No. 2, the equivalent size of the largest service-entrance conductor shall be determined by the largest sum of the areas of the corresponding conductors of each set.

2. Where there are no service-entrance conductors, the grounding electrode conductor size shall be determined by the equivalent size of the largest service-entrance conductor required for the load to be served.

[a] This table also applies to the derived conductors of separately derived ac systems.

[b] See installation restrictions in 250.64(A).

Table 20-1.

Analysis of Table 250.66 *Grounding Electrode Conductor and Service Conductor Compared As To Relative Size				
Size of Copper Service Conductor AWG	Continuous Rating of Service Conductor 75°C Wire Ampacity	Size of Copper Grounding Electrode Conductor AWG	Continuous Rating Grounding Electrode Conductor 75°C Wire Ampacity	Grounding Electrode Conductor Expressed as a percent of Service Conductor
2 or smaller	115	8	50	43%
0	150	6	65	43%
3/0	200	4	85	43%
350 kcmil	310	2	115	37%
600 kcmil	420	1/0	150	36%
1000 kcmil	545	2/0	175	32%
1250 kcmil (over 1100 kcmil)	590	3/0	200	34%

* This grounding Electrode Conductor is that part of the Grounding System which is the sole connection between the grounding electrode and the grounded system conductor.

Table 20-2.

Copper Wire Size. AWG	DC Resistance of Bare Wire. Ohms/100'	Trade Size of conduit to enclose conductor. Inch	*Approximate DC Resistance of conduit. Ohms/100'	*Approximate Impedance of Conductor in Conduit. Ohms/100' at 500 Amps Sq. In. density	
				Conduit bonded at both ends	Conduit NOT bonded
8	0.07780	1/2	0.0321	0.105	0.210
6	0.04910	3/4	0.0242	0.078	0.156
4	0.03080	3/4	0.0242	0.078	0.156
2	0.01940	1	0.0154	0.056	0.112
0	0.01220	1	0.0154	0.056	0.112
00	0.00967	1	0.0154	0.056	0.112
000	0.00766	1 1/4	0.0120	0.044	0.088

Comparison of Copper Grounding Electrode Conductors per Table 250.66 When Used With and Without Steel Conduit for Physical Protection.

* These values are approximate but sufficiently accurate for all practical purposes. Manufacturing tolerances for conduit, although quite satisfactory for its physical requirements, can vary the resistance up to about 15 percent compared to standard dimensions.

Table 20-3.

		Comparison of Aluminum Grounding Electrode Conductors per Table 250.66 When Used With and Without Aluminum Conduit for Physical Protection.		
Aluminum Wire Size	DC Resistance Ohms/100'	Trade Size of Conduit to Enclose Conductor In.	Approximate DC Resistance of Aluminum Conduit Ohms/100'
6	0.08080	$^1/_2$	0.00800
4	0.05080	$^1/_2$	0.00800
2	0.03190	$^3/_4$	0.00605
1/0	0.02010	1	0.00385
3/0	0.01260	1	0.00385
4/0	0.00100	$1^1/_4$	0.00300
250 kcmil	0.00847	$1^1/_4$	0.00300

*Must be installed to comply with 250.64, that is, made electrically continuous from the point of attachment to cabinets or equipment to the grounding electrode and shall be securely fastened to the ground clamp.

Table 20-4.

Rating of Grounding Electrode Conductors Specified in Table 250.66 and Voltage Drop under Maximum Short Time Rating.

Grounding Conductor Size AWG	DC Resistance per 100 ft. Copper Conductor	Ampacity 75° C Wire Continuous Rating	Circular Mils	*Short-Time Rating Amps	Ratio Short time Rating to Continuous Rating	Voltage Drop per 30 m (100 ft.) at Short time Rating
8	0.077800	50	16510	391	7.8	30.4
6	0.049100	65	26240	621	9.6	30.5
4	0.030800	85	41740	988	11.6	30.4
3**	0.024500	100	52620	1245	12.5	30.5
2	0.019400	115	66360	1571	13.7	30.5
1**	0.015400	130	83690	1981	15.2	30.5
1/0	0.012200	150	105600	2499	16.7	30.5
2/0	0.009670	175	133100	3150	18.0	30.5
3/0	0.007660	200	167800	3972	19.9	30.4
4/0** kcmil	0.006080	230	211600	5008	21.8	30.5
250**	0.005150	255	250000	5917	23.2	30.5
300**	0.004290	285	300000	7101	24.9	30.5
350**	0.003670	310	350000	8284	26.7	30.4
400**	0.003210	335	400000	9467	28.3	30.4
500**	0.002580	380	500000	11,834	31.1	30.5

* Based on 1 ampere per 42.25 of circular mil area for five seconds.

** Not shown in Table 250.66 but listed here to assist in sizing grounding conductor lengths over 30 m (100 ft.) long.

If length grounding conductor exceeds 30 m (100 ft.), the conductor should be increased in size so that voltage drop based on short-time current rating of the conductor specified should not exceed 40 volts. In other words, the resistance of the grounding conductor used when over 30 m (100 ft) long should not exceed the resistance of 30 m (100 ft) of the conductor specified in the table.

Table 20-5.

340

Table 250.122. Minimum Size Equipment Grounding Conductors for Grounding Raceway and Equipment.

Rating or Setting of Automatic Overcurrent Device in Circuit Ahead of Equipment, Conduit, etc., Not Exceeding (Amperes)	Size (AWG or kcmil)	
	Copper	Aluminum or Copper-Clad Aluminum*
15	14	12
20	12	10
30	10	8
40	10	8
60	10	8
100	8	6
200	6	4
300	4	2
400	3	1
500	2	1/0
600	1	2/0
800	1/0	3/0
1000	2/0	4/0
1200	3/0	250
1600	4/0	350
2000	250	400
2500	350	600
3000	400	600
4000	500	800
5000	700	1200
6000	800	1200

Note: Where necessary to comply with 250.4(A)(5) or 250.4(B)(4), the equipment grounding conductor shall be sized larger than this table.

* See installation restrictions in Section 250.120.

Table 20-6.

Rating of O. C. Device in Amperes	EGC Copper Conductor		*Short-Time Rating of EGC Conductor in Amps	** K-factor	Percent Rating of EGC Conductor to OC Device	*** Ampacity per Table 310.16
	AWG	Cir. Mils.				
20	12	6530	155	7.7	125%	25
30	10	10,380	246	8.2	117%	35
40	10	10,380	246	6.1	88%	35
60	10	10,380	246	4.1	58%	35
100	8	16,510	391	3.9	50%	50
200	6	26,240	621	3.1	33%	65
400	3	52,620	1,245	3.1	25%	100
600	1	83,690	1,981	3.3	22%	130
800	1/0	105,600	2,499	3.1	19%	150
1000	2/0	133,100	3,150	3.2	18%	175
1200	3/0	167,800	3,972	3.3	17%	200
1600	4/0	211,600	5,008	3.1	14%	230
2000	250 kcmil	250,000	5,914	3.0	13%	255
2500	350 kcmil	350,000	8,285	3.3	12%	310
3000	400 kcmil	400,000	9,467	3.2	11%	335
4000	500 kcmil	500,000	11,834	3.0	10%	380
5000	700 kcmil	700,000	16,568	3.3	9%	460
6000	800 kcmil	800,000	18,935	3.2	8%	490

Analysis of Table 250.122

*   One ampere per 42.25 circular mils for five seconds.

**   K factor. Ampere rating of overcurrent device times K equals short-time rating of E.G.C. conductor in amperes.

***   Based on 75°C copper wire, Table 310.16.

Table 20-7.

Conductor Properties

Size (AWG/kcmil)	Area mm²	Circular mils	Stranding Quantity	Diameter mm	in.	Overall Diameter mm	in.	Area mm²	in.²	Copper Uncoated ohm/km	ohm/kFT	Coated ohm/km	ohm/kFT	Aluminum km	kFT
18	0.823	1620	1	—	—	1.02	0.040	.823	0.001	25.5	7.77	26.5	8.08	42.0	12.8
18	0.823	1620	7	0.39	0.015	1.16	0.046	1.06	0.002	26.1	7.95	27.7	8.45	42.8	13.1
16	1.31	2580	1	—	—	1.29	0.051	1.31	0.002	16.0	4.89	16.7	5.08	26.4	8.05
16	1.31	2580	7	0.49	0.019	1.46	0.058	1.68	0.003	16.4	4.99	17.3	5.29	26.9	8.21
14	2.08	4110	1	—	—	1.63	0.064	2.08	0.003	10.1	3.07	10.4	3.19	16.6	5.06
14	2.08	4110	7		0.024	1.85	0.073	2.68	0.004	10.3	3.14	10.7	3.26	16.9	5.17
12	3.31	6530	1	—	—	2.05	0.081	3.31	0.005	6.34	1.93	6.57	2.01	10.45	3.18
12	3.31	6530	7	0.78	0.030	2.32	0.092	4.25	0.006	6.50	1.98	6.73	2.05	10.69	3.25
10	5.261	10,380	1	—	—	2.588	0.102	5.26	0.008	3.984	1.21	4.148	1.26	6.561	2.00
10	5.261	10,380	7	0.98	0.038	2.95	0.116	6.76	0.011	4.070	1.24	4.226	1.29	6.679	2.04
8	8.367	16,510	1	—	—	3.264	0.128	8.37	0.013	2.506	0.764	2.579	0.786	4.125	1.26
8	8.367	16,510	7	1.23	0.049	3.71	0.146	10.76	0.017	2.551	0.778	2.653	0.809	4.204	1.28
6	13.30	26,240	7	1.56	0.061	4.67	0.184	17.09	0.027	1.608	0.491	1.671	0.510	2.652	0.808
4	21.15	41,740	7	1.96	0.077	5.89	0.232	27.19	0.042	1.010	0.308	1.053	0.321	1.666	0.508
3	26.67	52,620	7	2.20	0.087	6.60	0.260	34.28	0.053	0.802	0.245	0.833	0.254	1.320	0.403
2	33.62	66,360	7	2.47	0.097	7.42	0.292	43.23	0.067	0.634	0.194	0.661	0.201	1.045	0.319
1	42.41	83,690	19	1.69	0.066	8.43	0.332	55.80	0.087	0.505	0.154	0.524	0.160	0.829	0.253
1/0	53.49	105,600	19	1.89	0.074	9.45	0.373	70.41	0.109	0.399	0.122	0.415	0.127	0.660	0.201
2/0	67.43	133,100	19	2.13	0.084	10.62	0.419	88.74	0.138	0.3170	0.0967	0.329	0.101	0.523	0.159
3/0	85.01	167,800	19	2.39	0.094	11.94	0.470	111.9	0.173	0.2512	0.0766	0.2610	0.0797	0.413	0.126
4/0	107.2	211,600	19	2.68	0.106	13.41	0.528	141.1	0.219	0.1996	0.0608	0.2050	0.0626	0.328	0.100
250	—		37	2.09	0.082	14.61	0.575	168	0.260	0.1687	0.0515	0.1753	0.0535	0.2778	0.0847
300	—		37	2.29	0.090	16.00	0.630	201	0.312	0.1409	0.0429	0.1463	0.0446	0.2318	0.0707
350	—		37	2.47	0.097	17.30	0.681	235	0.364	0.1205	0.0367	0.1252	0.0382	0.1984	0.0605
400	—		37	2.64	0.104	18.49	0.728	268	0.416	0.1053	0.0321	0.1084	0.0331	0.1737	0.0529
500	—		37	2.95	0.116	20.65	0.813	336	0.519	0.0845	0.0258	0.0869	0.0265	0.1391	0.0424
600	—		61	2.52	0.099	22.68	0.893	404	0.626	0.0704	0.0214	0.0732	0.0223	0.1159	0.0353
700	—		61	2.72	0.107	24.49	0.964	471	0.730	0.0603	0.0184	0.0622	0.0189	0.0994	0.0303
750	—		61	2.82	0.111	25.35	0.998	505	0.782	0.0563	0.0171	0.0579	0.0176	0.0927	0.0282
800	—		61	2.91	0.114	26.16	1.030	538	0.834	0.0528	0.0161	0.0544	0.0166	0.0868	0.0265
900	—		61	3.09	0.122	27.79	1.090	606	0.940	0.0470	0.0143	0.0481	0.0147	0.0770	0.0235
1000	—		61	3.25	0.128	29.26	1.150	673	1.040	0.0423	0.0129	0.0434	0.0132	0.0695	0.0212
1250	—		91	2.98	0.117	32.74	1.290	842	1.300	0.0338	0.0103	0.0347	0.0106	0.0554	0.0169
1500	—		91	3.26	0.128	35.86	1.410	1011	1.570	0.02814	0.00858	0.02814	0.00883	0.0464	0.0141
1750	—		127	2.98	0.117	38.76	1.520	1180	1.830	0.02410	0.00735	0.02410	0.00756	0.0397	0.0121
2000	—		127	3.19	0.126	41.45	1.630	1349	2.090	0.02109	0.00643	0.02109	0.00662	0.0348	0.0106

DC Resistance at 75°C (167°F)

1. These resistance values are valid ONLY for the parameters as given. Using conductors having coated strands, different stranding type, and especially, other temperatures, change the resistance.

2. Formula for temperature change: $R_2 = R_1 [1+a(T_2-75)]$ where: $a_{cu} = 0.00323, a_{AL} = 0.00330$.

3. Conductors with compact and compressed stranding have about 9 percent and 3 percent, respectively, smaller bare conductor diameters than those shown. See Table 5A for actual compact cable dimensions.

4. The IACS conductivities used: bare copper = 100% aluminum = 61%.

5. Class B stranding is listed as well as solid for some sizes. Its overall diameter and area is that of its circumscribing circle. (FPN): The construction information is per NEMA WC8-1992. The resistance is calculated per National Bureau of Standards Handbook 100, dated 1966, and Handbook 109, dated 1972.

Table 20-8.

Maximum length of steel conduit or tubing that may safely be used as an equipment grounding circuit conductor.						
Conduit Trade Size In.	AWG/ kcmil	OC 75°C	Fault 500%	EMT Length	IMC Length	RGC Length
½	14	15	75	231	232	227
	12	20	100	246	255	247
	10	30	150	226	250	240
¾	12	20	100	266	262	257
	10	30	150	255	259	250
	8	40	200	261	275	264
	8	50	250	211	229	222
1	8	40	200	288	284	275
	8	50	250	236	237	229
	6	60	300	263	270	257
	6	70	350	228	240	228
	4	80	400	254	274	259
	4	90	450	228	251	238
	3	100	500	227	256	244
1¼	3	100	500	268	274	258
	2	110	550	273	283	265
	2	125	625	244	257	243
	1	125	625	266	281	263
	1	150	750	226	245	234
1½	1	125	625	285	292	273
	1	150	750	243	256	241
	1/0	150	750	264	279	260
2	2/0	175	875	280	287	268
	3/0	200	1000	268	281	261
	4/0	225	1125	257	277	257
	4/0	250	1250	234	257	240
2½	250	250	1250	289	269	261
	250	300	1500	249	238	229
	300	250	1250	302	280	271

Based on a clearing ground-fault current of 500 percent of overcurrent device rating; circuit 120 volts to ground; 50 volts drop at the arc; 30° C ambient temperature; 75°C conductor temperature. Calculated with "Steel Conduit Analysis Vs 1.2" by Georgia Institute of Technology.

Table 20-11.

Conduit Trade Size In.	AWG/ kcmil	OC 75°C	Fault 500%	EMT Length	IMC Length	RGC Length
2¹/₂	400	300	1500	278	263	253
	400	350	1750	244	239	227
	500	350	1750	255	249	236
	500	400	2000	227	230	216
3	500	350	1750	280	262	253
	500	400	2000	251	240	231
	600	400	2000	260	248	238
	600	450	2250	235	231	220
	700	450	2250	242	237	225
	700	500	2500	221	222	209
3¹/₂	600	400	2000	280	256	248
	600	450	2250	257	237	229
	700	450	2250	264	244	235
	700	500	2500	243	228	218
	750	450	2250	267	246	237
	750	500	2500	246	230	221
	900	500	2500	254	237	227
	900	600	3000	220	212	200
4	750	450	2250	277	253	245
	750	500	2500	255	236	228
	900	500	2500	264	243	234
	900	600	3000	229	216	206
	1000	500	2500	269	247	238
	1000	600	3000	233	219	210

Maximum length of steel conduit or tubing that may safely be used as an equipment grounding circuit conductor.

Based on a clearing ground-fault current of 500 percent of overcurrent device rating; circuit 120 volts to ground; 50 volts drop at the arc; 30° C ambient temperature; 75°C conductor temperature. Calculated with "Steel Conduit Analysis Vs 1.2" by Georgia Institute of Technology.

Table 20-11(continued).

Size of Trade Conduit In.	Area of Conduit Wall Sq. In.	Area of Conduit Wall Based on NEMA Standards Sq. In.	D C Resistance Ohms/M ft.	Largest EGC Conductor as specified in Table 250.122 Aluminum Wire	Nearest Wire AWG Equivalent to Conduit	
					Aluminum Wire	Copper Wire
1/2	0.254	0.254	0.08000	12 AWG	4/0 AWG	2/0 AWG
3/4	0.337	0.337	0.06050	8	300 kcmil	4/0
1	*0.530	0.500	0.03850	6	500 kcmil	300 kcmil
1 1/4	*0.680	0.670	0.03000	4	600 kcmil	350 kcmil
1 1/2	*0.790	0.790	0.02580	4	700 kcmil	500 kcmil
2	*1.030	1.080	0.01980	1	900 kcmil	600 kcmil
2 1/2	1.710	1.710	0.01190	1	1500 kcmil	1000 kcmil
3	*2.310	2.250	0.00880	2/0	2000 kcmil	1250 kcmil
3 1/2	2.700	2.700	0.00752	2/0	over	1500 kcmil
4	*3.400	3.180	0.00598	2/0	2000 kcmil	2000 kcmil
5	4.300	4.300	0.00473	3/0	over 2000 kcmil	over 2000 kcmil

Aluminum Conduit used as an equipment grounding circuit conductor (EGC) compared with equipment grounding circuit (EGC) conductors as specified in Table 250.122 of the National Electrical Code.

*Values as measured from a sample.

Column 3 gives values as calculated on basis of NEMA standards.

Table 20-14.

				**** Percent			

Sizes of conductor (Copper Bar) for use in the equipment grounding circuit and for the Equipment Grounding Strap for overcurrent devices from 1600 amperes through 5000 amperes.

Rating of OC Device in Amps.	Copper EG Conductor Copper Bar - Inch	Cir. Mils.	* Conductor Rating Amps	** Short-Time Rating of EGC Conductor in Amps	*** K factor	**** Percent Rating of EGC Conductor to OC Device
1600	$^1/_4$ x 1	318,300	366	10,610	5.3 (6.6)	22.9
2000	$^1/_4$ x $1^1/_4$	397,900	457	13,260	6.0 (6.6)	22.9
2500	$^1/_4$ x $1^1/_2$	477,450	549	15,920	6.0 (6.4)	21.9
3000	$^1/_4$ x 2	636,600	647	21,220	6.0 (7.1)	21.6
4000	$^1/_4$ x $2^1/_2$	795,750	809	26,525	6.3 (6.6)	20.2
5000	$^1/_4$ x 4	1,273,000	1,220	42,430	6.8 (8.5)	24.4

* Based on 30°C rise, 40°C ambient.

** One ampere per 30 circular mils for 5 seconds. Use this value ONLY where insulated, current-carrying conductors will NOT come in contact with the bare equipment grounding bus bars. Where there is likelihood of contact, then recalculate the 5 second withstand rating on the basis of one ampere...for five seconds...for every 42.25 circular mils of copper conductor. This is discussed in chapter eleven of this text.

*** K-factor. Ampere rating of O. C. device times K-factor equals short-time rating of EGC conductor in am peres. Figures in brackets are based on the actual EGC conductor used.

**** Based on 30°C rise, 40°C ambient for EGC conductor.

Table 20-15.

The History and Mystery of Grounding

I n 1941, Howard S. Warren completed work on the manuscript "How The *Code* Came." This material was turned over to then president, Alvah Small, Underwriters Laboratories, to be made available to anyone should it serve a useful purpose. Alvah Small was chairman of the NFPA Electrical Committee from 1925 to 1950. This tome on the history of the *Code* was dedicated to Alvah for his work as chairman in keeping the *Code* abreast of developments in the electrical industry and always in the public interest. Warren represented the telephone group on the Electrical Committee from 1907 until 1939. Warren was preceded by C.J.H. Woodbury and followed by A.H. Schirmer, L.S. Inskip and L.H. Sessler.

The subject of grounding has always been confusing to many, and as such, precipitates much interest wherever sessions are held on *Code* discussions. Grounding has always had its controversial aspects. Part of this stems from the fact that the early sponsors of the *Code*, the Underwriters' National Electric Association, represented the insurance interests and the grounding of electrical circuits had much to offer for personnel safety, but little fire safety for the building could be obtained.

Many of the associations' names have changed today. The Underwriters National Electric Association was later to become the National Board of Fire Underwriters. The National Conference on Standard Electrical Rules was later to become the *National Electrical Code* Committee. AIEE was later to become the Institute of Electrical and Electronics Engineers (IEEE). The National Electric Light Association was later to become the Electric Light and Power Group.

The applications of grounding and the details of accomplishing it are many. The grounding section of the 1940 *Code* (Article 250) occupies over 18 pages. Curiously enough, grounding has always had a controversial aspect.

In the first (1897) *National Electrical Code*, as in the previous "Rules and Requirements" issued by the Underwriters' National Electric Association, the only mandatory grounding requirement was that for lightning arresters or protectors. There couldn't be much argument about that, since from its very nature a protector or lightning arrester must be provided with a path to earth.

In the July 1, 1894, issue of "Rules for Safe Wiring," there were included directions for installing communication circuit protectors, which covered the following: (1) the location of protector; (2) type of wiring to be used from the last outside support to the protector; (3) conditions which grounding conductor must comply with and (4) means for connecting grounding conductor to earth.

These provisions were carried over into the first (1897) edition of the *National Electrical Code*. In addition, the frames of generators and motors were required to be grounded when, and only when, it was impracticable to insulate them.

The fireworks began about three years later, when the question of grounding secondary distribution circuits was precipitated.

Circuit Grounding

Consider for a moment what the grounding of a secondary circuit involves. Take the simple case of a two-wire, 120-volt circuit fed by a transformer. The transformer keeps the difference of potential between the two wires of the circuit at 120 volts, but, in the absence of grounding, the potential of the circuit with respect to earth is indefinite. If there is no foreign electrical influence, the average potential of the two conductors will hover around zero or earth potential, but if one of the conductors becomes crossed with a foreign wire of high potential to earth, it will immediately become charged to the same high potential (assuming that no breakdown of insulation occurs). This brings about a hazardous state of affairs, for if a person in contact with earth or with any grounded metal should touch the charged conductor, he would be liable to receive a dangerous shock. Such hazardous condition might be produced by an accidental cross with a foreign wire or by breakdown of insulation of the transformer in such way as to impress potential from the primary circuit upon a secondary circuit conductor.

As a safeguard against such hazard, it was proposed that secondary circuits be grounded, i.e., that one conductor, in case of a two-wire circuit, or the neutral conductor, in case of a three-wire circuit, be connected to earth, thus assuring that no conductor of such grounded secondary circuit could have a potential to earth greater than the normal circuit voltage.

Such grounding seemed all right for cases where the normal secondary voltage was not in itself hazardous, but as grounding increases the liability of shocks at normal circuit voltage, it did not look so good for circuits of more than about 150 volts. For example, a 220/440-volt secondary, if grounded, would increase the chance of a 220-volt shock, which of itself would involve a serious hazard. On the other hand, the grounding would protect against the probably much higher voltages due to accidental crosses with foreign wires or to transformer breakdowns. So, in this case, it was a question of balancing the increase in hazard from 220-volt shocks against the decrease in hazard from shocks at higher voltages. The 220/440-volt secondary circuit was a borderline case; some experts favored grounding while others opposed it.

As this was a technical electrical question, the Underwriters, who at this period were in control of *Code* revisions, wisely declined to put a mandatory circuit-grounding rule into the *Code* until the electrical people could agree on what the rule should be, all the more so as such a rule was considered to be more concerned with life hazard than with fire hazard. Differences of opinion on the part of electrical experts as to the circuit voltage below which grounding ought to be made mandatory produced a stalemate that persisted for several years.

Discussions of electrical circuit grounding go back to the beginnings of the ac distribution system. In England, as well as in this country, such grounding (our British cousins call it *earthing*) was proposed. Thus Killingsworth Hedges said, in 1889:[1]

"One precaution is to earth the secondary circuit. Another is to connect one or both leads to a safety appliance which would automatically divert any excess current to earth and at the same time shut off the supply." Professor Elihu Thomson, in 1890, suggested grounding the transformer secondary winding or surrounding it with a grounded sheath. As alternatives, he suggested cutting off the secondary by automatic means or grounding it automatically by film cutouts when excessive voltage appeared. The placing of a grounded metal sheath between the primary and secondary windings of a transformer was called by the English, "Kent's sheath," although the idea had its origin in America. Thomson regarded this device as very effective for securing a safety grounding connection for the secondary circuit when there was a leak from the primary. Discussing this matter at an NELA convention, he said:

"In the ac system with low voltage secondary, good workmanship is all that is necessary to abolish risks if we leave out the high voltage primary which must be thoroughly insulated. Even some static capacity, which may give severe if not fatal shock, is to be looked out for. The writer early recognized this danger and felt it must be provided against. Hence came the expedients of grounding the secondary or surrounding it with a grounded sheath or cutting off the secondary absolutely by automatic means."

But the grounding of secondary circuits had plenty of opposition in the early days. Woodbury, in giving his suggestions about methods for reducing the electrical fire hazard, considered "it undesirable to permanently ground one wire of the secondary circuit."[3]

On March 10, 1892, the New York Board of Fire Underwriters issued a pamphlet report on "Grounding of Electric Wires," prepared by its Committee on Police and Origin of Fires. At a special meeting of the board, held on that date, it was voted that this report be printed and circulated to the members, together with certain parts of a report by Professor Henry Morton, and a reply thereto by the Edison Electric Illuminating Company of New York.

The Electrical World Announced
"The New York Board of Fire Underwriters has condemned the practice of grounding the neutral as dangerous and orders it to be stopped.

"There can be no doubt that the practice of grounding the neutral is not as safe as a completely insulated system. To change it now, however, may require much work and, therefore, may interfere somewhat with the lighting. It is a case in which prevention would have been better than cure."

The following resolutions were also adopted by the board and circulated as aforesaid:

"RESOLVED, that the Committee on Police and Origin of Fires be and they are hereby directed to require all electric companies furnishing current for power and lights (to parties that have received certificates of approval of electric equipments from this Board), to make regular weekly reports of the tests of their currents, as called for in the requirements of this Board, and in the event of failure on the part of the Electric Companies so to do, then the Superintendent of the Survey Department is hereby directed to decline to grant certificates, or make inspections of equipments supplied with current by said delinquent companies, and further.

"RESOLVED, that said Committee give notice to all Electric Companies receiving certificates from this Board that the intentional grounding of any portion of the equipment is a violation of the rules of this Board and such a practice must be discontinued.

"This Board will refuse certificates of approval to all companies who do not comply with this rule, and further

"RESOLVED that the committee require of Electric Companies that all grounds now existing on electric circuits shall be removed on or before October 1st, 1892."

These resolutions were promulgated over the name of William Del. Boughton, chairman of the board.

The extracts from Professor Morton's report on fire hazards from grounding of electric wires, particularly the neutral wire in the Edison system, included the following: "The grounding of the middle wire, in my opinion, decidedly increases the fire risk, for the following reasons:

"First. If all the wires are insulated, then two ground contacts must occur in order that fire should be produced by the contact of a conductor with the gas, water or steam pipe, or other conducting substance connected with the ground. If, on the other hand, the middle wire is "grounded," then every gas, water or steam pipe becomes in fact a 'live wire,' contact with which results in a current, only limited in amount by the capacity of the conductors (including fusible catches and the like) between the point of contact and the general network of supply wires.

"Second. It renders entirely impossible any testing by which from time to time the insulation of the systems can be watched and measured.

"As a result when a ground occurs, unless it attracts attention in its own neighborhood or is enormous in amount, it can go on indefinitely without being discovered, while it may be doing mischief all the time, whereas with an insulated system, a ground will instantly show itself by the proper instruments at the central station or elsewhere, and at the same time will be harmless until another ground, on the wire of opposite polarity, is developed.

"If the middle wire of the Edison system is grounded the entire system is then in such connection with the earth that the contact of either positive or negative conductors with water, gas or steam pipe, or iron parts of a building or other conductor, will establish a connection through which a large current will flow."

Commenting on the foregoing, R. R. Bowker, first vice president of the Edison Electric Illuminating Company of New York, said:[4]

"In view of the fact that many of the questions at issue are comparatively new to members of your board, we have thought it best to make a general explanation of the nature of the three-wire system, and of the practice involved in 'grounding the neutral,' as is done in the accompanying statement, which while it does not take up the leading questions of the report in regular order, does, it will be found from the references, present a direct and sufficient answer to each of these questions. The careful and specific attention of each member of your Board is requested for this statement, as the question involved is one of such large and increasing importance.

"We would emphasize to you that our street system is in no sense dependent on the practice of grounding the neutral and that contrary to the inference of Professor Morton's report, no copper or other element of cost is saved with that practice in view; the motive of the practice is not commercial but precautionary. We remind you that the practice is in vogue in other cities with the approval of Underwriters. We would also beg to correct the misapprehension that there is any grounding of the neutral wire in houses, or any decrease of insulating precautions because of that practice.

"We ask your Board for such modifications of your rules as will recognize the features of the three-wire system and provide for the practice thus adopted. We beg herewith, however, to say that should the Board, in full view of the protest of this company, decide to take the responsibility, with the insuring community and before the public, of directing this company to discontinue the practice, the superintendents of the company will have instructions to prepare to disconnect the neutral grounds at the junction boxes throughout this city, which would require, however, considerable time, and would not, we believe, produce the result desired of giving opportunities for insulating tests on lines of laboratory practice and of special circuit systems. We would, moreover, enter definite protest against the overruling of this company by your board in a question of practice which we believe can best be determined by working electrical engineers familiar with the needs and conditions of actual station practice."

A statement submitted to the New York Board of Fire Underwriters by the Edison Electric Illuminating Company contained the following:

"In the Edison 3-wire system, two dynamos

are coupled together both in series by a short copper conductor connecting the negative side of one with the positive side of the other. If each is running at 100 volts the product of the two thus in series is at 200 volts. The positive conductor runs from the outer side of one dynamo; the difference of potential between the positive and negative wire is therefore 200 volts. The middle or neutral wire runs out from the short connecting link between the two dynamos and like the positive and negative wire, is absolutely continuous from the dynamo in the station to the extreme point of every house installation. Incandescent lamps are placed 'in multiple' between the neutral wire and the positive conductor and between the neutral wire and the negative conductor. In each house and throughout the system lamps are 'balanced' as evenly as possible between the two sides of the system - that is, when a hundred lamps are placed in a house, it is the endeavor to put fifty of these on the positive side and fifty on the negative side of the system, so placed that the actual burning is about the same on either side. When one thousand lamps are burning on the positive side and one thousand lamps on the negative side of the system, the neutral wire brings back no current whatever to the dynamo at the station. The current at 200 volts flows through the lamp between the positive conductor and the neutral wire, giving this 100 volts, and thence through the lamp between the middle wire and the negative conductor giving that the other 100 volts, and the return is all on the negative side. If five of these lamps are taken out or turned off on the positive side, the neutral wire would become a positive conductor to the extent of the current for these five lamps, or if five lamps were taken out or turned off on the negative side, the neutral wire would become a negative conductor to the extent of the current for those five lamps.

"On the New York system the neutral is grounded at the junction boxes in the streets, and not at the stations; on the Brooklyn system the grounding is at the station. Authorities differ as to which is the better course to obtain the advantages of grounding the neutral. In New York the neutral is grounded at about 250 junction boxes, something less than 1/3 of the number of junction boxes on the whole system, at a distance from the station varying from 1/4 to 1-1/2 miles. The neutral is grounded, as it were, on the other side of the junction box from the station, so that if in any possible event the current should flow through the ground it would seek the shortest course to the nearest junction box grounded and thence through the regular neutral wire to the station. That the ground is practically not used as the neutral conductor is sufficiently evidenced by the fact that no electrolysis has yet been discovered by our electricians at the points where the neutral wire is grounded, as would certainly be the case were the ground connection used to any extent.

"The neutral wire is not in any case grounded within house installations or in any way in connection with houses.

"It should be clearly recognized moreover that it is not practicable on any large system of electrical conductors, whether overhead or underground, especially so large as is required for the city of New York, that absolute insulation, such as is possible in laboratory tests, can be obtained or maintained. There is no known material which is an absolute insulator and even electrical appliances and adjoining woodwork are themselves conductors to an infinitesimal degree. In any known system of considerable aggregate length, while no single part can be definitely criticized, the mean level of insulation of the system is necessarily brought into comparatively close association with the ground by numerous small leaks, each of individual importance but all together making a considerable aggregate.

"I desire to make perfectly clear that if any given system could be made and kept absolutely free from leakage and best electrical authorities, including Edison himself, would agree that the balance of advantage would be found in keeping the system in that condition and not in grounding the neutral. But the difference between laboratory tests and actual working conditions is such as to make their practical conclusions almost contradictory and we have the authority of Edison's chief electrician for the assertion that 'in endeavoring to attain the perfection of safety' the report 'advocates a course of procedure which would introduce greater risk and defeat its own purpose.'

"It would seem scarcely necessary after these remarks to reply more specifically to Professor Morton's position that there should be no difference between the rules made to apply to the two-wire and three-wire systems and no reason to alter previous rules of the Board to recognize or apply to the present situation. If Rule 40, which has often been quoted in this connection, is understood by your board to apply to the grounded neutral of the Edison three-wire system, it should certainly be modified. As a matter of fact, having been adopted before the three-wire system came into operation, this company asserts that it does not so apply and in this respect no alteration would then be needed.

"It is the general opinion of the Edison interest that, while absolute insulation, if it can be had and where it can be had, is preferable, the advantages of grounding the neutral under certain conditions and particularly on large systems, are such as to make that practice in those cases the best working method particularly as a precaution against fire risk."

Discussion at National Conference
At the meeting of the National Conference on Standard Electrical Rules in March 1896, Professor Kennelly suggested a rule requiring "that the secondary coil of a converter should be grounded at its center." This led to the following discussion:

James I. Ayer:
"Relative to the question of Mr. Kennelly as to the grounding of the secondary coil of the converter at its center, it seems to me that is putting a severe strain on the insulation. I assume a low resistance ground is meant, so that in event of a ground occurring on the line and a breaking down of the insulation of the transformer the current would pass through so as to destroy the converter and cut out the primary current from the secondary. There is no doubt about the advisability of getting some sort of transformer protection, especially on the primary lines where 2000 volts and upwards are transformed directly to the house service low potential.

"In a discussion of this question some time last spring, Professor Puffer recommended the same thing. I suggested carrying out a practice which I adopted years ago in a good many cases of using a magnetic cutout. This magnetic cutout was made with two solenoids, each of which have one connection to one of the secondary lines, one on one side and one on the other, the other terminal being connected to ground. These solenoids are of very high resistance so that the leakage of a 50-volt current amounts to 8 watts and for 110 volts to about 11 watts. This gives absolute protection. Any rise of pressure on the secondary mains above 500 volts would operate the circuit breaker. It is very positive. It rather straddles the question of grounding secondaries or grounding house mains by grounding them through a fixed high resistance ground connection. It gives an absolute, positive-acting device, not necessarily one that burns up transformers. In fact, it leaves the transformers to take care of itself but cuts out the primary from the secondary main.

"This is a matter to which I have given a great deal of thought. I found in the early days of the use of transformers the necessity of some such protection where there are many of them connected with the line. Thunderstorms would readily break them down and create punctures which would give a connection from one side of the transformer to one side of the secondary circuit. The lights would burn and things continue on apparently all right. Transformers do break down with the high potential being used and I believe the right thing to recommend, and in time to demand, is the proper protection for isolated transformers, to give you protection between the secondary and primary, but the question of grounding secondary mains and low resistances IS a very serious one and should receive a very full discussion before action is taken."

Professor Kennelly:
"I think the matter is so important that it is worth while to give it a few moments. I should be sorry to advocate anything which would unduly harass the manufacturers, but I believe that I make this suggested recommendation in the interests of the manufacturers just as much as in the interests of the parties who receive lighting from electric systems.

"If we exclude lightning, there can be no objection to grounding the secondary because it is obvious that the system should be capable of standing the working pressure. Then no one can fairly object to grounding the center of the secondary on the score of danger of puncture because, if the transformers are going to be so weakly insulated that grounding the secondary is going to injure them, those transformers had better be replaced. It may not be necessary, however, to actually ground the secondaries. A simple protective device that will ground the secondary if a cross develops would be preferable, and if Mr. Ayer's device is simple and practical, its adoption would seem worthy of consideration. A cross between the primary and coils of a transformer is, we know, of very rare occurrence, yet we all know that no one would be justified in permitting a person standing, say on a cellar floor, to handle the socket of a lamp if the operation involved making contact with one of the secondary conductors, unless the transformer had just been tested free from leakage or cross. If, however, the secondary coil of the transformer were grounded permanently, or protected in an equivalent degree there would be no such danger from accidental contact."

Mr. Ayer:

"In considering this question of the protection of the transformers, the practical conditions to be met with have got to be well understood. You take the case of distribution with the overhead system, where there are any material number of transformers in a network covering a large area, there is no such thing as getting an insulated circuit. You have got a grounded circuit, always.

"There has been a great deal of discussion especially among the Edison Companies and Underwriters, as to the wisdom of grounding the third wire of the Edison system. That has been a very fruitful topic of discussion. I think it is useless to try to enumerate the details or discuss the pros and cons of this here.

"In a resume of the English rules, you see that they contemplate at all times an automatic device in the secondary circuit, or in the primary wire, if not in the secondary, which will automatically cut off the primary current from the transformer or the secondary from the house. That is the English practice. They call for some intervening device which shall open the secondary circuit in the event of the primary forming a connection with high tension. It is a simple thing to make. I have one in my residence today and have had for five years, which has been in service all the time. The thing is operated with a leakage of 500 volts from the primary line to the secondary. It introduces

a high resistance ground on the secondary and a leakage on the secondary of about 8 watts. I believe the solution of the problem of the protection of the secondary coil is the introduction of this intermediate automatic device. Devices have been made in England operated by clock work, fuses, etc., but not similar to this to my knowledge."

The protective device for secondary circuits, which Ayer thus commended, did not meet with much favor and the solution of the problem was long delayed; indeed, it has not been solved even yet in all its aspects.

The Underwriters were slow to reverse their stand against allowing any permanent grounds on electric circuits.

The electric companies, on the other hand, were generally in favor of grounding the neutral wire. W.L.R. Emmet of the General Electric Company said, in 1899:

"The permanent grounding of transformer secondaries is prohibited by the fire underwriters' rules, which pro prohibition is generally respected throughout the country, so that most secondary circuits are entirely unprotected."[5]

AIEE Talks it Over

In 1899, Dr. Cary T. Hutchinson presented an AIEE paper on the protection of secondary circuits, in the course of which he heartily endorsed the practice of grounding and reproved the Underwriters for their stand against it saying,[6]

"The need of some effective method for protecting secondary circuits is well recognized yet there is at present probably no device for the purpose that is thoroughly trustworthy and efficient in all circumstances. The means that have been employed for this object may be classified under three general heads: (l) Devices intended to ground, short-circuit or open-circuit the secondary circuit when subjected to abnormal difference of potential; (2) grounded metallic shields imposed between the primary and secondary coils of the transformer; (3) permanent grounding of the secondary system.

"The third method, grounding the secondary permanently, is the only sure way to prevent the potential above earth of the secondary system rising above the voltage of the circuit.

"The grounding of the secondary circuit thus absolutely insures the safety of the circuit as regards abnormal pressure.

"It would seem that a remedy as simple as this, one generally applicable, would have been applied in all cases but the fact is that it has attained a limited use only. The reason for this state of affairs is the refusal of the Board of Fire Underwriters to authorize the practice of grounding any part of a circuit carrying current. The Underwriters take the position that to ground one side of a current brings about an increased liability to fire, because the full voltage of the circuit continually acts upon the insulation of the circuit, instead of one-half of this voltage as in an insulated system, and because one accidental ground on the insulated side may cause a fire.

"The Association of Edison Illuminating Companies has had a committee on grounding the neutral since 1890. The recommendations of this committee being based on extensive experience are entitled to great weight. Their recommendation of uniformity being the greatest safety was assured by the practice of grounding the neutral. This is the practice of nearly all the large Edison Companies as is well known. In some cases the companies have been forced to this position because they were not able to free the neutral from ground. In other cases they have deliberately adopted this as the best remedy for many troubles due to operation dangers from fire and to abnormal pressures.

"It is notorious fact that the systems in nearly all of the large cities, particularly the Edison system, are grounded. The neutrals in most cases are permanently grounded at the junction boxes. It is equally well known that in several of the large ac distributing systems the neutral wires are grounded. The authorities of the Board of Underwriters certainly should, and probably do know, these facts. To keep rules in print prohibiting such practices under the circumstances must weaken their authority and lessen the respect for all their rules.

"I therefore offer the following resolutions:

"RESOLVED that the AIEE favors a rule permitting the grounding of one wire of every low potential consumption system, and

"RESOLVED that the committee of the AIEE on the *National Electrical Code* is hereby requested to confer with the Electrical Committee of the Underwriters National Electric Association in relation to recommending to the National Board of Fire Underwriters the passage of a rule permitting the permanent grounding of one wire of such systems under suitable restrictions."

Hutchinson's resolution was passed, after discussion, some of which is noted below: Professor Elihu Thomson: "I endorse everything that is said and I hope some action will be taken by a body so influential as this, toward getting some such thing adopted."

C. M. Goddard:

"I should like to second the resolution and for several reasons; not, however, because the Underwriters are fully convinced as yet that one of their rules, almost as unchangeable as the laws of the Medes and Persians, that no circuit should be permanently grounded, has been wrong all the time. I am not convinced of that fact but because we are delighted to have the institute take up any subject in relation to our rules and try to help us arrive at a right solution.

"I think it very probable that this grounding of the secondary may be the best solution but the Underwriters are always conservative in such matters.

"We would be delighted to have you help us in making our rules. We modified this rule last year at the request of the Institute and we will certainly modify it further if you will only come to us and show us that you are right, for we have a great deal of respect for the opinions of the Institute."

Professor W. E. Goldsborough:

"I hardly feel like saying anything in opposition but I am very much interested in the subject and all my experience has tended to lead me to feel it is not best to ground the secondary."

Professor W. L. Puffer:

"I would carry grounding to such an extent that I would compel the absolute grounding immediately at the point of entrance in buildings of every system (except series arcs which ought not to be allowed in buildings) and including such things as water pipes and gas pipes."

While the consensus of this discussion was decidedly in favor of grounding the secondaries, there was some adverse opinion and the Underwriters seem to have continued other opposition to the practice, although the *Code*, i.e., the 1897 and 1899 editions, did not specifically prohibit it.

Father Defers to Mother

The Electrical World and Engineer, reporting the December, 1900, meeting of the Electrical Committee, said:[7]

"The motion to change the limit of low tension circuits from 300 to 550 volts was carried at the first session. A long discussion took place on the advisability of grounding one of the wires in low tension circuits in which numerous authorities were quoted and prominent electrical engineers spoke in its favor. The resolution was finally adopted to allow the grounding of one wire in every low potential circuit but making it optional with the installing engineer."

In accordance with this decision, the 1901 edition contained the following section:

"13A. Grounding Low Potential Circuits.

The grounding of low potential circuits under the following regulations is only allowed when such circuits are so arranged that under normal conditions of service there will be no passage of current over the ground wire.

"Direct-Current 3-Wire Systems:

a. Neutral wire may be grounded and when grounded the following rules must be complied with:

1. Must be grounded at the central station on a metal plate buried in coke beneath permanent moisture level and also through all available underground water and gas-pipe systems.
2. In underground systems the neutral wire must also be grounded at each distributing box through the box.
3. In overhead systems the neutral wire must be grounded every 500 feet, as provided in Sections c, e, f, and g.

Inspection departments having jurisdiction may require grounding if they deem it necessary.

Two-wire direct-current systems having no accessible neutral point are not to be grounded.

"Alternating-Current Secondary Systems

b. The neutral point of transformers or the neutral wire of distributing systems may be grounded and when grounded, the following rules must be complied with:

"1. Transformers feeding 2-wire systems must be grounded at the center of the secondary coils, as provided in sections d, e, f and g.

"2. Transformers feeding systems with a neutral wire must have the neutral wire grounded as provided in sections d, e, f and g, at the transformer and at least every 250 feet for overhead systems and every 500 feet for underground systems.

"Inspection departments having jurisdiction may *require* grounding if they deem it necessary."

Thus, by making the practice permissive, the Underwriters began a strategic retreat from their early position against the grounding of electrical circuits.

At the Electrical Committee meeting in December, 1901, the following proposal was considered:

"A letter was also referred for conference with the AIEE Committee, asking for modification of the rules with regard to grounding low potential circuits so that the ground might be on one side of the circuit instead of at the neutral point of the transformer."[8]

The Underwriters held to their position that the grounding of secondaries was a question for the electrical people to decide. A special subcommittee on the subject reported to the Electrical Committee meeting of December 9, 1902:[9]

"It was also agreed that Rule 13A-b should be amended to read as follows:

"(b) Transformer secondaries of distributing systems should preferably be grounded and when grounded the following rules must be applied:

"1. The grounding must be made at the neutral point or wire, whenever a neutral point or wire is accessible.

"2. When no neutral point or wire is accessible, the maximum difference of potential from the grounded point of the secondary circuit to any other point there of must not exceed 250 volts.

"3. To be identical with the present clause marked (2)." Here was a proposal to go further and take a position favoring circuit grounding. The meeting declined to either approve or reject this recommendation but voted to submit it to the Board of Directors of the AIEE and to the Underwriters' National Electric Association.

This recommendation to change the permissive "may be grounded" to "should preferably be grounded," was evidently accepted by the AIEE, for the rule in the 1903 *Code* was substantially the same as the subcommittee had recommended, namely:

"b. Transformer secondaries of distributing systems should preferably be grounded, and when grounded, the following rules must be complied with:

1. The grounding must be made at the neutral point or wire, whenever a neutral point or wire is accessible.
2. When no neutral point or wire is accessible, one side of the secondary circuit may be grounded, provided the maximum difference of potential between the grounded point and any other point in the circuit does not exceed 250 volts.
3. The ground connection must be at the transformer as provided in sections d, e, f, g and when transformers feed systems with a neutral wire, the neutral wire must also be grounded at least every 250 feet for overhead systems, and every 500 feet for underground systems. "Inspection Departments having jurisdiction may require grounding if they deem it necessary."

This rule in the 1903 *Code* represented quite a step away from the old position prohibiting circuit grounding, but still it was unsatisfactory in that it left the question open as to whether secondaries should be grounded or not. There was a substantial body of opinion in favor of making the grounding mandatory under certain conditions, but there were differences of opinion as to what the conditions were which warranted making grounding mandatory. It was generally agreed that where the voltage between conductor and ground did not exceed 150 volts, safety was promoted by grounding, but where the voltage was from 150 volts to 250 volts, the advantage of grounding was in a doubtful region. For voltages to ground over 250, the consensus of expert opinion was against grounding.

At the Electrical Committee meeting on December 2-3, 1903, a suggestion that the grounding of secondary neutrals be made compulsory was referred to a committee of three to be appointed by the chair, to report back.

More Discussion

At an AIEE meeting on December 19, 1903, H. G. Stott took up the discussion, saying:

"Some years ago a committee of the Institute met with the Board of Fire Underwriters and a rule was passed making it permissible to ground the neutral at the central point of the transformer or at one side of the transformer. The grounding of any part of the secondary network, especially the neutral point, is the best form of protection obtainable because in case of a cross with an electric light wire carrying perhaps 8,000 volts, the pressure of the wire is immediately reduced to the pressure of the circuit as the carrying capacity of the secondary network is greatly in excess of the amount of current passing through the average arc light circuit."

Discussion of this question continued. It was one of the subjects considered at the meeting of the National Conference on April 21, 1905. On that occasion, Dr. F. A. C. Perrine (formerly Professor of Electrical Engineering at Stanford University) stressed—perhaps overstressed—the opposing effects of grounding on the fire and life hazards, saying:

"For example, we have heard a good deal of the rules for grounding secondaries. We know the Underwriters have admitted that, at the request of the electrical interests, in spite of the fact that if anything it increases the fire hazard, decreases the life hazard. Now there is a condition in which the Underwriters have acceded to the request of the electrical interests but it is a condition in which the fire hazard risk is distinctly opposed to the life hazard risk."

Benallack confirmed the influence exerted on the Underwriters by the electrical people:[10]

"Other speakers had called attention to the importance of grounding secondary circuits as a good way, and in fact the only way, of making these circuits safe from the leakage of currents from the primary circuits. The Underwriters had been slow to permit this at first and it was largely due to the central station men who appreciated the danger that the practice was permitted and finally positively recommended by the Underwriters."[11]

At the National meeting on December 4, 1905, the subject of grounding secondaries was discussed further. H.C. Wirt remarked:

"The most important rule here is the rule on grounding of the secondary to make it compulsory." After discussion, the following resolutions were adopted:

"RESOLVED, that it is the sense of this Conference that the grounding of low tension secondary ac systems be strongly recommended in all cases wherever reliable grounding connections can be secured.

"RESOLVED, that in cases of three-wire systems with grounded neutral, solid connections without fuse be permitted on the neutral wire.

"RESOLVED, that it is the sense of this Conference that the grounding of secondary ac systems inside of buildings to water pipes, provided such connection is made at the nearest point to cellar wall on water pipes and outside of meter, is not only safe but places no additional burden or menace on such water pipes."

Woodbury referred to this meeting in his Annual Report for 1905, to H. V. Hayes:

"At the second meeting of the Conference on December 4, results of importance to the electrical interests were adopted in regard to grounding of secondaries on ac systems and also the resolution that grounding did not furnish any hazard or menace to water pipes to which they were attached. Although these results were filed by others yet I was concerned in their operation, particularly in relation to the last clause, on account of difficulties which some of the operating companies had experienced through claims of town officials for reimbursement for damage to water mains by lightning, alleging that the grounding of telephone lines to water pipes was the cause."

However, the Electrical Committee at its December 1905, meeting did not adopt the National Conference resolutions, as no formal endorsement of them by the AIEE or NELA had been received. The grounding of low-potential circuits was discussed but no changes in the rule were made and the matter was referred back to the subcommittee for further consideration. Nor was any such endorsement, or any further recommendation as to the grounding rule, received from the electrical people in 1906. Accordingly, in the 1907 edition, rule 13A remained practically unchanged from the previous edition. The year 1908, however, saw some progress toward a mandatory grounding rule for the *Code*. At the Electrical Committee meeting on March 25, the Committee on Rule 13A reported:

"The sense of the meeting was that, when suitable rules are prepared, the grounding of secondary systems shall be made mandatory."

At the meeting of the National Conference on March 27, 1908, C.M. Goddard made these remarks: "The Underwriters have all along felt that

it (grounding) was a rule which did not very seriously affect the fire hazard but I think the more it has been looked into, that the Underwriters have perhaps decided that there are certain fire hazards guarded against by the grounding although, of course, we all know that it is principally for guarding against life hazard."

"It is apparent that when the rule has been whipped into shape so that it will be satisfactory to all, it shall be made mandatory by the Underwriters instead of being simply permissive.

"Now the Underwriters want to be very sure that it is the desire of the electrical interests as a whole, both central station men and the engineering branch, also the municipal departments. We want to be very sure that we are not making the rule mandatory against their wishes, I should think the Underwriters would be perfectly satisfied to leave the rule as it stands."

It was voted to be the sense of the meeting that the grounding rule be made mandatory; also that instructions be given for making the grounds.

At the 17th annual meeting of the Electrical Committee, held in New York on March 24 and 25, 1909, the Committee on Rule 13A reported:[12]

"Your committee appointed to consider changes in grounding rule, and question as to whether it should be made mandatory, begs to report as follows:

"First—As we are asked only to make this rule mandatory up to and including 150 volts for alternating current secondary circuits, and to prohibit grounding of such circuits carrying in excess of 150 volts, it is evident that this suggestion cannot be based on the question of fire hazard because the fire hazard must certainly be as great from an ungrounded circuit in excess of 150 volts as from such circuit carrying not over 150 volts.

"Second—Your committee believes that the Underwriters have no right to enforce a rule which they cannot defend on the ground that it is adopted for the purpose of lessening the fire hazard.

"Third—Your committee believes that the Underwriters should use all of their moral influence in supporting any rule which may seem advisable in order to lessen the life hazard.

"Your committee would, therefore, recommend no change in the present rule but suggest that the rule, if made mandatory, might be followed in the rules of the National Board of Fire Underwriters for the government of insurance inspectors by a note saying, that in order to lessen the life hazard, compliance with the rule is recommended, but as the rule is not adopted for the purpose of lessening the fire hazard, no penalty could be enforced by the Underwriters for failure to observe its provisions.

"Your committee regrets that they have received no information as to the position of the American Institute of Electrical Engineers or the Associated Edison Illuminating Companies in regard to the subject."[13]

This report was adopted by the meeting and the grounding section was again left unchanged in the 1909 *Code*.

In its annual report for 1909, the Board of Directors of AIEE included this passage:[14]

"The Code Committee has held several meetings, the most important matter considered being in connection with Rule 13A of the Board of Fire Underwriters. This rule deals with the grounding of secondary circuits. The committee agreed that the rules should be made mandatory up to and including 150 volts between any wire and ground, and optional up to, and including, 250 volts between any wire and ground. This resolution was handed to the president of the National Conference on Standard Electrical Rules. Final action on the resolution was deferred pending an arrangement for a meeting of the *Code* Committee with a committee of the National Board of Fire Underwriters."[15]

So, in the 1911 *Code* the grounding rule, which in revision became Rule 15, was kept about as it had been, to wit:

"15. Grounding Low Potential Circuits

"Alternating-Current Secondary Systems

b. Transformer secondaries of distributing systems should preferably be grounded, and when grounded the following rules must be complied with:

1. The grounding must be made at the neutral point or wire, whenever a neutral point or wire is accessible.

2. When no neutral point or wire is accessible, one side of the secondary circuit may be grounded, provided the maximum difference of potential between the grounded point and any other point in the circuit does not exceed 250 volts.

3. The ground connection must be at the transformers or on the individual service as provided in sections c to g, and when transformers feed systems with a neutral wire, the neutral wire must also be grounded at least every 500 feet."

"Inspection Departments having jurisdiction may require grounding if they deem it necessary."

But the time was now approaching when the grounding of low-tension secondaries was to be made mandatory.

In its report for 1911, the Board of Directors of the AIEE included the following passage:[16]

"The Code Committee through its chairman, represented the institute at the annual meeting of the National Board of Fire Underwriters held at New York on March 20-21, 1911. The only matter of interest to the institute taken up at this meeting was the grounding of secondaries and the work of the institute's representative resulted in the passing of a resolution by the underwriters conference enforcing the practice of the grounding of secondaries and recommending that municipalities and lighting companies make such a rule mandatory with the further resolution that the institute use its efforts to bring about an agreement with the NELA in the matter of grounding secondaries up to 250 volts instead of at 150 volts, the present adopted standard of the Association."[17]

In March 1911, the Electrical Committee held its first meeting as the adopted child of the National Fire Protection Association. The following is an extract from a report of that meeting:[18]

"The following resolution, presented by the Committee on Rule 13A, was adopted:

"WHEREAS—The grounding of secondary alternating-current circuits having a normal difference of potential of not over 150 volts does effectually eliminate the life hazard of such circuits due to their accidental contact with circuits of a dangerous potential without introducing any hazard due to grounding, and

"WHEREAS—There is a difference of opinion as to whether the grounding of such circuits having a normal difference of potential of over 150 volts introduces hazards due to such grounding, and

"WHEREAS—Such grounding does not increase but rather tends to decrease, the fire hazard, therefore be it

"RESOLVED—That municipal departments are urged to make the grounding of secondary circuits up to 150 volts mandatory as a necessary safeguard to life; and

"RESOLVED—That Underwriters Inspection Departments be urged to recommend at all times such grounding as a proper and desirable precaution which introduces no fire hazard, and

"RESOLVED—That the National Electric Light Association be urged to see that all of its member companies be brought to realize the necessity of such grounding for the protection of their customers, and

"RESOLVED—That departments in charge of water works be urged to allow the attaching of such ground wires to their piping system in the full confidence that the integrity of such piping systems will in no way be affected, whatever may be the normal voltage, and

"RESOLVED—That all concerned give careful thought to other methods of obtaining satisfactory grounds wherever metallic water pipes are not available, to the end that this most necessary precaution of grounded secondary circuits may be available in all localities and under all conditions, and

"RESOLVED—That the American Institute of Electrical Engineers be urged to use its best endeavors to harmonize the present difference of opinion as to the limit of voltage at which grounding ceases to be desirable by determining a limit which shall meet with general approval."[19]

During the year following, the American Institute of Electrical Engineers joined with other prominent national electrical societies in a final effort to secure a mandatory requirement for the grounding of low-potential secondary circuits. The AIEE Board of Directors' report for 1912, contained this passage:

"The Code Committee held a meeting on March 12, 1912, with representatives of the National Electric Light Association, Association of Edison Illuminating Companies, the National Inspectors Association, and concurred in a joint recommendation to the NFPA in regard to the grounding of secondaries."

With this substantial backing, the Electrical Committee was encouraged to approve, at its meeting on March 26 and 27, 1913, the report of its Committee on Rule 15, which made "the grounding of transformer secondary circuits obligatory."

Circuit Grounding Made Compulsory
So, at last, in the 1913 Code, a mandatory grounding rule was included, reading as follows:

"15. Grounding Low-Potential Circuits

"Alternating-Current Secondary Systems b. Transformer secondaries of distributing systems (except when supplied from private industrial power or lighting plants when the primary voltage does not exceed 550 volts) must be grounded, provided the maximum difference of potential between the grounded point and any other point in the circuit does not exceed 150 volts and may be grounded when the maximum difference of potential between the grounded point and any other point in the circuit exceeds 150 volts. In either case the following rules must be complied with:

"1. The grounding must be made at the neutral point or wire, whenever a neutral point or wire is accessible.

2. When no neutral point or wire is accessible, one side of the secondary circuit must be grounded.

3. The ground connection must be at the transformers or on the individual service as provided in sections c to g, and when transformers feed systems with a neutral wire, the neutral wire must also be grounded at least every 500 feet."

This rule has not been changed much, in substance, since that time. The main difference is that in the 1940 Code, grounding for voltages from 150 to 300 is recommended, instead of being merely permitted. This change first appeared in the 1923 edition.

Specifications to Assure Safety
Certain conclusions can thus be established to assist in writing specifications to cover Service and Equipment Grounding to assure safety in installations. They are as follows:

1. All ungrounded systems should have a ground detector and proper maintenance applied to avoid, as far as practicable, the occurrence of grounds on opposite phases at the same time. If proper maintenance cannot be assured for an ungrounded system then a grounded system should be installed.

2. The use of grounding electrodes at a transformer bank and at the building service provide by themselves virtually no protection in the event of a ground fault, as far as overcurrent protection is concerned. However, the lowest practical resistance for such grounds is desirable to serve the second purpose of holding the equipment as near to ground potential as possible.

3. The service ground and the equipment ground must have their common point of connection within the service equipment enclosure.

If the service equipment is in the switchboard, then the common point must be within the switchboard section which houses the service equipment. A common grounding conductor to the grounding electrode must be run from that common point of connection.

4. When a transformer bank is grounded, that is, when a service is run to a building from a system which has any point grounded, the grounded conductor must be run to the service equipment whether that grounded conductor is to be used for voltage requirements or not, and that grounded conductor must be connected to the equipment grounding conductor within the service equipment enclosure.

5. When a fault occurs on any grounded system we must not assume it began as a short circuit. (phase/to/phase fault) until we have clearly indisputable evidence to the contrary, if we are to analyze the fault correctly. It is almost self-evident, when the overcurrent devices do not operate promptly that the trouble was owing to a ground fault.

Unless *effective grounding* is followed and carefully observed, we cannot expect a ground fault to clear unless the circuit is manually opened or the ground fault continues and develops into a phase-to-phase fault. The latter condition may require many minutes during which time the equipment grounding conductor will develop a potential to ground, which may become high enough to be dangerous. Further, we may have a disastrous fire loss. Some method of automatically clearing a ground fault in limited time is highly desirable.

Engineering specifications must be carefully written to cover service and equipment grounding and not left to chance interpretation of how to accomplish safety. It is not an uncommon practice for specifications to read: "Ground in accordance with the provisions of Article 250 of the *National Electrical Code*." The engineer is not assuming his full responsibility for the safety of the installation when he writes such specifications. It is the engineer's responsibility to outline in his specifications how the job should be installed to fully comply with all three of the objectives of "effective grounding" and not leave it to chance, as is now often the case.

Ufer Ground

H.G. Ufer, in an IEEE Conference Paper, CP-61-978, describes an installation of made ground electrodes on 24 buildings in 1942, in Arizona, to meet a 5-ohm maximum value. The resistance values were checked bimonthly over an 18-year period, during which time no servicing was required.

In 1960, the maximum reading was 4.8 ohms and the minimum 2.1 ohms. The average value of the 24 installations was 3.57 ohms.

The installations used $1/2$-inch steel reinforcing rods set in a concrete footing. They were at two locations in Arizona. The first was near Tucson, Arizona, which is normally hot and dry during most of the year and has an average annual rainfall of 10.91 inches. The soil is sand and gravel. The second location was near Flagstaff, Arizona, where the soil is clay, shale gumbo, and loam with small area stratas of soft limestone. The made electrodes were used, as no water piping system was available.

As a result of these installations and the 18-year test period, Mr. Ufer suggested that a No. 4 or larger copper wire be embedded in the concrete footing of a building and that test data be compiled further to verify the effectiveness. Based on this data, CMP-5 accepted a concrete-encased electrode commonly referred to as an *Ufer Ground*. The concrete-encased electrode shall consist of at least 20 feet of bare copper not smaller than No. 4 AWG encased in 2 inches of concrete near the bottom of the footing or foundation.

---

[1] "The Fire Risks of Electric Lighting" by Killingsworth Hedges, *Electrical World*, October 6, 1889.

[2] "Safety and Safety Devices in Electrical Installations" by Professor Elihu Thomson, Electrical World, Feb. 22, 1890.

[3] Address at Cornell University by C. J. H. Woodbury, Electrical Engineer, November 18, 1891.

[4] *Electrical World*, March 26, 1892.

[5] "Means of Attaining Safety in Electrical Distribution" by W. L. R. Emmet, *Electrical Review*, June 7, 1899.

[6] "The Protection of Secondary Circuits from Fire Risk" by Dr. Cary T. Hutchinson, *Trans. AIEE*, Vol. XVI, 1899.

[7] *Electrical World and Engineer*, December 15, 1900.

[8] *Electrical World and Engineer*, December 14, 1901.

[9] *National Electrical Contractor*, January 1903.

[10] "The Electrical Fire Hazard" by William T. Bennallack, Journal Western Society of Engineers, Chicago IL., March 10, 1905.

[11] *National Electrical Contractor*, January 1906.

[12] *National Electrical Contractor*, April 1909.

[13] *National Electrical Contractor*, April 1909.

[14] Trans. AIEE 1909, Part 2, p. 1513.

[15] "The Fire Risks of Electric Lighting" by Killingsworth Hedges, *Electrical World*, October 6, 1889.

[16] Trans. AIEE, 1911, Part 3, P. 2591.

[17] "The Protection of Secondary Circuits from Fire Risk" by Dr. Cary T. Hutchinson, *Trans. AIEE*, Vol. XVI, 1899.

[18] *National Electrical Contractor*, April 1911.

[19] *National Electrical Contractor*, January 1903.

Why Measure Soil Resistivity?

Soil resistivity measurements have a threefold purpose. First, such data are used to make sub-surface geophysical surveys as an aid in identifying ore locations, depth to bedrock and other geological phenomena. Second, resistivity has a direct impact on the degree of corrosion in underground pipelines. A decrease in resistivity relates to an increase in corrosion activity and therefore dictates the protective treatment to be used. Third, soil resistivity directly affects the design of a grounding system, and it is to that task that this discussion is directed. When designing an extensive grounding system, it is advisable to locate the area of lowest soil resistivity in order to achieve the most economical grounding installation.

Effects of Soil Resistivity on Ground Electrode Resistance

Soil resistivity is the key factor that determines what the resistance of a grounding electrode will be, and to what depth it must be driven to obtain low ground resistance. The resistivity of the soil varies widely throughout the world and changes seasonally. Soil resistivity is determined largely by its content of electrolytes, which consist of moisture, minerals and dissolved salts. A dry soil has high resistivity if it contains no soluble salts (figure B-1).

Factors affecting soil resistivity

Two samples of soil, when thoroughly dried, may in fact become very good insulators having a resistivity in excess of 109 ohm-centimeters. The resistivity of the soil sample is seen to change quite rapidly until approximately 20 percent or greater moisture content is reached (figure B-2).

The resistivity of the soil is also influenced by temperature. Figure B-2 shows the variation of the resistivity of sandy loam, containing 15.2 percent moisture, with temperature changes from 20° to -15°C. In this temperature range the resistivity is seen to vary from 7200 to 330,000 ohm-centimeters.

Because soil resistivity directly relates to moisture content and temperature, it is reasonable to assume that the resistance of any grounding system will vary throughout the different seasons of the year. Such variations are shown in figure B-4. Since both temperature and moisture content become more stable at greater distances below the surface of the earth, it follows that a grounding system — to be most effective at all times — should be constructed with the ground rod driven down a considerable distance below the surface of the earth. Best results are obtained if the ground rod reaches the water table.

In some locations, the resistivity of the earth is so high that low-resistance grounding can be obtained only at considerable expense and with an elaborate grounding system. In such situations, it may be economical to use a ground rod system of limited size and to reduce the ground resistivity by periodically increasing the soluble chemical content of the soil. Figure B-5 shows the substantial reduction in resistivity of sandy loam brought about by an increase in chemical salt content.

Soil	Resistivity (approx.), V-cm		
	Min.	Average	Max.
Ashes, cinders, brine, waste	590	2.370	7,000
Clay, shale, gumbo, loam	340	4,060	16,300
Same, with varying proportions of sand and gravel	1,020	15,800	135,000
Gravel, sand, stones with little clay or loam	59,000	94,000	458,000

Figure B-1.

Moisture content, % by weight	Resistivity, V-cm	
	Top soil	Sandy loam
0	>10⁹	>10⁹
2.5	250,000	150,000
5	165,000	43,000
10	53,000	18,500
15	19,000	10,500
20	12,000	6,300
30	6,400	4,200

Figure B-2.

Temperature		Resistivity, Ohm-cm
C	F	
20	68	7,200
10	50	9,900
0	32 (water)	13,800
0	32 (ice)	30,000
-5	23	79,000
-15	14	330,000

Figure B-3.

Seasonal variation of earth resistance with an electrode of 3/4 inch pipe in rather stony clay soil. Depth of electrode in earth is 3 ft for Curve 1, and 10 ft for Curve 2.

Figure B-4.

THE EFFECT OF SALT* CONTENT ON THE RESISTIVITY OF SOIL
(Sandy loam, Moisture content, 15% by weight, Temperature, 17°C)

Added Salt (% by weight of moisture)	Resistivity (Ohm-centimeters)
0	10,700
0.1	1,800
1.0	460
5	190
10	130
20	100

Figure B-5.

THE EFFECT OF TEMPERATURE ON THE RESISTIVITY OF SOIL CONTAINING SALT*

(Sandy loam, 20% moisture. Salt 5% of weight of moisture)

Temperature (Degrees C)	Resistivity (Ohm-centimeters)
20	110
10	142
0	190
-5	312
-13	1,440

*Such as copper sulfate, sodium carbonate, and others. Salts must be EPA or local ordinance approved prior to use.

Figure B-6.

Chemically treated soil is also subject to considerable variation of resistivity with temperature changes, as shown in figure B-6. If salt treatment is employed, it is necessary to use ground rods which will resist chemical corrosion.

Soil Resistivity Measurements (4-Point Measurement)

Resistivity measurements are of two types; the 2-point and the 4-point method. The 2-point method is simply the resistance measured between two points. For most applications the most accurate method is the 4-point method. The 4-point method (figures B-7 and B-8), as the name implies, requires the insertion of four equally spaced and in-line electrodes into the test area. A known current from a constant current generator is passed between the outer electrodes. The potential drop (a function of the resistance) is then measured across the two inner electrodes. Most earth resistance meters read directly in ohms.

$$\rho = \frac{4\pi AR}{1 + \frac{2A}{\sqrt{(A^2+4B^2)}} - \frac{2A}{\sqrt{(4A^2+4B^2)}}}$$

Where: A = distance between the electrodes in centimeters

B = electrode depth in centimeters

If A > 20 B, the formula becomes:

$\rho = 2\pi$ AR (with A in cm)

ρ = 191.5 AR (with A in feet)

ρ = Soil resistivity (ohm-cm)

This value is the average resistivity of the ground at a depth equivalent to the distance "A" between two electrodes.

Soil Resistivity Measurements

Given a sizable tract of land in which to determine the optimum soil resistivity some intuition is in order. Assuming that the objective is low resistivity, preference should be given to an area containing moist loam as opposed to a dry sandy area. Consideration must also be given to the depth at which resistivity is required.

Example

After inspection, the area investigated has been narrowed down to a plot of ground approximately 75 square feet (7 m^2). Assume that you need to determine the resistivity at a depth of 15 feet (450 cm). The distance "A" between the electrodes must then be equivalent to the depth at which average resistivity is to be determined (15 ft, or 450 cm). Using the more simplified Wenner formula ($\rho = 2\pi$ AR), the electrode depth must then be 1/20th of the electrode spacing or 8-7/8" (22.5 cm).

Lay out the electrodes in a grid pattern and connect to the meter as shown in figures B-7 and B-8. Proceed as follows:

- Remove the shorting link between X and Xv (C1, P1)
- Connect all four auxiliary rods (figure B-7)

For example, if the reading is R = 15

ρ (resistivity) = 2π A x R

A (distance between electrodes) = 450 cm

ρ = 6.28 x 15 x 450 = 42,390 Ω-cm

Figure B-7.

Figure B-8.

Ground electrodes

The term *ground* is defined as a conducting connection by which a circuit or equipment is connected to the earth. The connection is used to establish and maintain as closely as possible the potential of the earth on the circuit or equipment connected to it. A *ground* consists of a grounding conductor, a bonding connector, its grounding electrode(s), and the soil in contact with the electrode.

Grounds have several protection applications. For natural phenomena, such as lightning, grounds are used to discharge the system of current before personnel can be injured or system components damaged. For foreign potentials due to faults in electric power systems with ground returns, grounds help ensure rapid operation of the protection relays by providing low-resistance fault current paths. This provides for the removal of the foreign potential as quickly as possible. The ground should drain the foreign potential before personnel are injured and the power or communications system is damaged.

Ideally, to maintain a reference potential for instrument safety, protect against static electricity, and limit the system to frame voltage for operator safety, a ground resistance should be zero ohms. In reality, as we describe further in the text, this value cannot be obtained.

Last, but not least, low ground resistance is essential to meet *NEC*, OSHA and other electrical safety standards.

Ground Rod and Clamp

Contact Resistance Between Rod and Soil

Concentric Shells of Earth

Figure B-9.

Figure B-9 illustrates a grounding rod. The resistance of the electrode has the following components:
(A) the resistance of the metal and that of the connection to it.
(B) the contact resistance of the surrounding earth to the electrode.
(C) the resistance in the surrounding earth to current flow or earth resistivity which is often the most significant factor.
More specifically:
(A) Grounding electrodes are usually made of a very conductive metal (copper or copper clad) with adequate cross sections so that the overall resistance is negligible.
(B) The National Institute of Standards and Technology has demonstrated that the resistance between the electrode and the surrounding earth is negligible if the electrode is free of paint, grease or other coating, and if the earth is firmly packed.
(C) The only component remaining is the resistance of the surrounding earth. The electrode can be thought of as being surrounded by concentric shells of earth or soil, all of the same thickness. The closer the shell to the electrode, the smaller its surface; hence, the greater its resistance. The farther away the shells are from the electrode, the greater the surface of the shell; hence, the lower the resistance. Eventually, adding shells at a distance from the grounding electrode will no longer noticeably affect the overall earth resistance surrounding the electrode. The distance at which this effect occurs is referred to as the effective resistance area and is directly dependent on the depth of the grounding electrode.
In theory, the ground resistance may be derived from the general formula:

$$R = \frac{\rho\, L}{A} \quad \text{Resistance = Resistivity} \ \times\ \frac{\text{Length}}{\text{Area}}$$

This formula illustrates why the shells of concentric earth decrease in resistance the farther they are from the ground rod:

In the case of ground resistance, uniform earth (or soil) resistivity throughout the volume is assumed, although this is seldom the case in nature. The equations for systems of electrodes are very complex and often expressed only as approximations. The most commonly used formula for single ground electrode systems, developed by Professor H. R. Dwight of the Massachusetts Institute of Technology, is the following:

$$R = \frac{\rho}{2 \pi L} \quad \frac{\{(\ln 4L) - 1\}}{r}$$

R = resistance in ohms of the ground rod to the earth (or soil)

L = grounding electrode length

r = grounding electrode radius

ρ = average resistivity in ohms-cm

Effect of Ground Electrode Size and Depth on Resistance

Size: Increasing the diameter of the rod does not materially reduce its resistance. Doubling the diameter reduces resistance by less than 10% (figure B-10).

Depth: As a ground rod is driven deeper into the earth, its resistance is substantially reduced. In general, doubling the rod length reduces the resistance by an additional 40% (figure B-11). The

Figure B-10.

Driven Depth in Feet
Ground Resistance Versus Ground Rod Depth

Figure B-11.

NEC [2002, 250.53(G)] requires a minimum of 8 ft (2.4 m) to be in contact with the soil. The most common is a 3 m (10 ft) cylindrical rod which meets the *Code*. A minimum diameter of 15.87 mm ($^5/_8$ in.) (1.59 cm) is required for steel rods and 13 mm ($^1/_2$ in.) (1.27 cm) for copper or copper clad steel rods [*NEC* 2002 250.52(A)(5)]. Minimum practical diameters for driving limitations for 3 m (10 ft) rods are:

- $^1/_2$ inch (1.27 cm) in average soil
- $^5/_8$ inch (1.59 cm) in moist soil
- $^3/_4$ inch (1.91 cm) in hard soil or more than 10 ft driving depths

Ground Resistance Values

NEC 250.56 2002: Resistance of man-made electrodes:

"A single electrode consisting of a rod, pipe, or plate that does not have a resistance to ground of 25 ohms or less shall be augmented by one additional of any of the types specified in 250.52(A)(2) through (A)(7). Where multiple rod, pipe or plate electrodes are installed to meet the requirements of this section, they shall be not less than 1.8 m (6 ft) apart."

NEC states that the resistance to ground shall not exceed 25 ohms. This is an upper limit and guideline, since much lower resistance is required in many instances.

"How low in resistance should a ground be?" An arbitrary answer to this in ohms is difficult. The lower the ground resistance, the safer, and for positive protection of personnel and equipment, it is worth the effort to aim for less than one ohm. It is generally impractical to reach such a low resistance along a distribution system or a transmission line or in small substations. In some regions, resistances of 5 ohms or less may be obtained without much trouble.

In other regions, it may be difficult to bring resistance of driven grounds below 100 ohms.

Accepted industry standards stipulate that transmission substations should be designed not to exceed one ohm resistance. In distribution substations, the maximum recommended resistance is for 5 ohms or even 1 ohm. In most cases, the buried grid system of any substation will provide the desired resistance.

In light industrial or in telecommunication central offices, 5V is often the accepted value. For lightning protection, the arrestors should be coupled with a maximum ground resistance of 1V.

Ground Rod Resistance-Ohms	Soil Resistivity (Ohm-centimeters)	Rod Depth Feet	Rod Diameter Inches

1. Select required resistance on R scale.
2. Select apparent resistivity on P scale.
3. Lay straightedge on R and P scale, and allow to intersect with K scale.
4. Mark K scale point.
5. Lay straightedge on K scale point & DIA scale, and allow to intersect with D scale.
6. Point on D scale will be rod depth required for resistance on R scale.

Figure B-12.

These parameters can usually be met with the proper application of basic grounding theory. There will always exist circumstances which will make it difficult to obtain the ground resistance required by the *NEC* or other safety standards. When these situations develop, several methods of lowering the ground resistance can be employed. These include parallel rod systems, deep driven rod systems utilizing sectional rods and chemical treatment of the soil. Additional methods, discussed in other published data, are buried plates, buried conductors (counterpoise), electrically connected building steel, and electrically connected concrete reinforced steel.

Electrically connecting to existing water and gas distribution systems was often considered to yield low ground resistance; however, recent design changes utilizing non-metallic pipes and insulating joints have made this method of obtaining a low resistance ground questionable and in many instances unreliable.

The measurement of ground resistances may only be accomplished with specially designed test equipment. Most instruments use the fall-of-potential principle of alternating current (AC) circulating between an auxiliary electrode and the ground electrode under test; the reading will be given in ohms, and represents the resistance of the ground electrode to the surrounding earth. AEMC has also recently introduced clamp-on ground resistance testers.

Ground Resistance Testing Principle
(Fall of Potential — 3-Point Measurement)

The potential difference between rods X and Y is measured by a voltmeter, and the current flow between rods X and Z is measured by an ammeter. (Note: X, Y and Z may be referred to as X, P

Figure B-13.

Figure B-14.

Figure B-15.

and C in a 3-point tester or C1, P2 and C2 in a 4-point tester.) (See figure B-13.)

By Ohm's Law, E = RI or R = E/I, we may obtain the ground electrode resistance R. If E = 20 V and I = 1A, then

$$R = \frac{E}{1} = \frac{20}{1} = 20$$

It is not necessary to carry out all the measurements when using a ground tester. The ground tester will measure directly by generating its own current and displaying the resistance of the ground electrode.

Position of the Auxiliary Electrodes on Measurements

The goal in precisely measuring the resistance to ground is to place the auxiliary current electrode Z far enough from the ground electrode under test so that the auxiliary potential electrode Y will be outside of the effective resistance areas of both the ground electrode and the auxiliary current electrode. The best way to find out if the auxiliary potential rod Y is outside the effective resistance areas is to move it between X and Z and to take a reading at each location. If the auxiliary potential rod Y is in an effective resistance area (or in both if they overlap, as in figure B-14), by displacing it, the readings taken will vary noticeably in value. Under these conditions, no exact value for the resistance to ground may be determined.

On the other hand, if the auxiliary potential rod Y is located outside of the effective resis-

Figure B-16.

tance areas (figure B-15), as Y is moved back and forth the reading variation is minimal. The readings taken should be relatively close to each other, and are the best values for the resistance to ground of the ground X. The readings should be plotted to ensure that they lie in a "plateau" region as shown in figure B-15. The region is often referred to as the "62% area."

Measuring Resistance of Ground Electrodes (62 percent method)

The 62 percent method has been adopted after graphical consideration and after actual test. It is the most accurate method but is limited by the fact that the ground tested is a single unit.

This method applies only when all three electrodes are in a straight line and the ground is a single electrode, pipe, or plate, etc., as in figure B-16.

Consider figure B-17, which shows the effective resistance areas (concentric shells) of the ground electrode X and of the auxiliary current electrode Z. The resistance areas overlap. If readings were taken by moving the auxiliary potential electrode Y towards either X or Z, the read-

ing differentials would be great and one could not obtain a reading within a reasonable band of tolerance. The sensitive areas overlap and act constantly to increase resistance as Y is moved away from X.

Now consider figure B-18, where the X and Z electrodes are sufficiently spaced so that the areas of effective resistance do not overlap. If we plot the resistance measured we find that the measurements level off when Y is placed at 62 percent of the distance from X to Z, and that the readings on either side of the initial Y setting are most likely to be within the established tolerance band. This tolerance band is defined by the user and expressed as a percent of the initial reading: ± 2%, ± 5%, ± 10%, and so forth.

Auxiliary Electrode Spacing

No definite distance between X and Z can be given, since this distance is relative to the diameter of the electrode tested, its length, the homogeneity of the soil tested, and particularly, the effective resistance areas. However, an approximate distance may be determined from the following chart which is given for a homogeneous soil and an electrode of 1" in diameter. (For a diameter of $1/2$", reduce the distance by 10 percent; for a diameter of 2" increase the distance by 10 percent.)

Multiple Electrode System

A single driven ground electrode is an economical and simple means of making a good ground system. But sometimes a single rod will not provide sufficient low resistance, and several ground electrodes will be driven and connected in parallel by a cable. Very often when two, three or four ground electrodes are being used, they are driven in a straight line; when

Figure B-17.

Figure B-18.

Approximate distance to auxiliary electrodes using the 62% method

Depth Driven	Distance to Y	Distance to Z
6 ft	45 ft	72 ft
8 ft	50 ft	80 ft
10 ft	55 ft	88 ft
12 ft	60 ft	96 ft
18 ft	71 ft	115 ft
20 ft	74 ft	120 ft
30 ft	86 ft	140 ft

Figure B-19.

four or more are being used, a hollow square configuration is used and the ground electrodes are still connected in parallel and are equally spaced (figure B-20).

In multiple electrode systems, the 62 percent method electrode spacing may no longer be applied directly. The distance of the auxiliary electrodes is now based on the maximum grid distance (i.e., in a square, the diagonal; in a line, the total length. For example, a square having a side of 20 ft will have a diagonal of approximately 28 ft).

Listed Grounding Electrode Systems

In addition to ground rods and plates, grounding electrode systems consisting of 2.125 in. diameter copper pipe have been developed. These pipe electrodes are available in straight or L-shaped versions and in various lengths from 10 to 20 feet. Longer models are available on special order.

These grounding electrode systems are filled with a mixture of nonhazardous earth salts. Weep holes are provided at the top and various locations toward the bottom or end of the pipe. This allows the circulation of moisture through the pipe to enhance performance. In addition, a specially formulated backfill material is available. This material improves the earth-to-grounding electrode performance.

Grounding Electrode Study

The Southern Nevada Chapter of the IAEI has embarked on an ambitious study of the effectiveness of several types of grounding electrodes. The testing program began in 1992. It is planned that the five different sites with 18 electrodes will be monitored over a ten year period. (See the figure B-23 for a description of electrodes being studied.) Data on electrode resistance, soil moisture, area temperature and rainfall is being collected on a regular basis.

Soil conditions at the five different sites are reported to be as follows:

"Balboa" – predominately gravel rock loosely compacted and is normally dry.

"Lone Mountain" – normally dry with sandy/silty soil.

"Pecos" – generally dry sand and silt down to a depth of about six feet and has clay and is wet down to about 12 feet.

"Pawnee" – a high water table with silt and clay.

Figure B-22 gives results of actual on-site resistance measurements made of the various grounding electrodes at each location. The results are an average of the readings taken over the six-year period of 1993 through 1998. For ex-

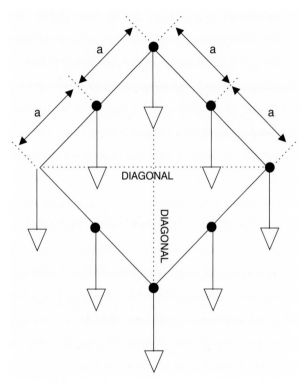

Figure B-20.

Multiple Electrode System

Max Grid Distance	Distance to Y	Distance to Z
6 ft	78 ft	125 ft
8 ft	87 ft	140 ft
10 ft	100 ft	160 ft
12 ft	105 ft	170 ft
14 ft	118 ft	190 ft
16 ft	124 ft	200 ft
18 ft	130 ft	210 ft
20 ft	136 ft	220 ft
30 ft	161 ft	260 ft
40 ft	186 ft	300 ft
50 ft	211 ft	340 ft
60 ft	230 ft	370 ft
80 ft	273 ft	440 ft
100 ft	310 ft	500 ft
120 ft	341 ft	550 ft
140 ft	372 ft	600 ft
160 ft	390 ft	630 ft
180 ft	434 ft	700 ft
200 ft	453 ft	730 ft

Figure B-21.

Description of Electrodes

A. No. 2 AWG stranded copper wire 50' in length centered in 12" of sand at 36" measured to the conductor.

B. No. 4 AWG steel reinforcing bar located within and near the bottom of a concrete foundation consisting of 12" by 12" of concrete 2500 psi. Top of concrete located approx. 6" below grade.

C. No. 4 AWG solid copper wire 25' in length centered in 6" by 6" of Erico ground enhancement material. Located at 20" to the bottom of the concrete.

D. No. 4 AWG solid copper wire 25' in length centered in 6" by 6" of concrete 2500 psi; located at 20" to the bottom of the concrete. This electrode is designed to imitate a thickened edge of a post tensioned concrete construction.

E. 8' by 5/8" copper clad ground rod centered in a pre-drilled 9" dia. by 9' vertical hole, encased by Erico GEM.

F. 8' by 5/8" copper clad ground rod installed horizontally centered in a 6" Erico GEM located 20" to the bottom of the GEM.

G. 8' by 5/8" copper clad ground rod installed horizontally directly buried at 30".

H. 8' by 5/8" copper clad ground rod vertically driven.

I. 10' by 3/4" galvanized ground rod driven vertically.

J. 10' by 3/4" galvanized ground rod directly buried horizontally at a depth of 30".

K. Grounding plate directly buried at 30".

L. Lyncole XIT 10' vertical electrode in pre-drilled 9" hole.

M. Arrangement of steel and concrete designed to imitate a light pole base; six, 2 ft. No. 4 steel vertical rods tied with three No. 2 steel hoops separated 12" vertically in 2' dia. by 30" concrete 2500 psi.

N. No. 4 AWG solid copper wire 20' roll shaped as a coil installed in 2' by 2' concrete 2500 psi.

O. No. 4 AWG solid copper wire 20' roll shaped as a coil installed in 2' by 2' Erico GEM.

P. Wood pole 18" dia. by 8' with pole plate attached at bottom with No. 6 solid copper wire wrapped in a spiral for 6' at 6" spacing between wraps.

Q. 8' by 1/2" copper clad ground rod installed horizontally directly buried at 30".

R. Lightning Eliminators Chemrod vertical installed per manufacturer's recommendations.

S. Lyncole XIT horizontal rod installed per manufacturer's recommendations.

Figure B-22.

ample, electrode "E" (which happens to have the lowest average resistance over the period) has an average resistance of 16.84 ohms at the Balboa site and an average resistance of 2.46 ohms at the Pawnee site. The average for all sites for the six-year period is 6.57 ohms.

The averages are calculated from data furnished. The data is believed to be accurate and is subject to errors and omissions in collection and recording. In some cases, the data is incomplete as indicated for electrodes "R" and "S."

This is an active study and while valuable data is being collected on a regular basis, it may be too early to make conclusions. The sponsorship of the grounding electrode study has been assumed by the National Fire Protection Association Research Foundation. Under their sponsorship, the study has been expanded to four other locations including Illinois, Texas, Florida, and Virginia. As a result, additional and significant data from diverse locations with different soil conditions is being collected.

Also, the study now includes information on soil temperature and moisture content.

Lightning Protection System

Any comprehensive discussion of lightning protection systems is beyond the scope of this text. Lightning protection systems should be installed in accordance with the *Standard for the Installation of Lightning Protection Systems*, NFPA-780.

The *Code* prohibits, in 250.60, the use of air terminal conductors and grounding electrodes such as driven pipes, rods, or other made electrodes in place of the grounding electrodes required by the *Code* for a wiring system and for equipment. Note that where two grounding electrodes are installed, they are required to be bonded together. See 250.106 for the requirement that lightning protection ground terminals (grounding electrodes) be bonded to the building or structure grounding electrode system.

The fine print notes following 250.106 provide valuable information regarding lightning protection systems. The *National Electrical Code* no longer has specific spacing requirements for separation of lightning protection system conductors from metal raceways and other metal enclosures of the building electrical system. The required spacing is typically 6 ft. through air and 3 ft. through dense construction materials such as concrete, brick or wood. Specific requirements are given in the *Lightning Protection Code*.

Conclusion

No potential exists between a conductor or equipment enclosure and earth if the system is properly and adequately grounded, except during a fault. By careful and intelligent design we must reduce the time of clearing a fault to a minimum.

While it is desirable to obtain a grounding electrode having a resistance as low as practical, it is more important that we provide a path of low impedance for the return of ground-fault current which will clear a ground fault when it occurs.

No hazard can exist in the distribution system if there is no insulation failure to create a ground fault. Further, the hazard can exist only for the period of time it takes to clear the fault. If a ground fault clears promptly, it is extremely unlikely that any loss of life would occur and property damage would be kept to a minimum.

For maximum safety, one rule that must be followed is; **ONLY ONE** grounding electrode system must be used in or on a building, with everything that should or is required to be grounded connected to that same grounding electrode system.

If more than one grounding electrode is required, they may be used providing **ALL** the grounding electrodes used are bonded together to form a grounding electrode system which in effect becomes **ONE** common grounding electrode.

Grounding electrodes should never be relied upon to provide the ground-fault return path for equipment. That is not their intended function in the electrical system. They cannot be relied upon to provide that path.[1]

---

[1] Information provided by Travis Lindsey and updated to 2002 NEC.

	Balboa	Lone Mountain	Pecos	Pawnee	Average
Electrode A	45.15	11.48	22.69	4.38	20.93
Electrode B	31.94	9.17	10.08	5.79	14.24
Electrode C	30.53	7.21	9.60	29.44	19.19
Electrode D	34.00	10.11	24.45	8.46	19.26
Electrode E	16.84	3.50	3.49	2.46	6.57
Electrode F	30.98	4.21	5.84	2.84	10.97
Electrode G	222.08	74.47	27.44	11.41	83.85
Electrode H	52.38	6.35	6.73	9.03	18.62
Electrode I	32.80	6.86	14.15	3.50	14.33
Electrode J	142.46	10.72	11.67	3.50	42.08
Electrode K	589.74	92.40	114.35	36.88	208.34
Electrode L	19.43	17.05	6.83	5.11	12.11
Electrode M	82.00	32.05	19.50	15.25	37.20
Electrode N	72.59	14.18	41.02	15.46	35.81
Electrode O	55.32	8.59	21.36	6.63	22.98
Electrode P	73.94	15.79	6.46	11.94	27.03
Electrode Q	184.98	26.15	20.63	19.62	62.84
Electrode R	15.41	---	---	---	15.41
Electrode S	---	---	13.70	3.60	8.65

Figure B-23

Key Notes

ANSWERS TO CHAPTERS ONE THROUGH NINETEEN

The references included can be found on the pages in this textbook or in the 2002 *National Electrical Code* under the sections given.

Chapter 1

1. c. In textbook
2. d. 250.4(A)
3. d. 250.4(A)(5)
4. d. In textbook
5. b. In textbook
6. a. In textbook
7. b. In textbook
8. b. In textbook
9. d. In textbook
10. d. In textbook
11. c. In textbook
12. c. In textbook
13. b. In textbook
14. b. In textbook
15. c. In textbook
16. a. In textbook
17. b. In textbook
18. d. In textbook, 250.110
19. b. Watt's wheel/ Ohms Law - textbook

Chap. 1 - 7500 ÷ 480 x 1.73 = 9.03

Chapter 2

1. d. 250.20(A)(1)
2. b. 250.20(A)(2)
3. a. 250.20(A)(3)
4. b. 250.20(B)(1)
5. b. 250.20(B)(2)
6. c. 250.20(B)(3)
7. d. 250.21(1)
8. b. 250.21(2)
9. d. 250.21(3)
10. b. 250.21(4)
11. d. 250.21(5)
12. b. 250.20(C)
13. a. 250.20(C)
14. d. In textbook
15. d. Article 100
16. c. 250.20(D)
17. b. In textbook
18. c. In textbook
19. d. In textbook
20. d. In textbook
21. a. In textbook
22. c. In textbook
23. b. In textbook
24. d. In textbook
25. d. In textbook Sections 250.22(4), 411.5(A)

Chapter 3

1. b. Article 100
2. a. Article 100
3. c. Article 100
4. a. 200.6(A)
5. c. 200.6(E), Exception No. 1
6. d. 200.6(B)
7. d. 200.6(A)
8. c. Sections 215.8, 230.56 and 110.15
9. d. In textbook
10. a. In textbook
11. d. In textbook
12. b. In textbook
13. d. In textbook
14. c. In textbook
15. b. In textbook
16. d. In textbook
17. c. In textbook
18. b. In textbook
19. c. 240.85
20. d. 430.36

Chapter 4

1. c. 250.24(B)
2. d. 250.24(B)
3. a. 250.24(B)
4. b. 250.24(B)
5. a. 250.24(B)
6. c. 250.24(B)
7. c. 250.24(B)
8. a. In textbook
9. d. In textbook
10. a. 250.24(A)(2)
11. c. Table 250.66
12. c. In textbook
13. d. In textbook
14. d. 250.36
15. a. 250.170
16. b. 250.170, Exception
17. d. 250.172, Exception

Chapter 5

1. b. Article 100
2. a. Article 100
3. d. 250.28(A)
4. b. 250.28B)
5. d. Table 250.66, 250.28(D)
6. d. 250.28(D)
7. a. 250.28(D)
8. d. 250.71
9. b. 250.142(A)

10. c. In textbook
11. a. Table 250.102(C)
12. d. 250.92(B)

Chapter 6

1. b. In textbook
2. a. Article 100
3. d. 250.52(C)
4. c. 250.52(C)
5. b. 250.52(D)
6. a. 250.52(A)(2)
7. d. 250.52(A)(2)
8. c. 250.50
9. d. 250.52
10. a. 250.58
11. d. 250.6(B)
12. b. 250.6(C)
13. d. 250.56 (In text)
14. a. In textbook
15. d. In textbook
16. b. 250.60
17. b. 250.54

Chapter 7

1. b. Article 100
2. c. Table 250.66
3. c. Table 250.66
4. a. In textbook
5. c. Table 250.66
6. c. Table 250.66
7. d. In textbook
8. d. Table 250.66
9. c. 230.71(A)
10. b. 250.66(A)
11. b. 250.66(B)
12. d. 250.70
13. b. 250.70
14. a. 250.64(B)
15. b. 250.64(B)
16. b. 250.64(A)

Chapter 8

1. d. 250.97
2. d. 250.102(B), 250-8
3. d. 314.3, Exception Nos. 1 and 2
4. c. 250.148
5. b. 250.146(D)
6. c. 250.102(E)
7. d. 250.104(A)(1)
8. d. In textbook
9. c. 250.104(B)

10. b. 250.98
11. a. 250.100, 250.92(B)
12. c. 250.122
13. c. 250.104(A)
14. c. In textbook
15. b. 250.104(A)
16. d. 250.104(C)

Chapter 9
1. a. Article 100
2. d. In textbook
3. c. In textbook
4. a. Table 250.122
5. d. 250.118
6. c. 250.118
7. c. 250.118(7)
8. c. 250.120(A)
9. b. 250.122(E)
10. b. 250.136(A)
11. c. 250.122(F)(1)
12. a. 250.122(F)(2)

Chapter 10
1. c. Article 100
2. a. Article 100
3. b. 250.110
4. d. 250.110
5. d. 250.86 Exception No. 1
6. c. 250.110, Exception No. 1
7. b. 250.110, Exception No. 2
8. d. 250.110, Exception No. 3
9. b. 250.86, Exception No. 1
10. a. 250.112(L)
11. d. 250.8
12. b. 250.8, 250.70
13. c. 250.80, Exception
14. b. 250.24(A)(5), 250.140, 250.142
15. b. 250.24(A), 215.6 and 408.20

Chapter 11
1. c. In textbook
2. b. In textbook
3. c. In textbook
4. b. In textbook
5. a. In textbook
6. b. In textbook
7. d. In textbook
8. b. In textbook
9. a. In textbook
10. b. In textbook
11. a. 6 AWG, In textbook

Chapter 12
1. c. Article 100
2. b. 250.30(A)(2) and Table 250.66
3. a. 250.30(B)
4. c. 250.30(A)(3)
5. b. 250.30(A)(3)
6. c. 250.30(A)(3)
7. a. 250.58
8. d. In textbook
9. a. In textbook
10. b. In textbook
11. d. In textbook
12. a. 250.30(A)(5), 250.28
13. c. 250.30(A)(6), 220.22, Table 250.66

Chapter 13
1. b. Article 100
2. c. 250.32(A)
3. c. In textbook
4. a. 250.32(A)
5. c. 250.32(B)(1)
6. a. 250.52
7. c. In textbook
8. c. 250.32(A)
9. a. 250.142(A)
10. a. 250.6(D)
11. d. 547.9(C)

Chapter 14
1. b. 210.7(D)
2. a. 210.7(D), Exception
3. b. 210.8(A)(2), Exception No. 2
4. c. 210.8(A)(4)
5. c. 210.8(A)(6)
6. c. 210.63
7. d. 210.8(B)(1)
8. d. 210.8(A)(3)
9. b. In textbook
10. a. In textbook
11. a. In textbook
12. b. In textbook
13. a. In textbook
14. c. 210.8(B)(3) and textbook

Chapter 15
1. b. Article 100
2. b. 230.95
3. c. 230.95, Exception Nos. 1 and 2
4. b. 230.95(A)
5. a. In textbook, Section 215.10
6. c. In textbook

7. a. In textbook
8. d. 230.95(C)
9. b. 517.17(A)
10. c. 215.10

Chapter 16
1. c. In textbook
2. a. 250.106
3. c. 501.16(A), 502.16(A) and 503.16(A)
4. d. In textbook
5. a. 250.102(E)
6. b. 547.4(F)
7. a. 547.10(C)
8. c. 547.9
9. d. In textbook
10. d. 517.13(A)
11. d. 517.19(D)
12. c. 680.26(B)(5)
13. a. 680.26(C)(3)
14. a. 680.23(C)(1)
15. d. 680.22(A)(5)
16. c. 680.23(F)(3)(4)
17. a. 680.7(A)
18. a. 680.5
19. d. 680.23(A)(3)
20. d. 680.23(A)(3)
21. b. 680.23(B)(2)(A)
22. a. 680.23(B)(5)
23. b. 680.23(A)(2)
24. b. 680.23(A)(2)
25. c. 600.7
26. c. 600.7(C)
27. d. 600.7(A)

Chapter 17 (New)
1. b. Article 645
2. d.
3. a.
4. c.
5. c.
6. b.
7. b.
8. c.
9. a.
10. d.
11. a.
12. d.
13. d.
14. b.
15. d.
16. d.

17. a.
18. a.
19. c.

Chapter 18 (New)
1. d. Article 100 definition of grounding conductor
2. b. 250.94
3. b. 810.21(H)
4. c. 820.40(A)(4)
5. c. 820.40(A)(2) and (3)
6. c. 820.40(D)
7. d. 820.40(B)(1)
8. c. 830.42(B)
9. b. 800.40(A)(6)
10. c. 830.40(D)
11. b. 830.11(I)(3)
12. b 250.20(A)(3)
13. c. 504.50(B)
14. a. 504.50(C) Exception
15. b. 800.41(B)

Chapter 19
1. b. In textbook
2. a. In textbook
3. b. In textbook
4. c. In textbook
5. c. In textbook
6. a. In textbook
7. c. In textbook
8. b. In textbook
9. c. In textbook
10. d. In textbook
11. a. In textbook
12. b. In textbook

Index

Soares Book on Grounding

Eighth Edition

Designed by: Melody Dalton
Michael J. Johnston

Project Manager: Michael J. Johnston

Cover design: Melody Dalton

Contributors: Michael J. Johnston
J. Philip Simmons
Paul Dobrowski
Chuck Mello
Travis Lindsey

Technical Review: Michael J. Johnston
J. Philip Simmons

Technical Editing: Michael J. Johnston
J. Philip Simmons
Philip H. Cox

Technical Drawings: Michael J. Johnston
Brady Davis

Graphic Artists: Bob Sanford
Laura Hildreth

Photos / Illustration Reprints:

AFC Cable Systems, New Bedford, MA
AVO International, Biddle Instruments, Dallas, TX
Electro-Test Inc., Milwaukie, OR
Cooper Industries, Bussman Division, St. Louis, MO
Eagle Electric Company, Long Island, NY
Insulated Cable Engineers Association, South Yarmouth, MA
International Electronic and Electrical Engineers, Piscataway, PA
National Electrical Manufacturers Association, Rosslyn, VA
National Fire Protection Association, Quincy, MA
NFPA Research Foundation, Quincy, MA
Post Glover Resistors, Inc., Erlanger, KY
Square D Company Palatine, IL
Steel Conduit Section, National Electrical Manufacturers Association, Rosslyn, VA
Thomas & Betts, Memphis, TN
Underwriters Laboratories, Northbrook, IL
Greaves Corporation, Guilford, CT
Lyncole XIT, Torrance, CA
Harger, Grayslake, IL

Digital Photos: Michael J. Johnston
IAEI Archives
Chuck Mello, Electro-test Inc.

Managing Editor: Kathryn Ingley

Composed at: International Association of Electrical Inspectors in Book Antiqua and Swiss 721BT

Printed: Affiliated Graphics on Luna Matte text and bound in C1S White Tango cover